T0183943

Lecture Notes in Computer Science 9685

Commenced Publication in 1973
Founding and Former Series Editors:
Gerhard Goos, Juris Hartmanis, and Jan van Leeuwen

More information about this series at http://www.springer.com/series/7407

Andrew V. Goldberg · Alexander S. Kulikov (Eds.)

Experimental Algorithms

15th International Symposium, SEA 2016
St. Petersburg, Russia, June 5–8, 2016
Proceedings

 Springer

Editors
Andrew V. Goldberg
Amazon.com, Inc.
Palo Alto, CA
USA

Alexander S. Kulikov
St. Petersburg Department of Steklov
   Institute of Mathematics
Russian Academy of Sciences
St. Petersburg
Russia

ISSN 0302-9743   ISSN 1611-3349   (electronic)
Lecture Notes in Computer Science
ISBN 978-3-319-38850-2   ISBN 978-3-319-38851-9   (eBook)
DOI 10.1007/978-3-319-38851-9

Library of Congress Control Number: 2016939104

LNCS Sublibrary: SL1 – Theoretical Computer Science and General Issues

Printed on acid-free paper

This Springer imprint is published by Springer Nature
The registered company is Springer International Publishing AG Switzerland

Preface

This volume contains the 25 papers presented at SEA 2016, the 15th International Symposium on Experimental Algorithms, held during June 5–8, 2016, in St. Petersburg, Russia. The symposium was organized by the Steklov Mathematical Institute at St. Petersburg of the Russian Academy of Sciences (PDMI). SEA covers a wide range of topics in experimental algorithmics, bringing together researchers from algorithm engineering, mathematical programming, and combinatorial optimization communities. In addition to the papers, three invited lectures were given by Juliana Freire (New York University, USA), Haim Kaplan (Tel Aviv University, Israel), and Yurii Nesterov (Ecole Polytechnique de Louvain, Belgium).

The Program Committee selected the 25 papers presented at SEA 2016 and published in these proceedings from the 54 submitted papers. Each submission was reviewed by at least three Program Committee members, some with the help of qualified subreferees. We expect the full versions of most of the papers contained in these proceedings to be submitted for publication in refereed journals.

Many people and organizations contributed to the smooth running and the success of SEA 2016. In particular our thanks go to:

- All authors who submitted their current research to SEA
- Our reviewers and subreferees who gave input into the decision process
- The members of the Program Committee, who graciously gave their time and expertise
- The members of the local Organizing Committee, who made the conference possible
- The EasyChair conference management system for hosting the evaluation process
- Yandex
- The Government of the Russian Federation (Grant 14.Z50.31.0030)
- Steklov Mathematical Institute at St. Petersburg of the Russian Academy of Sciences
- Monomax Congresses & Incentives

June 2016

Andrew V. Goldberg
Alexander S. Kulikov

Organization

Program Committee

Ittai Abraham	VMware Research, USA
Maxim Babenko	Moscow State University, Russia
Daniel Bienstock	Columbia University, USA
Daniel Delling	Sunnyvale, CA, USA
Paola Festa	University of Naples Federico II, Italy
Stefan Funke	Universität Stuttgart, Germany
Andrew V. Goldberg	Amazon.com, Inc., USA
Dan Halperin	Tel Aviv University, Israel
Michael Juenger	Universität zu Köln, Germany
Alexander S. Kulikov	St. Petersburg Department of Steklov Institute of Mathematics, Russian Academy of Sciences, Russia
Alberto Marchetti Spaccamela	Sapienza University of Rome, Italy
Petra Mutzel	University of Dortmund, Germany
Tomasz Radzik	King's College London, UK
Rajeev Raman	University of Leicester, UK
Ilya Razenshteyn	CSAIL, MIT, USA
Mauricio Resende	Amazon.com, Inc., USA
Peter Sanders	Karlsruhe Institute of Technology, Germany
David Shmoys	Cornell University, USA
Daniele Vigo	Università di Bologna, Italy
Neal Young	University of California, Riverside, USA

Organizing Committee

Asya Gilmanova	Monomax Congresses & Incentives, Russia
Ekaterina Ipatova	Monomax Congresses & Incentives, Russia
Alexandra Novikova	St. Petersburg Department of Steklov Institute of Mathematics, Russian Academy of Sciences, Russia
Alexander Smal	St. Petersburg Department of Steklov Institute of Mathematics, Russian Academy of Sciences, Russia
Alexander S. Kulikov	St. Petersburg Department of Steklov Institute of Mathematics, Russian Academy of Sciences, Russia

Additional Reviewers

Akhmedov, Maxim
Artamonov, Stepan
Atias, Aviel
Becchetti, Luca
Birgin, Ernesto G.
Bonifaci, Vincenzo
Bökler, Fritz
Botez, Ruxandra
Cetinkaya, Elcin
Ciardo, Gianfranco
de Andrade, Carlos
Dietzfelbinger, Martin
Fischer, Johannes
Fleischman, Daniel
Fogel, Efi
Gog, Simon
Gomez Ravetti, Martin
Gondzio, Jacek
Gonçalves, José
Gopalan, Parikshit
Gronemann, Martin
Halperin, Eran
Harchol, Yotam
Hatami, Pooya
Hatano, Kohei
Hübschle-Schneider, Lorenz
Johnson, David
Karypis, George
Kashin, Andrei
Kleinbort, Michal
Kolesnichenko, Ignat
Kärkkäinen, Juha

Lattanzi, Silvio
Luo, Haipeng
Mallach, Sven
Miyazawa, Flavio K.
Mu, Cun
Narodytska, Nina
Pajor, Thomas
Pardalos, Panos
Pascoal, Marta
Pferschy, Ulrich
Pouzyrevsky, Ivan
Prezza, Nicola
Ribeiro, Celso
Rice, Michael
Roytman, Alan
Sagot, Marie-France
Salzman, Oren
Savchenko, Ruslan
Schlechte, Thomas
Schmidt, Daniel
Schöbel, Anita
Shamai, Shahar
Solovey, Kiril
Sommer, Christian
Spisla, Christiane
Starikovskaya, Tatiana
Storandt, Sabine
Suk, Tomáš
Valladao, Davi
Vatolkin, Igor
Wieder, Udi
Zey, Bernd

Abstracts of Invited Talks

Provenance for Computational Reproducibility and Beyond

Juliana Freire

New York University, New York, USA

The need to reproduce and verify experiments is not new in science. While result verification is crucial for science to be self-correcting, improving these results helps science to move forward. Revisiting and reusing past results — or as Newton once said, "standing on the shoulders of giants" — is a common practice that leads to practical progress. The ability to reproduce computational experiments brings a range of benefits to science, notably it: enables reviewers to test the outcomes presented in papers; allows new methods to be objectively compared against methods presented in reproducible publications; researchers are able to build on top of previous work directly; and last but not least, recent studies indicate that reproducibility increases impact, visibility, and research quality and helps defeat self-deception.

Although a standard in natural science and in Math, where results are accompanied by formal proofs, reproducibility has not been widely applied for results backed by computational experiments. Scientific papers published in conferences and journals often include tables, plots and beautiful pictures that summarize the obtained results, but that only loosely describe the steps taken to derive them. Not only can the methods and implementation be complex, but their configuration may require setting many parameters. Consequently, reproducing the results from scratch is both time-consuming and error-prone, and sometimes impossible. This has led to a credibility crisis in many scientific domains. In this talk, we discuss the importance of maintaining detailed provenance (also referred to as lineage and pedigree) for both data and computations, and present methods and systems for capturing, managing and using provenance for reproducibility. We also explore benefits of provenance that go beyond reproducibility and present emerging applications that leverage provenance to support reflective reasoning, collaborative data exploration and visualization, and teaching.

This work was supported in part by the National Science Foundation, a Google Faculty Research award, the Moore-Sloan Data Science Environment at NYU, IBM Faculty Awards, NYU School of Engineering and Center for Urban Science and Progress.

Minimum Cost Flows in Graphs with Unit Capacities

Haim Kaplan

Tel Aviv University, Tel Aviv, Israel

We consider the minimum cost flow problem on graphs with unit capacities and its special cases. In previous studies, special purpose algorithms exploiting the fact that capacities are one have been developed. In contrast, for maximum flow with unit capacities, the best bounds are proven for slight modifications of classical blocking flow and push-relabel algorithms.

We show that the classical cost scaling algorithms of Goldberg and Tarjan (for general integer capacities) applied to a problem with unit capacities achieve or improve the best known bounds. For weighted bipartite matching we establish a bound of $O(\sqrt{r}m \log C)$ on a slight variation of this algorithm. Here r is the size of the smaller side of the bipartite graph, m is the number of edges, and C is the largest absolute value of an arc-cost. This simplifies a result of Duan et al. and improves the bound, answering an open question of Tarjan and Ramshaw. For graphs with unit vertex capacities we establish a novel $O(\sqrt{n}m \log (nC))$ bound.

This better theoretical understanding of minimum cost flow on one hand, and recent extensive experimental work on algorithms for maximum flow on the other hand, calls for further engineering and experimental work on algorithms for minimum cost flow. I will discuss possible future research along these lines.

Complexity Bounds for Primal-Dual Methods Minimizing the Model of Objective Function

Yurii Nesterov

CORE/INMA, UCL, Louvain-la-Neuve, Belgium

We provide Frank–Wolfe (Conditional Gradients) method with a convergence analysis allowing to approach a primal-dual solution of convex optimization problem with composite objective function. Additional properties of complementary part of the objective (strong convexity) significantly accelerate the scheme. We also justify a new variant of this method, which can be seen as a trust-region scheme applying the linear model of objective function. Our analysis works also for a quadratic model, allowing to justify the global rate of convergence for a new second-order method. To the best of our knowledge, this is the first trust-region scheme supported by the worst-case complexity bound.

Contents

Practical Variable Length Gap Pattern Matching

Johannes Bader[1], Simon Gog[1(✉)], and Matthias Petri[2]

[1] Institute of Theoretical Informatics, Karlsruhe Institute of Technology,
76131 Karlsruhe, Germany
gog@kit.edu

[2] Department of Computing and Information Systems,
The University of Melbourne, VIC 3010, Australia

Abstract. Solving the problem of reporting all occurrences of patterns containing variable length gaps in an input text T efficiently is important for various applications in a broad range of domains such as Bioinformatics or Natural Language Processing. In this paper we present an efficient solution for static inputs which utilizes the wavelet tree of the suffix array. The algorithm partially traverses the wavelet tree to find matches and can be easily adapted to several variants of the problem. We explore the practical properties of our solution in an experimental study where we compare to online and semi-indexed solutions using standard datasets. The experiments show that our approach is the best choice for searching patterns with many gaps in large texts.

1 Introduction

The classical *pattern matching problem* is to find all occurrences of a pattern $\mathcal{P}$ (of length m) in a text $\mathcal{T}$ (of length n both drawn from an alphabet Σ of size σ). The *online* algorithm of Knuth, Morris and Pratt [10] utilizes a precomputed table over the pattern to solve the problem in $\mathcal{O}(n + m)$ time. Precomputed indexes over the input text such as *suffix arrays* [14] or *suffix trees* allow matching in $\mathcal{O}(m \times \log n)$ or $\mathcal{O}(m)$ time respectively. In this paper we consider the more general *variable length gap pattern matching problem* in which a pattern does not only consist of characters but also length constrained *gaps* which match any character in the text. We formally define the problem as:

Problem 1 Variable Length Gap (VLG) Pattern Matching [5]. Let $\mathcal{P}$ be a pattern consisting of $k \geq 2$ subpatterns $p_0 \ldots p_{k-1}$, of lengths $m = m_0 \ldots m_{k-1}$ drawn from Σ and $k-1$ gap constraints $C_0 \ldots C_{k-2}$ such that $C_i = \langle \delta_i, \Delta_i \rangle$ with $0 \leq \delta_i \leq \Delta_i < n$ specifies the smallest (δ_i) and largest (Δ_i) distance between a match of p_i and p_{i+1} in $\mathcal{T}$. Find all matches – given as k-tuples $\langle i_0 \ldots i_{k-1} \rangle$ where i_j is the starting position for subpattern p_j in $\mathcal{T}$ – such that all gap constraints are satisfied.

If overlaps between matches are permitted, the number of matching positions can be polynomial in $k + 1$. We refer to this problem type as *all*. In standard

A.V. Goldberg and A.S. Kulikov (Eds.): SEA 2016, LNCS 9685, pp. 1–16, 2016.
DOI: 10.1007/978-3-319-38851-9_1

implementations of *regular expressions* overlaps are not permitted[1] and two types for variable gap pattern matching are supported. The *greedy* type (a gap constraint is written as $C_i = .\{\delta_i, \Delta_i\}$) maximizes while the *lazy* type (a gap constraint is written as $C_i = .\{\delta_i, \Delta_i\}?$) minimizes the characters matched by gaps. The following example illustrates the three VLG problem types.

Example 1. *Given a text* $\mathcal{T} = aaabbbbaaabbbb$ *and a pattern* $\mathcal{P} = ab\langle 1, 6 \rangle b$ *consisting of two subpatterns* $p_0 =$ "ab" *and* $p_1 =$ "b" *and a gap constraint* $C_0 = \langle 1, 6 \rangle$. *Type "all" returns* $S_{all} = \{\langle 2, 5 \rangle, \langle 2, 6 \rangle, \langle 2, 10 \rangle, \langle 9, 12 \rangle, \langle 9, 13 \rangle\}$, *greedy matching results in* $S_{greedy} = \{\langle 2, 10 \rangle\}$ *and lazy evaluation* $S_{lazy} = \{\langle 2, 5 \rangle, \langle 9, 12 \rangle\}$.

VLG pattern matching is an important problem which has numerous practical applications. Traditional Unix utilities such as `grep` or `mutt` support VLG pattern matching on small inputs using regular expression engines. Many areas of computer science use VLG matching on potentially very large data sets. In Bioinformatics, Navarro and Raffinot [18] investigate performing VLG pattern matching in protein databases such as PROSITE [9] where individual protein site descriptions are expressed as patterns containing variable length gaps. In Information Retrieval (IR), the proximity of query tokens within a document can be an indicator of relevance. Metzler and Croft [15] define a language model which requires finding query terms to occur within a certain window of each other in documents. In Natural Language Processing (NLP), this concept is often referred to as collocations of words. Collocations model syntactic elements or semantic links between words in tasks such as word sense disambiguation [16]. In Machine Translation (MT) systems VLG pattern matching is employed to find translation rule sets in large text corpora to improve the quality of automated language translation systems [13].

In this paper, we focus on the *offline* version of the VLG pattern matching problem. Here, a static input is preprocessed to generate an *index* which facilitates faster query processing. Our contributions are as follows:

1. We build an index consisting of the wavelet tree over the suffix array and propose different algorithms to efficiently answer VLG matching queries. The core algorithms is conceptionally simple and can be easily adjusted to the three different matching modes outlined above.
2. In essence our WT algorithm is faster than other intersection based approaches as it allows to combine the sorting and filtering step and does not require copying of data. Therefore our approach is specially suited for a large number of subpatterns.
3. We provide a thorough empirical evaluation of our method including a comparison to different practical baselines including other index based approaches like qgram indexes and suffix arrays.

---

[1] I.e. any two match tuples $\langle i_0 \ldots i_{k-1} \rangle$ and $\langle i'_0 \ldots i'_{k-1} \rangle$ spanning the intervals $[i_0, i_{k-1} + m_{k-1} - 1]$ and $[i'_0, i'_{k-1} + m_{k-1} - 1]$ do not overlap.

2   Background and Related Work

Existing solutions to solving VLG can be categorized into three classes of algorithms. In general the algorithms discussed here perform lazy evaluation, but can be implemented to also support greedy evaluation. The first category of algorithms build on the classical algorithm of Thompson [20] to construct a finite automaton to solve the VLG problem used by many regular expression engines. Let $L = \sum_{i=0}^{k-2} \Delta_i$. The classical automaton requires $\mathcal{O}(n(L\sigma + m))$ time which can not be reduced much further [4,5]. The matching process scans $\mathcal{T}$, transitioning between states in the automaton to find occurrences of $\mathcal{P}$. Algorithms engineered to solve the VLG problem specifically can achieve better runtime performance by utilizing bit-parallelism or placing constraints on individual gap constraints in $\mathcal{P}$ [4,6,18]. For example the runtime of Bille and Thorup [4] is $\mathcal{O}(n(k\frac{\log w}{w} \times \log k))$ time after prepocessing $\mathcal{P}$ (w is the word size).

A second class of algorithms take into account the occurrences of each subpattern in $\mathcal{P}$ [5,17,19]. The algorithms operate in two stages. First, all occurrences of each subpattern $p \in \mathcal{P}$ in $\mathcal{T}$ are determined. Let α_i be the number of occurrences of p_i in $\mathcal{T}$ and $\alpha = \sum_{i=0}^{k-1} \alpha_i$ be the total number of occurrences of all p_i in $\mathcal{T}$. The occurrences of each subpattern can be obtained via a classical online algorithm such as Aho and Corasick [1] (AC), or using an index such as the suffix array (SA) of $\mathcal{T}$. The algorithms of Morgante et al. [17,19] require additional $\mathcal{O}(\alpha)$ space to store the occurrences, whereas Bille et al. [5] only requires $\mathcal{O}(S)$ extra space where $S = \sum_{i=0}^{k-2} \delta_i$. The algorithms keep track of, for each p_{i+1} the occurrences of p_i for which C_i is satisfied. Similarly to Rahman et al. [19], the occurrences $X_i = [x_0, \ldots, x_{\alpha_i-1}]$ of p_i are used to satisfy $C_i = \langle \delta_i, \Delta_i \rangle$ by searching for the next larger (or equal) value for all $x_j + \delta_i$ in the sorted list of occurrences of p_{i+1}. Performing this process for all p_i and gap constraints C_i can be used to perform lazy evaluation of VGP. While Rahman et al. [19] consider only AC or SA to identify occurrences of subpatterns, many different ways to store or determine and search positions exist. For example, the folklore q-gram index, which explicitly stores the occurrences of all q-grams in $\mathcal{T}$, can be used to obtain the occurrences of all subpatterns by performing intersection of the positional lists of the q-grams each subpattern in $\mathcal{P}$. List compression affects the performance of the required list intersections and thus provides different time space-trade-offs [11]. Similarly, different list intersection algorithms can also affect the performance of such a scheme [2].

A combination of schemata one and two use occurrences of subpatterns to restrict the segments within $\mathcal{T}$ where efficient *online* algorithms are used to *verify* potential matches. For example, q-gram lists can be intersected until the number of possible occurrences of $\mathcal{P}$ is below a certain threshold. Then an automaton based algorithm can match $\mathcal{P}$ at these locations.

The third category are suffix tree based indexes. In its simplest form, Lewenstein [12] augments each node of a suffix tree over $\mathcal{T}$ with multiple gap-r-tree for all $1 \leq r < G$, where G is the longest gap length which has to be specified at construction time. If k subpatterns are to be supported, the nodes in each gap-r-tree have to be recursively augmented with additional gap-r-trees at a total space cost of $\mathcal{O}(n^k G^{k-1})$ space. Queries are answered in $\mathcal{O}(\sum_0^{k-1} m_i)$ time by traversing the suffix tree from the root, branching into the gapped-r-trees after the locus of p_0 is found. Lewenstein [12] further propose different time-space trade-offs, reducing the space to $\mathcal{O}(nG^{2k-1}\log^{k-1} n)$ by performing centroid path decomposition which increases query time to $\mathcal{O}(\sum_0^{k-1} m_i + 2^{k-1}\log\log n)$. Bille and Gørtz [3] propose a suffix tree based index which requires two ST over $\mathcal{T}$ ($ST(\mathcal{T})$) and the reverse text ($ST(\mathcal{T}^R)$) plus a range reporting structure. Single fixed length gap queries can then be answered by matching p_0 in $ST(\mathcal{T})$ and the reverse of p_1 in $ST(\mathcal{T}^R)$. Then a range reporting structure is used to find the matching positions in $\mathcal{T}$.

In practice, Lopez [13] use a combination of (1) intersection precomputation (2) fast list intersection and (3) enhanced version of Rahman et al. [19] to solve a restricted version of VGP.

3   VLG Pattern Matching Using the Wavelet Tree over SA

We first introduce notation which is necessary to describe our algorithms. Let range I from index ℓ to r be denoted by $[\ell, r]$. A range is considered empty ($I = \emptyset$) if $\ell > r$. We denote the intersection of two ranges $I_0 = [\ell_0, r_0]$ and $I_1 = [\ell_1, r_1]$ as $I_0 \cap I_1 = [\max\{\ell_0, \ell_1\}, \min\{r_0, r_1\}]$. We further define the addition $I_0 + I_1$ of two ranges to be $[\ell_0 + \ell_1, r_0 + r_1]$. Shorthands for the left and right border of a non-empty range I are $lb(I)$ and $rb(I)$.

Let $X_i = x_{i,0}, \ldots, x_{i,\alpha_i - 1}$ be the list of starting positions of subpattern p_i in $\mathcal{T}$. Then for $k = 2$ the solution to the VLG pattern matching problem for $\mathcal{P} = p_0\langle\delta, \Delta\rangle p_1$ are pairs $\langle x_{0,i}, x_{1,j}\rangle$ such that $([x_{0,i}, x_{0,i}] + [m_0 + \delta, m_0 + \Delta]) \cap [x_{1,j}, x_{1,j}] \neq \emptyset$. The generalization to $k > 2$ subpatterns is straightforward by checking all $k - 1$ constraints. For ease of presentation we will restrict the following explanation to $k = 2$. Assuming all X_i are present in sorted order all matches can be found in $\mathcal{O}(\alpha_0 + \alpha_1 + z)$ time, where z refers to the number of matches of $\mathcal{P}$ in $\mathcal{T}$. Unfortunately, memory restrictions prohibit the storage of all possible $\mathcal{O}(n^2)$ sorted subpattern lists, but we will see next that the linear space suffix array can be used to retrieve any unsorted subpattern list.

A suffix $\mathcal{T}[i, n-1]$ is identified by its starting position i in $\mathcal{T}$. The suffix array (SA) contains all suffixes in lexicographically sorted order, i.e. SA[0] points to the smallest suffix in the text, SA[1] to the second smallest and so on. Figure 1 depicts the SA for an example text of size $n = 32$. Using SA and $\mathcal{T}$ it is easy to determine all suffixes which start with a certain prefix p by performing binary search. For example, $p_0 = \mathtt{gt}$ corresponds to the SA-interval $[17, 21]$ which contains suffixes $16, 13, 21, 4, 28$. Note that the occurrences of the p in SA are not stored sorted order. Answering a VLG pattern query using SA can be achieved by first determining the SA-intervals of all subpatterns and next, filtering out all occurrence tuples which fulfill the gap constraints [19].

Let $\mathcal{P} = \mathtt{gc}\langle 1, 2\rangle\mathtt{c}$ containing $p_0 = \mathtt{gc}$ and $p_1 = \mathtt{c}$. In Example Fig. 1, the SA-interval of c (SA[9, 16]) contains suffixes $26, 10, 11, 19, 12, 20, 1$ and 8.

$i =$	0	1	2	3	4	5	6	7	8	9	10	11	12	13	14	15	16	17	18	19	20	21	22	23	24	25	26	27	28	29	30	31
$\mathcal{T} =$	a	c	t	a	g	t	a	t	c	t	c	c	c	g	t	a	g	t	a	c	c	g	t	a	t	a	c	a	g	t	t	$
$SA(\mathcal{T}) =$	31	25	18	0	15	3	27	23	6	26	10	11	19	12	20	1	8	16	13	21	4	28	30	24	17	14	2	22	5	9	7	29

Fig. 1. Sample text $\mathcal{T} = $ `actagtatctcccgtagtaccgtatacagtt$` and suffix array (SA) of $\mathcal{T}$.

Sorting the occurrences of both subpatterns returns in $X_0 = 4, 13, 16, 21, 28$ and $X_1 = 1, 8, 10, 11, 12, 19, 20, 26$. Filtering X_0 and X_1 based on $C_0 = \langle 1, 2 \rangle$ produces tuples $\langle 4, 8 \rangle, \langle 16, 19 \rangle$ and $\langle 16, 20 \rangle$. The time complexity of this process is $\mathcal{O}(\sum_{i=0}^{k-1} \alpha_i \log \alpha_i + z)$, where the first term (sorting all X_i) is independent of z (the output size) and can dominate if subpatterns occur frequently.

Using a wavelet tree (WT) [8] allows *combining the sorting and filtering process*. This enables *early termination* for text regions which do not contain all required subpatterns in correct order within the specified gap constraints. A wavelet tree $WT(X)$ of a sequence $X[0, n-1]$ over an alphabet $\Sigma[0, \sigma - 1]$ is defined as a perfectly balanced binary tree of height $H = \lceil \log \sigma \rceil$. Conceptually the root node v represents the whole sequence $X_v = X$. The left (right) child of the root represents the subsequence X_0 (X_1) which is formed by only considering symbols of X which are prefixed by a 0-bit(1-bit). In general the i-th node on level L represents the subsequence $X_{i_{(2)}}$ of X which consists of all symbols which are prefixed by the length L binary string $i_{(2)}$. More precisely the symbols in the range $R(v_{i_{(2)}}) = [i \cdot 2^{H-L}, (i+1) \cdot 2^{H-L} - 1]$. Figure 2 depicts an example for $X = SA(\mathcal{T})$. Instead of actually storing $X_{i_{(2)}}$ it is sufficient to store the bitvector $B_{i_{(2)}}$ which consists of the ℓ-th bits of $X_{i_{(2)}}$. In connection with a *rank structure*, which can answer how many 1-bits occur in a prefix $B[0, j-1]$ of bitvector $B[0, n-1]$ in constant time using only $o(n)$ extra bits, one is able to reconstruct all elements in an arbitrary interval $[\ell, r]$: The number of 0-bits (1-bits) left to ℓ corresponds to ℓ' in the left (right) child and the number of 0-bits (1-bits) left to r corresponds to $r' + 1$ in the left (right) child. Figure 2 shows this *expand* method. The red interval $[17, 21]$ in the root node v is expanded to $[9, 10]$ in node v_0 and $[8, 10]$ in node v_1. Then to $[4, 4]$ in node v_{00} and $[5, 5]$ in node v_{01} and so on. Note that WT nodes are only traversed if the interval is not empty (i.e. $\ell \leq r$). E.g. $[4, 4]$ at v_{00} is split into $[3, 2]$ and $[1, 1]$. So the left child v_{000} is omitted and the traversal continues with node v_{001}. Once a leaf is reached we can output the element corresponding to its root to leaf path. The wavelet tree $WT(X)$ uses just $n \cdot h + o(n \cdot h)$ bits of space.

In our application the initial intervals correspond to the SA-intervals of all subpatterns p_i in $\mathcal{P}$. However, our traversal algorithm only considers the existence of a SA-interval at a given node and not its size. A non-empty SA-interval of subpattern p_i in a node v_x at level L means that p_i occurs somewhere in the text range $R(v_x) = [x \cdot 2^{H-L}, (x+1) \cdot 2^{H-L} - 1]$. Figure 3 shows the text ranges for each WT node. A node v and its parent edge is marked red (resp. blue) if subpattern p_0's (resp. p_1's) occurs in the text range $R(v)$.

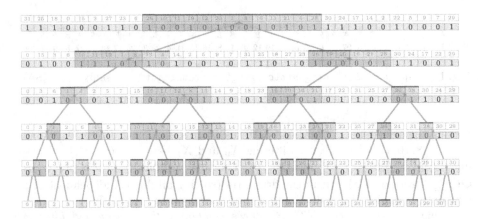

Fig. 2. Wavelet tree built for the suffix array of our example text. The SA-interval of gt (resp. c) in the root and its expanded intervals in the remaining WT nodes are marked red (resp. blue). (Color figure online)

3.1 Breadth-First Search Approach

For both subpatterns p_0 and p_1, at each level in the WT, j we iteratively materialize lists N_0^j and N_1^j of all WT nodes at level j in which the ranges corresponding to p_0 and p_1 occur by expanding the nodes in the lists N_0^{j-1} and N_1^{j-1} of the previous level. Next all nodes v_xs in N_0^j are removed if there is no node v_y in N_1^j such that $(R(v_x) + [m_0 + \delta, m_0 + \Delta]) \cap [R(v_y)] \neq \emptyset$ and vice versa. Each list N_i^{j-1} stores nodes in sorted order according to the beginning of their ranges. Thus, removing all "invalid" nodes can be performed in $\mathcal{O}(|N_0^j| + |N_1^j|)$ time. The following table shows the already filtered list for our running example.

WT level (j)	p_0 text ranges (N_0^j)	p_1 text ranges (N_1^j)
0	$[0, 31]$	$[0, 31]$
1	$[0, 15], [16, 31]$	$[0, 15], [16, 31]$
2	$[0, 7], [8, 15], [16, 23], [24, 31]$	$[0, 7], [8, 15], [16, 23], [24, 31]$
3	$[4, 7], [12, 15], [16, 19], [20, 23]$	$[8, 11], [12, 15], [16, 19], [20, 23], [24, 27]$
4	$[4, 5], [16, 17], [20, 21]$	$[8, 9], [18, 19], [20, 21], [24, 25]$
5	$[4, 4], [16, 16]$	$[8, 8], [19, 19], [20, 20]$

The WT nodes in the table are identified by their text range as shown in Fig. 3. One example of a removed node is $[0, 3]$ in lists N_0^3 and N_1^3 which was expanded from node $[0, 7]$ in N_0^2 and N_1^2. It was removed since there is no text range in N_1^3 which overlaps with $[0, 3] + [2 + 1, 2 + 2] = [3, 6]$. Figure 3 connects removed WT nodes with dashed instead of solid edges. Note that all text positions at the leaf level are the start of a subpattern which fulfills all gap constraints. For

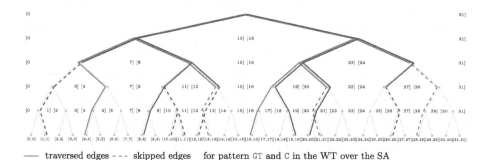

Fig. 3. Wavelet tree nodes with annotated text ranges and path of subpattern iterators. (Color figure online)

the *all* variant it just takes $\mathcal{O}(z)$ time to output the result. The disadvantage of this BFS approach is that the lists of a whole level have to be kept in memory, which takes up to n words of space. We will see next that a DFS approach lowers memory consumption to $\mathcal{O}(k \log n)$ words.

3.2 Depth-First Search Approach

For each subpattern p_i we create a depth-first search *iterator* it_i. The iterator consists of a stack of (WT node, SA-interval) pairs, which is initialized by the WT root node and the SA-interval of p_i, in case the SA-interval is not empty. The iterator is invalid, if the stack is empty – this can be checked in constant time using a method $valid(it_i)$. We refer with $it_i.v$ to the current WT node of a valid iterator (which is on top of the stack). A valid iterator can be incremented by operations $next\_down$ and $next\_right$. Method $next\_down$ pops pair $(v, [\ell, r])$, expands SA-interval $[\ell, r]$ and pushes the right child of v with its SA-interval and the left child of v with its SA-interval onto the stack, if the SA-interval is not empty. That is, we traverse to the leftmost child of $it.v$ which contains p_i. The $next\_right(it_i)$ operation pops one element from the stack, i.e. we traverse to the leftmost node in the WT which contains p_i and is right of $it_i.v$.

Using these iterators the VLG pattern matching process can be expressed succinctly in Algorithm 1, which reports the next match. The first line checks, if both iterators are still valid so that a further match can be reported. Lines 2 and 4 check if the gap constraints are met. If the text range of p_1's iterator is too far right (Line 2), the iterator of p_0 is moved right in Line 3. Analogously, the iterator of p_1 is moved right in Line 5 if the text range of p_0's iterator is too far right (Line 4). If the gap constrained is met and not both text ranges have size one (Line 7) we take the iterator which is closer to the root (and break ties by i) and refine its range. Finally, if both iterators reach the leaf level a match can be reported. Since the traversal finds the two leftmost leaf nodes – i.e. positions – which met the constraint the direct output of $\langle lb(R(it_0.v)), lb(R(it_1.v)) \rangle$ in Line 11 corresponds to the *lazy* problem type. For *lazy* Line 12 would move it_0 to the

Algorithm 1. dfs_next_match($it_0, it_1, m_0, \Delta_0, \delta_0$)

1: **while** $valid(it_0)$ **and** $valid(it_1)$ **do**
2:     **if** $rb(R(it_0.v)) + m_0 + \Delta_0 < lb(R(it_1.v))$ **then**   # gap constraint violated?
3:         $it_0 \leftarrow next\_right(it_0)$
4:     **else if** $rb(R(it_1.v)) < lb(R(it_0.v)) + m_0 + \delta_0$ **then**   # gap constraint violated?
5:         $it_1 \leftarrow next\_right(it_1)$
6:     **else**   # gap constrained fulfilled
7:         **if not** $(is\_leaf(it_0.v)$ **and** $is\_leaf(it_1.v))$ **then**
8:             $x \leftarrow \arg\min_{i \in \{0,1\}}\{\langle depth(it_i.v), i \rangle\}$ # select itr closest to the root
9:             $it_x \leftarrow next\_down(it_x)$ # refine range
10:        **else**
11:            report match according to VLG problem type   # found match
12:            move it_0 and it_1 according to VLG problem type and return $\langle it_0, it_1 \rangle$

---

right of it_1 by calling $it_0 \leftarrow next\_right(it_0)$ until $lb(R(it_0.v)) > lb(R(it_1.v))$ is true and no overlapped matches are possible. Type *greedy* can be implemented by moving it_1 in Line 11 as far right as possible within the gap constrains, output $\langle lb(R(it_0.v)), lb(R(it_1.v)) \rangle$, and again moving it_0 to the right of it_1. Type *all* reports the first match in Line 11, then iterates it_1 as long as it meets the gap constraint and reports a match if $it_1.v$ is a leaf. In Line 12 it_0 is move one step to the right and it_1 it reset to its state before line 11.

3.3   Implementation Details

Our representation of the WT requires two rank operations to retrieve the two child nodes of any tree node. In our DFS approach, k tree iterators partially traverse the WT. For higher values of k it is likely that the child nodes of a specific WT node are retrieved multiple times by different iterators. We therefore examined the effect of caching the child nodes of a tree node when they are retrieved for the first time, so any subsequent child retrieval operations can be answered without performing further rank operations. Unfortunately, this approach resulted in a slowdown of our algorithm by a factor of 3. We conjecture, that one reason for this slowdown is the additional memory management overhead (even when using custom allocators) of dynamically allocating and releasing the cached data. Also, critical portions of the algorithm (being called most frequently) contain more branching and were even inlined before we implemented the cache. Furthermore, we determined that more than 65 % of tree nodes traversed once were never traversed a second time, so caching children for these nodes will not yield any run time performance improvements. On average, each cache entry was accessed less than 2 times after creation. Thus, only very few rank operations are actually saved. Therefore we do not cache child nodes in our subsequent empirical evaluation.

4   Empirical Evaluation

In this section we study the practical impact of our proposals by comparing to standard baselines in different scenarios. Our source code – including baselines and dataset details – is publicly available at https://github.com/olydis/vlg_matching and implemented on top of SDSL [7] data structures. We use three datasets from different application domains:

- The CC data set is a 371 GiB prefix of a recent 145 TiB web crawl from commoncrawl.org.
- The Kernel data set is a 78 GiB file consisting of the source code of all (332) Linux kernel versions 2.2.X, 2.4.$X.Y$ and 2.6.$X.Y$ downloaded from kernel.org. This data set is very repetitive as only minor changes exist between subsequent kernel versions.
- The Dna-Hg38 data set data consisting of the 3.1 GiB Genome Reference Consortium Human Reference 38 in fasta format with all symbol $\notin \{A, C, G, T\}$ removed from the sequence.

We have implemented our BFS and DFS wavelet tree approaches. We omit the results of the BFS approach, as DFS dominated BFS in both query time and memory requirement. Our index is denoted by WT-DFS the following. We use a pointerless WT (`wt_int`) in combination with a fast rank enabled bitvector (`bit_vector_il`). We compare to three baseline implementations:

- RGXP: A "off-the-shelf" automaton based regular expression engine (BOOST library version 1.58; `ECMAScript` flag set) which scans the whole text.
- QGRAM-RGXP: A q-gram index ($q = 3$) which stores absolute positions of all unique 3-grams in the text using Elias-Fano encoding. List intersection is used to produce candidate positions in $\mathcal{T}$ and subsequently checked by the RGXP engine.
- SA-SCAN: The plain SA is used as index. The SA-intervals of the subpatterns are determined, sorted, and filtered as described in earlier. This approach is similar to that of Rahman et al. [19] while replacing the van Emde Boas tree by sorting ranges.

All baselines and indexes are implemented using `C++11` and compiled using `gcc` `4.9.1` using all optimizations. The experiments were performed on a machine with an Intel Xeon E4640 CPU and 148 GiB RAM. The default VLG matching type in our experiments is *lazy*, which is best suited for proximity search. Pattern were generated systematically for each data set. We fix the gap constraints $C_i = \langle \delta_i, \Delta_i \rangle$ between subpatterns to $\langle 100, 110 \rangle$ small (C_S), $\langle 1\,000, 1\,100 \rangle$ medium (C_M), or $\langle 10\,000, 11\,000 \rangle$ large (C_L). For each dataset we extract the 200 most common subpatterns of length 3, 5 and 7 (if possible). We form 20 regular expressions for each dataset, k, and gap constraint by selecting from the set of subpatterns.

Matching Performance for Different Gap Constraint Bands. In our first experiment we measure the impact of gap constraint size on query time. We

Table 1. Median query time in milliseconds for fixed $m_i = 3$ and text size $2\,\mathrm{GiB}$ for different gap constraints $\langle 100, 110 \rangle$ small (C_S), $\langle 1000, 1100 \rangle$ medium (C_M) or $\langle 10\,000, 11\,000 \rangle$ large (C_L) and three data sets.

Method	Kernel-2G			CC-2G			Dna-Hg38-2G		
	C_S	C_M	C_L	C_S	C_M	C_L	C_S	C_M	C_L
$k = 2$									
RGXP	6 383	7 891	18 592	2 533	4 148	17 394	24 363	26 664	9 849
QGRAM-RGXP	695	2 908	20 775	650	2 604	21 027	48 984	33 911	7 711
SA-SCAN	115	114	113	132	130	132	6 762	6 661	6 433
WT-DFS	279	244	347	180	211	277	17 041	10 978	8 350
$k = 4$									
RGXP	5 130	6 840	30 948	5 076	6 889	28 931	34 025	41 800	24 549
QGRAM-RGXP	1 336	8 992	$\geq 10^5$	1 284	9 187	$\geq 10^5$	$\geq 10^5$	$\geq 10^5$	91 137
SA-SCAN	247	249	250	284	284	289	14 667	14 971	14 191
WT-DFS	160	164	183	195	201	232	19 977	12 608	8 506
$k = 8$									
RGXP	3 243	5 089	31 796	2 426	4 215	28 943	33 126	$\geq 10^5$	$\geq 10^5$
QGRAM-RGXP	3 307	30 174	$\geq 10^5$	2 894	27 488	$\geq 10^5$	$\geq 10^5$	$\geq 10^5$	$\geq 10^5$
SA-SCAN	594	585	596	759	761	765	29 850	30 621	29 296
WT-DFS	263	282	228	184	185	179	28 343	16 707	8 843
$k = 16$									
RGXP	3 447	5 278	32 782	2 407	4 229	33 828	37 564	$\geq 10^5$	$\geq 10^5$
QGRAM-RGXP	6 843	61 787	$\geq 10^5$	5 967	65 722	$\geq 10^5$	$\geq 10^5$	$\geq 10^5$	$\geq 10^5$
SA-SCAN	1 400	1 402	1 416	1 714	1 711	1 690	56 558	62 423	55 017
WT-DFS	508	507	463	331	331	316	55 660	26 041	9 152
$k = 32$									
RGXP	3 446	5 237	32 979	3 673	6 041	33 957	24 040	$\geq 10^5$	$\geq 10^5$
QGRAM-RGXP	14 732	$\geq 10^5$	$\geq 10^5$	11 506	$\geq 10^5$	$\geq 10^5$	$\geq 10^5$	$\geq 10^5$	$\geq 10^5$
SA-SCAN	2 885	2 926	2 924	3 573	3 560	3 562	82 663	92 756	81 164
WT-DFS	1 183	1 083	965	614	609	594	35 495	35 212	5 501

fix the dataset size to $2\,\mathrm{GiB}$ and the subpattern length $|p_i| = m_i = 3$; Table 1 shows the results for pattern consisting of $k = 2^1, \ldots, 2^5$ subpatterns. For RGXP, the complete text is scanned for all bands. However, the size of the underlying automaton increases with the gap length. Thus, the performance decreases for larger gaps. The intersection process in QGRAM-RGXP reduces the search space of RGXP to a portion of the text. There are cases where the search space reduction is not significant enough to amortize the overhead of the intersection. For example, the large gaps or the small alphabet test case force QGRAM-RGXP to perform more work than RGXP. The two SA based solutions, SA-SCAN and WT-DFS, are

Table 2. Median query time in milliseconds for fixed gap constraint $\langle 100, 110 \rangle$ and text size 2 GiB for different subpattern lengths $m_i \in 3, 5, 7$ for three data sets.

Method	Kernel-2G			CC-2G			Dna-Hg38-2G		
	C_S	C_M	C_L	C_S	C_M	C_L	C_S	C_M	C_L
$k = 2$									
RGXP	6 383	7 891	18 592	2 533	4 148	17 394	24 363	26 664	9 849
QGRAM-RGXP	695	2 908	20 775	650	2 604	21 027	48 984	33 911	7 711
SA-SCAN	115	114	113	132	130	132	6 762	6 661	6 433
WT-DFS	279	244	347	180	211	277	17 041	10 978	8 350
$k = 4$									
RGXP	5 130	6 840	30 948	5 076	6 889	28 931	34 025	41 800	24 549
QGRAM-RGXP	1 336	8 992	$\geq 10^5$	1 284	9 187	$\geq 10^5$	$\geq 10^5$	$\geq 10^5$	91 137
SA-SCAN	247	249	250	284	284	289	14 667	14 971	14 191
WT-DFS	160	164	183	195	201	232	19 977	12 608	8 506
$k = 8$									
RGXP	3 243	5 089	31 796	2 426	4 215	28 943	33 126	$\geq 10^5$	$\geq 10^5$
QGRAM-RGXP	3 307	30 174	$\geq 10^5$	2 894	27 488	$\geq 10^5$	$\geq 10^5$	$\geq 10^5$	$\geq 10^5$
SA-SCAN	594	585	596	759	761	765	29 850	30 621	29 296
WT-DFS	263	282	228	184	185	179	28 343	16 707	8 843
$k = 16$									
RGXP	3 447	5 278	32 782	2 407	4 229	33 828	37 564	$\geq 10^5$	$\geq 10^5$
QGRAM-RGXP	6 843	61 787	$\geq 10^5$	5 967	65 722	$\geq 10^5$	$\geq 10^5$	$\geq 10^5$	$\geq 10^5$
SA-SCAN	1 400	1 402	1 416	1 714	1 711	1 690	56 558	62 423	55 017
WT-DFS	508	507	463	331	331	316	55 660	26 041	9 152
$k = 32$									
RGXP	3 446	5 237	32 979	3 673	6 041	33 957	24 040	$\geq 10^5$	$\geq 10^5$
QGRAM-RGXP	14 732	$\geq 10^5$	$\geq 10^5$	11 506	$\geq 10^5$	$\geq 10^5$	$\geq 10^5$	$\geq 10^5$	$\geq 10^5$
SA-SCAN	2 885	2 926	2 924	3 573	3 560	3 562	82 663	92 756	81 164
WT-DFS	1 183	1 083	965	614	609	594	35 495	35 212	5 501

considerably faster than scanning the whole text for Kernel and CC. We also observe the WT-DFS is less dependent on the number of subpatterns k than SA-SCAN, since no overhead for copying and explicitly sorting SA ranges is required. Also WT-DFS profits from larger minimum gap sizes as larger parts of the text are skipped when gap constraints are violated near the root of the WT. For Dna-Hg38, small subpattern length of $m_i = 3$ generate large SA intervals which in turn decrease query performance comparable to processing the complete text.

Matching Performance for Different Subpattern Lengths. In the second experiment, we measure the impact of subpattern lengths on query time. We fix

Table 3. Space usage relative to text size at query time of the different indexes for three data sets of size 2 GiB, different subpattern lengths $m_i \in 3, 5, 7$ and varying number of subpatterns $k \in 2, 4, 8, 16, 32$

Method	Kernel-2G			CC-2G			Dna-Hg38-2G		
	3	5	7	3	5	7	3	5	7
$k = 2$									
RGXP	1.00	1.00	1.00	1.00	1.00	1.00	1.00	1.00	1.00
QGRAM-RGXP	7.93	7.93	7.93	7.49	7.49	7.49	7.94	7.94	7.94
SA-SCAN	5.01	5.00	5.00	5.01	5.00	5.00	5.20	4.90	4.88
WT-DFS	5.50	5.50	5.50	5.50	5.50	5.50	5.41	5.38	5.38
$k = 4$									
RGXP	1.00	1.00	1.00	1.00	1.00	1.00	1.00	1.00	1.00
QGRAM-RGXP	7.93	7.93	7.93	7.49	7.49	7.49	7.94	7.94	7.94
SA-SCAN	5.01	5.01	5.00	5.02	5.01	5.00	5.89	4.94	4.89
WT-DFS	5.50	5.50	5.50	5.50	5.50	5.50	5.38	5.38	5.38
$k = 8$									
RGXP	1.00	1.00	1.00	1.00	1.00	1.00	1.00	1.00	1.00
QGRAM-RGXP	7.93	7.93	7.93	7.49	7.49	7.49	7.94	7.94	7.94
SA-SCAN	5.06	5.02	5.01	5.06	5.02	5.01	6.57	5.03	4.91
WT-DFS	5.50	5.50	5.50	5.50	5.50	5.50	5.38	5.38	5.38
$k = 16$									
RGXP	1.00	1.00	1.00	1.00	1.00	1.00	1.00	1.00	1.00
QGRAM-RGXP	7.93	7.93	7.93	7.49	7.49	7.49	7.94	7.94	7.94
SA-SCAN	5.22	5.09	5.02	5.18	5.09	5.03	7.86	5.25	4.94
WT-DFS	5.50	5.50	5.50	5.50	5.50	5.50	5.38	5.38	5.38
$k = 32$									
RGXP	1.00	1.00	1.00	1.00	1.00	1.00	1.00	1.00	1.00
QGRAM-RGXP	7.93	7.93	7.93	7.49	7.49	7.49	7.94	7.94	7.94
SA-SCAN	5.32	5.09	5.04	5.31	5.11	5.03	10.20	5.54	5.00
WT-DFS	5.50	5.50	5.50	5.50	5.50	5.50	5.38	5.38	5.38

the gap constraint to $\langle 100, 110 \rangle$ and the data sets size to 2 GiB. Table 2 shows the results. Larger subpattern length result in smaller SA ranges. Consequently, query time performance of SA-SCAN and WT-DFS improves. As expected RGXP performance does not change significantly, as the complete text is scanned irrespectively of the subpattern length.

Matching Performance for Different Text Sizes. In this experiment we explore the dependence of query time on text size. The results are depicted in Fig. 4. The boxplot summarizes query time for all ks and all gap constraints for a

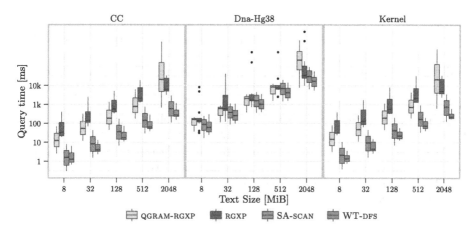

Fig. 4. Average query time dependent on input size for subpattern length $m_i = 3$.

fixed subpattern length $m = 3$. As expected, the performance of RGXP increases linearly with the text size for all datasets. The indexed solutions QGRAM-RGXP SA-SCAN and WT-DFS also show a linear increase with dataset size. We observe again that the SA based solutions are significantly faster than the RGXP base methods. For CC and Kernel the difference is one order of magnitude – even for small input sizes of 8 MiB. We also observe that WT-DFS is still the method of choice for the unfavorable case of a small alphabet text with small subpattern length of $m = 3$.

Space Usage at Query Time. In addition to run time performance, we evaluate the in memory space usage of the different indexes at query time. The space usage considered is the space of the underlying index structure in addition to the temporary space required to answer queries. For example, the SA-SCAN method requires additional space to sort the positions of each subinterval to perform efficient intersection. Similarly, the QGRAM-RGXP index requires additional space to store results of intersections for subpatterns larger than q. The space usage of the different index structures relative to the text size is shown in Table 3. Clearly RGXP requires only little extra space in addition to the text to store the regexp automaton. The QGRAM-RGXP requires storing the text, the compressed q-gram lists, the regexp automaton for verification, and during query time, q-gram intersection results. The SA-SCAN index requires storing the suffix array ($n \log n$ bits), which requires roughly $4n$ bytes of space for a text of size 2 GiB plus the text (n bytes) to determine the subpattern ranges in the suffix array. Additionally, SA-SCAN requires temporary space to sort subpattern ranges. Especially for frequent subpatterns, this can be substantial. Consider the Dna-Hg38 dataset for $k = 32$ and $m = 3$. Here the space usage of SA-SCAN is $9n$, which is roughly twice the size of the index structure. This implies that SA-SCAN potentially requires large amounts of additional space at query time which can be prohibitive. The WT-DFS index encodes the suffix array using a wavelet tree. The structure requires $n \log n$ bits of space plus $o(n \log n)$ bits to efficiently support rank operations. In our setup we use an rank structure which requires 12.5% of the space of the WT bitvector. In addition, we store the text to determine the suffix array ranges via forward search. This requires another $n \log \sigma$ bits which corresponds to n bytes for CC and CC. For this reason the WT-DFS index is slightly larger than SA-SCAN. We note that the index size of WT-DFS can be reduced from $5.5n$ to $4.5n$ by not including the text explicitly. The suffix array ranges can still be computed with a logarithmic slowdown if the WT over the suffix array is augmented with select structures. The select structure enables access to the inverse suffix array and we can therefore simulate Ψ and LF. This allows to apply backward search which does not require explicit access to the original text.

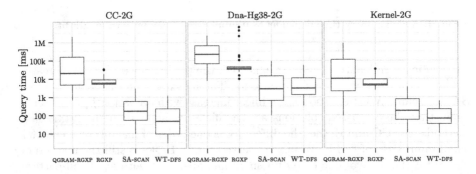

Fig. 5. Overall runtime performance of all methods for three data sets, accumulating the performance for all $m_i \in 3, 5, 7$ and C_S, C_M and C_L for text size 2 GiB.

Overall Runtime Performance. In a final experiment we explored the whole parameter space (i.e. $k \in \{2^1, \dots, 2^5\}$, $m_i \in \{3, 5, 7\}$, $C \in \{C_S, C_M, C_L\}$) and summarize the results in Fig. 5. Including also the large subpattern length $m_i = 5$ and $m_i = 7$ results in even bigger query time improvement compared to the RGXP based approaches: for CC and Kernel SA based method queries can be processed in about 100 ms while RGXP require 10 s on inputs of size 2 GiB. The average query time for Dna-Hg38 improves with SA based methods from 50 s to 5 s. The WT based approach significantly improves the average time for CC and Kernel and is still the method of choice for Dna-Hg38.

5   Conclusion

In this paper we have shown another virtue of the wavelet tree. Built over the suffix array its structure allows to speed up variable length gap pattern queries by combining the sorting and filtering process of suffix array based indexes. Compared to the traditional intersection process it does not require copying of data and enables skipping of list regions which can not satisfy the intersection criteria. We have shown empirically that this process outperforms competing approaches in many scenarios.

In future work we plan to reduce the space of our index by not storing the text explicitly and using the wavelet tree augmented with a select structure to determine the intervals of the subpatterns in the suffix array.

Acknowledgement. We are grateful to Timo Bingmann for profiling our initial implementation. This work was supported under the Australian Research Council's Discovery Projects scheme (project DP140103256) and Deutsche Forschungsgemeinschaft.

References

1. Aho, A.V., Corasick, M.J.: Efficient string matching: an aid to bibliographic search. Commun. ACM **18**(6), 333–340 (1975)
2. Baeza-Yates, R.: A fast set intersection algorithm for sorted sequences. In: Sahinalp, S.C., Muthukrishnan, S.M., Dogrusoz, U. (eds.) CPM 2004. LNCS, vol. 3109, pp. 400–408. Springer, Heidelberg (2004)
3. Bille, P., Gørtz, I.L.: Substring range reporting. Algorithmica **69**(2), 384–396 (2014)
4. Bille, P., Thorup, M.: Regular expression matching with multi-strings and intervals. In: Proceedings of SODA, pp. 1297–1308 (2010)
5. Bille, P., Gørtz, I.L., Vildhøj, H.W., Wind, D.K.: String matching with variable length gaps. Theor. Comput. Sci. **443**, 25–34 (2012)
6. Fredriksson, K., Grabowski, S.: Efficient algorithms for pattern matching with general gaps, character classes, and transposition invariance. Inf. Retrieval **11**(4), 335–357 (2008)
7. Gog, S., Beller, T., Moffat, A., Petri, M.: From theory to practice: plug and play with succinct data structures. In: Gudmundsson, J., Katajainen, J. (eds.) SEA 2014. LNCS, vol. 8504, pp. 326–337. Springer, Heidelberg (2014)
8. Grossi, R., Gupta, A., Vitter, J.S.: High-order entropy-compressed text indexes. In: Proceedings of SODA, pp. 841–850 (2003)
9. Hulo, N., Bairoch, A., Bulliard, V., Cerutti, L., De Castro, E., Langendijk-Genevaux, P.S., Pagni, M., Sigrist, C.J.A.: The PROSITE database. Nucleic Acids Res. **34**(suppl 1), D227–D230 (2006)
10. Knuth, D.E., Morris Jr., J.H., Pratt, V.R.: Fast pattern matching in strings. SIAM J. Comput. **6**(2), 323–350 (1977)
11. Lemire, D., Boytsov, L.: Decoding billions of integers per second through vectorization. Soft. Prac. Exp. **45**(1), 1–29 (2015)
12. Lewenstein, M.: Indexing with gaps. In: Grossi, R., Sebastiani, F., Silvestri, F. (eds.) SPIRE 2011. LNCS, vol. 7024, pp. 135–143. Springer, Heidelberg (2011)
13. Lopez, A.: Hierarchical phrase-based translation with suffix arrays. In: Proceedings of EMNLP-CoNLL, pp. 976–985 (2007)
14. Manber, U., Myers, E.W.: Suffix arrays: a new method for on-line string searches. SIAM J. Comput. **22**(5), 935–948 (1993)
15. Metzler, D., Croft, W.B.: A Markov random field model for term dependencies. In: Proceedings of SIGIR, pp. 472–479 (2005)
16. Mihalcea, R., Tarau, P., Figa, E.: Pagerank on semantic networks, with application to word sense disambiguation. In: Proceedings of COLING (2004)

17. Morgante, M., Policriti, A., Vitacolonna, N., Zuccolo, A.: Structured motifs search. J. Comput. Biol. **12**(8), 1065–1082 (2005)
18. Navarro, G., Raffinot, M.: Fast and simple character classes and bounded gaps pattern matching, with applications to protein searching. J. Comput. Biol. **10**(6), 903–923 (2003)
19. Rahman, M.S., Iliopoulos, C.S., Lee, I., Mohamed, M., Smyth, W.F.: Finding patterns with variable length gaps or don't cares. In: Chen, D.Z., Lee, D.T. (eds.) COCOON 2006. LNCS, vol. 4112, pp. 146–155. Springer, Heidelberg (2006)
20. Thompson, K.: Regular expression search algorithm. Commun. ACM **11**(6), 419–422 (1968)

Fast Exact Computation of Isochrones
in Road Networks

Moritz Baum, Valentin Buchhold$^{(\boxtimes)}$, Julian Dibbelt, and Dorothea Wagner

Karlsruhe Institute of Technology, Karlsruhe, Germany
{moritz.baum,valentin.buchhold,julian.dibbelt,dorothea.wagner}@kit.edu

Abstract. We study the problem of computing isochrones in static and dynamic road networks, where the objective is to identify the boundary of the region in range from a given source within a certain amount of time. While there is a wide range of practical applications for this problem (e. g., urban planning, geomarketing, visualizing the cruising range of a vehicle), there has been little research on fast algorithms for large, realistic inputs, and existing approaches tend to compute more information than necessary. Our contribution is twofold: (1) We propose a more compact but sufficient definition of isochrones, based on which, (2) we provide several easy-to-parallelize, scalable algorithmic approaches for faster computation. By extensive experimental analysis, we demonstrate that our techniques enable fast isochrone computation within milliseconds even on continental networks, significantly faster than the state-of-the-art.

1 Introduction

Online map services, navigation systems, and other route planning and location-based applications have gained wide usage, driven by significant advances [2] in shortest path algorithms for, e. g., location-to-location, many-to-many, POI, or kNN queries. Less attention has been given to the fast computation of *isochrones*, despite its relevance in urban planning [3,23,24,33,35], geomarketing [17], range visualization for (electric) vehicles [4,28], and other applications [30].

Interestingly, there is no canonical definition of isochrones in the literature. A unifying property, however, is the consideration of a range limit (time or some other limited resource), given only a source location for the query and no specific target. As a basic approach, a pruned variant of Dijkstra's algorithm [16] can be used to compute shortest path distances to all vertices within range. Newer approaches [18,23,24] still subscribe to the same model (computing distances). However, for the applications mentioned above it suffices to identify only the set of vertices or edges within range (and no distances). Moreover, for visualization [4] it serves to find just the vertices and edges on the boundary of the range. Exploiting these observations, we derive new approaches for faster computation of isochrones.

Supported by the EU FP7 under grant agreement no. 609026 (project MOVE-SMART).

A.V. Goldberg and A.S. Kulikov (Eds.): SEA 2016, LNCS 9685, pp. 17–32, 2016.
DOI: 10.1007/978-3-319-38851-9_2

Related Work. Despite its low asymptotic complexity, Dijkstra's algorithm [16] is too slow in practice. *Speedup techniques* [2] accelerate online shortest-path queries with data *preprocessed* in an offline phase. Many employ *overlay edges (shortcuts)* that maintain shortest path distances, allowing queries to skip parts of the graph. Contraction Hierarchies (CH) [27] contracts vertices in increasing order of importance, creating shortcuts between yet uncontracted neighbors. Customizable Route Planning (CRP) [7] adds shortcuts between separators of a multilevel partition [10,29,31]. As separators are independent of routing costs, CRP offers fast, dynamic *customization* of preprocessed data to a new cost metric (e. g., user preferences, traffic updates). Customizable CH (CCH) was evaluated in [14,15].

While proposed for point-to-point queries, both CH and CRP can be extended to other scenarios. Scanning the hierarchy (induced by a vertex order or multi-level partition, respectively) in a final top-down sweep enables *one-to-all* queries: PHAST [5] applies this to CH, GRASP [18] to CRP. For *one-to-many* queries, RPHAST [5] and reGRASP [18] restrict the downward search by initial target selection. POI, kNN, and similar queries are possible [1,12,19,22,25,26,32].

Since the boundary of an isochrone is not known in advance but part of the query output, target selection (as in one-to-many queries) or backward searches (as in [19]) are not directly applicable in our scenario. To the best of our knowledge, the only speedup technique extended to isochrone queries is GRASP.[1] However, isoGRASP [18] computes distances to all vertices in range, which is more than we require. MINE [23] and MINEX [24] consider multimodal networks (including road and public transit), however, due to the lack of preprocessing, running times are prohibitively slow, even on instances much smaller than ours.

Our Contribution. We give a compact definition of isochrones that serves the applications mentioned above, but requires no output of distances (Sect. 2). We propose several techniques that enable fast computation of isochrones and are easy to parallelize. First, we describe a new algorithm based on CRP (Sect. 3). Moreover, we present a faster variant of isoGRASP [18], exploiting that distances are not required (Sect. 4). Then, we introduce novel approaches that combine graph partitions with variants of (R)PHAST (Sect. 5). Our experimental evaluation (Sect. 7) on large, realistic input reveals that our techniques compute isochrones in a few milliseconds, clearly outperforming the state-of-the-art.

2   Problem Statement and Basic Approach

Let $G = (V, E, \text{len})$ be a directed, weighted graph, representing the road network, with *length function* $\text{len}\colon E \to \mathbb{R}_{\geq 0}$, representing, e. g., travel time. Denote

---

[1] Extension of CRP to isochrones is outlined in a patent (US Patent App. 13/649,114; http://www.google.com/patents/US20140107921), however, in a simpler than our intended scenario. Furthermore, the approach was neither implemented nor evaluated.

by $d\colon V \times V \mapsto \mathbb{R}_{\geq 0}$ the associated shortest path distance. We assume that G is *strongly connected*. Our *isochrone problem* takes as input a source $s \in V$ and a limit $\tau \in \mathbb{R}_{\geq 0}$. We say that a vertex $v \in V$ is *in range* if $d(s,v) \leq \tau$, else it is *out of range*. We define the output of the isochrone problem as the set of all *isochrone edges* that separate vertices in range from those out of range. Observe that these are the edges $(u,v) \in E$ with exactly one endpoint in range. To distinguish, we call e *outward (isochrone)* if and only if $d(s,u) \leq \tau, d(s,v) > \tau$ and *inward (isochrone)* if and only if $d(s,u) > \tau, d(s,v) \leq \tau$. This set of edges compactly represents the area in range [4]. However, all approaches presented below can be modified to serve other output definitions (requiring, e. g., the set of vertices in range); see Sect. 6. In what follows, we first describe a basic approach for the isochrone problem as specified above.[2] Afterwards, we propose speedup techniques that employ offline preprocessing on the graph G to quickly answer online queries consisting of a source $s \in V$ and a limit $\tau \in \mathbb{R}_{\geq 0}$. We distinguish *metric-independent* preprocessing (must be run when the topology of the input graph changes) and *metric-dependent* customization (only the length function changes).

Basic Approach. Dijkstra's algorithm [16] computes distances $d(s,v)$ from a source s to all $v \in V$. It maintains *distance labels* $d(\cdot)$ for each vertex, initially set to ∞ (except $d(s) = 0$). In each iteration, it extracts a vertex u with minimum $d(u)$ from a priority queue (initialized with s) and *settles* it. At this point, $d(u)$ is *final*, i. e., $d(u) = d(s,u)$. It then *scans* all edges (u,v): If $d(u) + \mathrm{len}(u,v) < d(v)$, it updates $d(v)$ accordingly and adds (or updates) v in the queue. For our problem setting, *isoDijkstra* can be stopped once the distance label of the minimum element in the queue exceeds the limit τ (*stopping criterion*). Then, outward isochrone edges are easily determined: We sweep over all vertices left in the queue, which must be out of range, outputting incident edges where the other endpoint is in range. Inward isochrone edges can be determined during the same sweep if we apply the following modification to the graph search. When settling a vertex u, we also scan incoming edges (v,u). If $d(v) = \infty$, we insert v into the queue with a key of infinity. Thereby, we guarantee that for both types of isochrone edges the unreachable endpoint is contained in the queue when the search terminates.

Partitions. Below, we propose speedup techniques based on graph partitions. Formally, a *(vertex) partition* is a family $\mathcal{V} = \{V_1, \dots, V_k\}$ of *cells* $V_i \subseteq V$, such that $V_i \cap V_j = \emptyset$ for $i \neq j$ and $\bigcup_{i=1}^{k} V_i = V$. A *(nested) multilevel partition* with L levels is a family $\Pi = \{\mathcal{V}^1, \dots, \mathcal{V}^L\}$ of partitions of nested cells, i. e., for each level $\ell \leq L$ and cell $V_i^\ell \in \mathcal{V}^\ell$, there is a cell $V_j^{\ell+1} \in \mathcal{V}^{\ell+1}$ at level $\ell + 1$ with $V_i^\ell \subseteq V_j^{\ell+1}$. For consistency, we define $\mathcal{V}^0 := \{\{v\} \mid v \in V\}$ (the trivial partition where each vertex has its own cell) and $\mathcal{V}^{L+1} := \{V\}$ (the trivial single-cell partition). An edge $(u,v) \in E$ is a *boundary edge* (u and v are *boundary*

---

[2] Strictly speaking, iso*chrone* implies time as a resource. While *isoline* or *isocontour* would be more precise, we have settled for the term most common in the literature.

vertices) on level ℓ, if u and v are in different cells of $\mathcal{V}^\ell$. Similar to vertex partitions, we define *edge partitions* $\mathcal{E} = \{E_1, \ldots, E_k\}$, with $E_i \cap E_j = \emptyset$ for $i \neq j$ and $\bigcup_{i=1}^k E_i = E$. A vertex $v \in V$ is *distinct* (wrt. $\mathcal{E}$) if all its incident edges belong to the same cell, else v is a *boundary vertex* or *ambiguous*.

3   IsoCRP

The three-phase workflow of CRP [7] distinguishes preprocessing and metric customization. Preprocessing finds a (multilevel) vertex partition of the road network, inducing for each level ℓ an *overlay graph* H^ℓ containing all boundary vertices and boundary edges wrt. $\mathcal{V}^\ell$, and *shortcut edges* between pairs of boundary vertices that belong to the same cell $V_i^\ell \in \mathcal{V}^\ell$. Metric customization computes the lengths of all shortcuts. The basic idea of *isoCRP* is to run isoDijkstra on the overlay graphs. Thus, we use shortcuts to skip cells that are entirely in range, but *descend* into lower levels in cells that intersect the isochrone frontier, to determine isochrone edges. There are two major challenges. First, descending into cells where shortcuts exceed the limit τ is not sufficient (we may miss isochrone edges that are part of no shortcut, but belong to shortest paths leading into the cell), so we have to precompute additional information. Second, descents into cells must be consistent for all boundary vertices (i. e., we have to descend at all vertices), motivating two-phase queries.

Customization. Along the lines of plain CRP, we obtain shortcut lengths by running Dijkstra's algorithm restricted to the respective cell. Additionally, we make use of the same searches to compute *eccentricities* for all boundary vertices. Given a boundary vertex u in a cell V_i^ℓ, its (level-ℓ) eccentricity, denoted $\mathrm{ecc}_\ell(u)$, is the maximum finite distance to some $v \in V_i^\ell$ in the subgraph induced by V_i^ℓ. This subgraph is not strongly connected in general (i. e., some vertices may be unreachable), but restricting eccentricities to cells allows fast customization.

At the lowest level, the eccentricity of a boundary vertex u is the distance label of the last vertex settled in the search from u. To accelerate customization, previously computed overlays are used to obtain shortcuts on higher levels. We compute *upper bounds* on eccentricities for those levels. When settling a vertex v, we check if the sum of the label $d(v)$ and $\mathrm{ecc}_{\ell-1}(v)$ exceeds the current bound on $\mathrm{ecc}_\ell(u)$ and update it if needed. Shortcuts of a cell are represented as a square matrix for efficiency, and storing eccentricities adds a single column to them.

To improve data locality and simplify index mapping, vertices are reordered in descending order of level during preprocessing, breaking ties by cell [7].

Queries. We say that a cell is *active* if its induced subgraph contains at least one isochrone edge. Given a source $s \in V$ and a limit τ, queries work in two phases. The first phase determines active cells, while the second phase descends into active cells to determine isochrone edges. The *upward phase* runs isoDijkstra on the search graph consisting of the union of the top-level overlay and all subgraphs induced by cells containing s. To determine active cells, we maintain two flags $i(\cdot)$ (initially false) and $o(\cdot)$ (initially true) per cell and level, to indicate whether the cell contains at least one vertex that is *in* or *out* of range, respectively. When settling a vertex $u \in V_i^\ell$, we set $i(V_i^\ell)$ to true if $d(u) \le \tau$. Next, we check whether $d(u) + \mathrm{ecc}_\ell(u) \le \tau$. Observe that this condition is not sufficient to unset $o(V_i^\ell)$, because $\mathrm{ecc}_\ell(u)$ was computed in the subgraph of V_i^ℓ. If this subgraph is not strongly connected, $d(u) + \mathrm{ecc}_\ell(u)$ is not an upper bound on the distance to any vertex in V_i^ℓ in general. Therefore, when scanning an outgoing shortcut (u, v) with length ∞ (such shortcuts exist due to the matrix representation), we also check whether $d(v) + \mathrm{ecc}_\ell(v) \le \tau$. If the condition holds for u and all boundary vertices v unreachable from u (wrt. V_i^ℓ), we can safely unset $o(V_i^\ell)$. Toggled flags are *final*, so we no longer need to perform any checks for them. After the upward phase finished, cells V_i^ℓ that have both $i(V_i^\ell)$ and $o(V_i^\ell)$ set are active (isochrone edges are only contained in cells with vertices both in and out of range).

The *downward phase* has L subphases. In descending order of level, and for every active cell at the current level ℓ, each subphase runs isoDijkstra restricted to the respective cell in $H_{\ell-1}$. Initially, all boundary vertices are inserted into the queue with their distance labels according to the previous phase as key. As before, we check eccentricities on-the-fly to mark active cells for the next subphase. Isochrone edges are determined at the end of each isoDijkstra search (see Sect. 2). On overlays, only boundary edges are reported.

Parallelization. For faster customization, cells of each level are processed in parallel [7]. During queries, the (much more expensive) downward phase is parallelized in a natural way, as cells at a certain level can be handled independently. We assign cells to threads and synchronize results between subphases. To reduce the risk of false sharing, we assign blocks of consecutive cells (wrt. vertex ordering) to the same thread. Moreover, to reduce synchronization overhead, we process cells on lower levels in a top-down fashion within the same thread.

4   Faster IsoGRASP

GRASP [18] extends CRP to batched query scenarios by storing for each level-ℓ boundary vertex, $0 \le \ell < L$, (incoming) *downward shortcuts* from boundary vertices of its supercell at level $\ell + 1$. Customization follows CRP, collecting downward shortcuts in a separate *downward graph* $H^{\downarrow}$. Original *isoGRASP* [18] runs Dijkstra's algorithm on the overlays (as in CRP), marks all in-range top-level cells, and propagates distances in marked cells from boundary vertices to those at the levels below in a sweep over the corresponding downward shortcuts. We accelerate isoGRASP significantly by making use of eccentricities.

Customization. Metric customization of our variant of isoGRASP is similar to isoCRP, computing shortcuts and eccentricities with Dijkstra's algorithm as in Sect. 3. We obtain downward shortcuts in the same Dijkstra searches. We apply edge reduction (removing shortcuts via other boundary vertices) [18] to downward shortcuts, but use the matrix representation for overlay shortcuts.

Queries. As in isoCRP, queries run two phases, with the *upward phase* being identical to the one described in Sect. 3. Then, the *scanning phase* handles levels from top to bottom in L subphases to process active cells. For an active level-ℓ cell V_i^ℓ, we sweep over its *internal* vertices (i. e., all vertices of the overlay $H_{\ell-1}$ that lie in V_i^ℓ and are no level-ℓ boundary vertices). For each internal vertex v, its incoming downward shortcuts are scanned, obtaining the distance to v. To determine active cells for the next subphase, we maintain flags $\mathrm{i}(\cdot)$ and $\mathrm{o}(\cdot)$ as in isoCRP. This requires checks at all boundary vertices that are unreachable from v within $V_i^{\ell-1}$. We achieve some speedup by precomputing these vertices, storing them in a separate adjacency array.

Similar to isoCRP, the upward phase reports all (original) isochrone edges contained in its search graph. For the remaining isochrone edges, we sweep over internal vertices and their incident edges a second time after processing a cell in the scanning phase. To avoid duplicates and to ensure that endpoints of examined edges have correct distances, we skip edges leading to vertices with higher indices. Both queries and customization are parallelized in the same fashion as isoCRP.

5   IsoPHAST

Preprocessing of PHAST [5] contracts vertices in increasing order of (heuristic) importance, as in the point-to-point technique CH [27]. To contract a vertex, shortcut edges are added between yet uncontracted neighbors to preserve distances, if necessary. Vertices are assigned *levels* $\ell(\cdot)$, initially set to zero. When contracting u, we set $\ell(v) = \max\{\ell(v), \ell(u) + 1\}$ for each uncontracted neighbor v. Given the set E^+ of all shortcuts added during preprocessing, PHAST handles one-to-all queries from some given source s as follows. During the *forward CH search*, it runs Dijkstra's algorithm on $G^\uparrow = (V, E^\uparrow)$, $E^\uparrow = \{(u,v) \in E \cup E^+ : \ell(u) < \ell(v)\}$. The subsequent *downward phase* is a linear sweep over all vertices in descending order of level, reordered accordingly during preprocessing. For each vertex, it scans its incoming edges in $E^\downarrow = \{(u,v) \in E \cup E^+ : \ell(u) > \ell(v)\}$ to update distances. Afterwards, distances from s to all $v \in V$ are known. RPHAST [9] is tailored to one-to-many queries with given target sets T. It first extracts the relevant subgraph $G_T^\downarrow$ that is reachable from vertices in T by a backward search in $G^\downarrow = (V, E^\downarrow)$. Then, it runs the linear sweep for $G_T^\downarrow$.

Our *isoPHAST* algorithm builds on (R)PHAST to compute isochrones. Since the targets are not part of the input, we use graph partitions to restrict the subgraph that is examined for isochrone edges. Queries work in three phases, in which we (1) run a forward CH search, (2) determine active cells, and (3) perform linear sweeps over all active cells as in PHAST. Below, we describe preprocessing of isoPHAST, before proposing different strategies to determine active cells.

First, we find a (single-level) partition $\mathcal{V} = \{V_1, \ldots, V_k\}$ of the road network and reorder vertices such that boundary vertices (or *core* vertices) are pushed to the front, breaking ties by cell (providing the same benefits as in CRP). Afterwards, we use CH to contract all cell-induced subgraphs, but leave core vertices *uncontracted*. Non-core vertices inside cells are reordered according to their CH levels to enable linear downward sweeps. The output of preprocessing consists of an *upward graph* $G^\uparrow$, containing for each cell all edges leading to vertices of higher level, added shortcuts between core vertices, and all boundary edges. We also obtain a *downward graph* $G^\downarrow$ that stores for each non-core vertex its incoming edges from vertices of higher level. Further steps of preprocessing depend on the query strategy and are described below.

IsoPHAST-CD. Our first strategy (Core-Dijkstra) performs isoDijkstra on the core graph to determine active cells. This requires eccentricities for core vertices, which are obtained during preprocessing as follows. To compute $\mathrm{ecc}(u)$ for some vertex u, we run (as last step of preprocessing) Dijkstra's algorithm on the subgraph induced by all core vertices of $G^\uparrow$ in the cell V_i of u, followed by a linear sweep over the internal vertices of V_i. When processing a vertex v during this sweep, we update the eccentricity of u to $\mathrm{ecc}(u) = \max\{\mathrm{ecc}(u), d(v)\}$.

Queries start by running isoDijkstra from the source in $G^\uparrow$. Within the source cell, this corresponds to a forward CH search. At core vertices, we maintain flags $\mathtt{i}(\cdot)$ and $\mathtt{o}(\cdot)$ to determine active cells (as described in Sect. 3, using an adjacency array to store unreachable core neighbors as in Sect. 4). If the core is not reached, only the source cell is set active. Next, we perform for each active cell a linear sweep over its internal vertices, obtaining distances to all vertices that are both in range and contained in an active cell.

Isochrone edges crossing cell boundaries are added to the output during the isoDijkstra search, whereas isochrone edges connecting non-core vertices are obtained in the linear sweeps as follows. When scanning incident edges of a vertex v, neighbors at higher levels have final distance labels. Moreover, the label of v is final after scanning incoming edges $(u, v) \in G^\downarrow$. Thus, looping through incoming original edges a second time suffices to find the remaining isochrone edges. Since original edges $(v, u) \in E$ to vertices u at higher levels are not contained in $G^\downarrow$ in general, we add dummy edges of length ∞ to $G^\downarrow$ to ensure that neighbors in G are also adjacent in $G^\downarrow$.

isoPHAST-CP. Instead of isoDijktra, our second strategy (Core-PHAST) performs a linear sweep over the core. Eccentricities are precomputed after generic preprocessing as described above. Next, we use CH preprocessing to contract vertices in the core, and reorder core vertices according to their levels. Finally, we update $G^\uparrow$ and $G^\downarrow$ by adding core shortcuts.

Queries strictly follow the three-phase pattern discussed above. We first run a forward CH search in $G^\uparrow$. Then, we determine active cells and compute distances for all core vertices in a linear sweep over the core. Again, we maintain flags $\mathtt{i}(\cdot)$ and $\mathtt{o}(\cdot)$ for core vertices (cf. Section 3) and use an adjacency array storing unreachable core neighbors (cf. Section 4). To find isochrone edges between core vertices, we insert dummy edges into the core to preserve adjacency. The third phase (linear sweeps over active cells) is identical to isoPHAST-CD.

isoPHAST-DT. Our third strategy (Distance Table) uses a *distance (bounds) table* to accelerate the second phase, determining active cells. Working with such tables instead of a dedicated core search benefits from *edge partitions*, since the unique assignment of edges to cells simplifies isochrone edge retrieval. Given a partition $\mathcal{E} = \{E_1, \ldots, E_k\}$ of the edges, the table stores for each pair E_i, E_j of cells a lower bound $\underline{d}(E_i, E_j)$ and an upper bound $\overline{d}(E_i, E_j)$ on the distance from E_i to E_j, i. e., $\underline{d}(E_i, E_j) \leq d(u, v) \leq \overline{d}(E_i, E_j)$ for all $u \in E_i$, $v \in E_j$ (we abuse notation, saying $u \in E_i$ if u is an endpoint of at least one edge $e \in E_i$). Given a source $s \in E_i$ (if s is ambiguous, pick any cell containing s) and a limit τ, cells E_j with $\underline{d}(E_i, E_j) \leq \tau < \overline{d}(E_i, E_j)$ are set active.

Preprocessing first follows isoPHAST-CP, with three differences: (1) We use an edge partition instead of a vertex partition; (2) Eccentricities are computed on the *reverse* graph, with Dijkstra searches that are not restricted to cells but stop when all boundary vertices of the current cell are reached; (3) After computing eccentricities, we recontract the whole graph using a CH order (i. e., contraction of core vertices is not delayed), leading to sparser graphs $G^\uparrow$ and $G^\downarrow$. Afterwards, to quickly compute (not necessarily tight) distance bounds, we run for each cell E_i a (multi-source) forward CH search in $G^\uparrow$ from all boundary vertices of E_i. Then, we perform a linear sweep over $G^\downarrow$, keeping track of the minimum and maximum distance label per target cell. This yields, for all cells, lower bounds $\underline{d}(E_i, \cdot)$, and upper bounds on the distance from *boundary* vertices of E_i to each cell. To obtain the desired bounds $\overline{d}(E_i, \cdot)$, we increase the latter values by the *(backward) boundary diameter* of E_i, i. e., the maximum distance from any vertex in E_i to a boundary vertex of E_i. This diameter equals the maximum eccentricity of the boundary vertices of E_i on the reverse graph (which we computed before). As last step of preprocessing, we extract and store the relevant search graph $G_i^\downarrow$ for each $E_i \in \mathcal{E}$. This requires a target selection phase as in RPHAST for each cell, using all (i. e., distinct and ambiguous) vertices of a cell as input.

Queries start with a forward CH search in $G^\uparrow$. Then, active cells are determined in a sweep over one row of the distance table. The third phase performs a linear sweep over $G_i^\downarrow$ for each active cell E_i, obtaining distances to all its vertices. Although vertices can be contained in multiple search graphs, distance labels do not need to be reinitialized between sweeps, since the source remains unchanged. To output isochrone edges, we proceed as before, looping through incoming downward edges twice (again, we add dummy edges to $G_i^\downarrow$ for correctness). To avoid duplicates (due to vertices contained in multiple search graphs), edges in $G_i^\downarrow$ have an additional flag to indicate whether the edge belongs to E_i.

Search graphs may share vertices, which increases memory consumption and slows down queries (e. g., the vertex with maximum level is contained in every search graph). We use *search graph compression*, i. e., we store the topmost vertices of the hierarchy (and their incoming edges) in a separate graph $G_c^\downarrow$ and remove them from all $G_i^{\downarrow}$. During queries, we first perform a linear sweep over $G_c^{\downarrow}$ (obtaining distances for all $v \in G_c^{\downarrow}$), before processing search graphs of active cells. The size of $G_c^{\downarrow}$ (more precisely, its number of vertices) is a tuning parameter.

Parallelization. The first preprocessing steps are executed in parallel, namely, building cell graphs, contracting non-core vertices, inserting dummy edges, and reordering non-core vertices by level. Afterwards, threads are synchronized, and $G^{\uparrow}$ and $G^{\downarrow}$ are built sequentially. Eccentricities are again computed in parallel. Since our CH preprocessing is sequential, the core graph is contracted in a single thread (if needed). Computation of distance bounds is parallelized (if needed).

Considering queries, the first two phases are run sequentially. Both isoDijkstra and the forward CH search are difficult to parallelize. Executing PHAST (on the core) in parallel does not pay off (the core is rather dense, resulting in many levels). Distance table operations, on the other hand, are very fast, and parallelization is not necessary. In the third phase, however, active cells can be assigned to different threads. We store a copy of the graph $G^{\downarrow}$ once per NUMA node for faster access during queries. Running the third phase in parallel can make the second phase of isoPHAST-CP a bottleneck. Therefore, we alter the way of computing flags $i(\cdot)$ and $o(\cdot)$. When settling a vertex $v \in V_i$, we set $i(V_i)$ if $d(v) \leq \tau$, and $o(V_i)$ if $d(v) + \text{ecc}(v) > \tau$. Note that these checks are less accurate (more flags are toggled), but we no longer have to check unreachable boundary vertices. Correctness of isoPHAST-CP is maintained, as no stopping criterion is applied and $\max_{v \in V_i}(d(v) + \text{ecc}(v))$ is a valid upper bound on the distance to each vertex in V_i. Hence, no active cells are missed.

6   Alternative Outputs

Driven by our primary application, visualizing the cruising range of a vehicle, we introduced a compact, yet sufficient representation of isochrones. However, all approaches can be adapted to produce a variety of other outputs, without increasing their running times significantly. As an example, we modify our algorithms to output a list of all vertices in range (instead of isochrone edges).

Even without further modifications, we can check in constant time if a vertex is in range after running the query. Consider the (top-level) cell V_i of a vertex v. If $i(V_i)$ is not set, the cell contains no in-range vertices and v must be out of range. Similarly, if $o(V_i)$ is not set, v is in range. If both flags are set, we run the same check for the cell containing v on the level below. If both flags are set for the cell on level 1 containing v, we check if the distance label of v exceeds the time limit (since all cells considered are active, the distance label of v is correct).

A simple approach to output the vertices in range performs a sweep over all vertices and finds those in range as described above. We can do better by collecting vertices in range on-the-fly. During isoDijkstra searches and when scanning active cells, we output each scanned vertex that is in range. In the scanning phase we also add all internal vertices for cells V_i where $o(V_i)$ is not set.

7   Experiments

Our code is written in C++ (using OpenMP) and compiled with g++ 4.8.3 -O3. Experiments were conducted on two 8-core Intel Xeon E5-2670 clocked at 2.6 Ghz, with 64 GiB of DDR3-1600 RAM, 20 MiB of L3 and 256 KiB of L2 cache. Results are checked against a reference implementation (isoDijkstra) for correctness.

Input and Methodology. Experiments were done on the European road network (with 18 million vertices and 42 million edges) made available for the 9th DIMACS Implementation Challenge [13], using travel times in seconds as edge lengths.

We implemented CRP following [7], with a matrix-based clique representation. Our GRASP implementation applies implicit initialization [18] and (downward) shortcut reduction [20]. The CH preprocessing routine follows [27], but takes priority terms and hop limits from [5]. We use PUNCH [8] to generate multilevel partitions for isoCRP/isoGRASP, and Buffoon [37] to find single-level partitions for isoPHAST. Edge partitions are computed following the approach in [36, 38].

We report parallel customization times, and both sequential and parallel query times. Parallel execution uses all available cores. Customization times for isoPHAST exclude partitioning, since it is metric-independent. For queries, reported figures are averages of 1 000 random queries (per individual time limit τ).

Tuning Parameters. We conducted preliminary studies to obtain reasonable parameters for partitions and search graph compression. For isoCRP/isoGRASP, we use the 4-level partition from [6], with maximum cell sizes of 2^8, 2^{12}, 2^{16}, 2^{20}, respectively. Although [18] uses 16 levels, resorting to a 4-level partition had only minor effects in preliminary experiments (similar observations are made in [19]).

For sequential isoPHAST-CD (CP) queries, a partition with $k = 2^{12}$ (2^{11}) cells yields best query times. For fewer cells (i. e., coarser partitions), the third query phase scans a large portion of the graph and becomes the bottleneck. Using more fine-grained partitions yields a larger core graph, slowing down the second query phase. Consequently, fewer cells ($k = 256$) become favorable when queries are executed in parallel (as the third phase becomes faster due to parallelization).

For isoPHAST-DT, similar effects occur for different values of k. Moreover, search graph compression has a major effect on query times (and space consumption). If there are few vertices in $G_c^\downarrow$, then vertices at high levels occur in search graphs of multiple cells, but large $G_c^\downarrow$ cause unnecessary vertex scans. Choosing $k = 2^{14}$ (2^{12}) and $|G_c^\downarrow| = 2^{16}$ (2^{13}) yields fastest sequential (parallel) queries.

Evaluation. Table 1 summarizes the performance of all algorithms discussed in this paper, showing figures on customization and queries. We report query times for medium-range ($\tau = 100$) and long-range time limits ($\tau = 500$, this is the hardest limit for most approaches, since it maximizes the number of isochrone edges).

Table 1. Performance of our algorithms. We report parallel customization time and space consumption (space per additional metric is given in brackets, if it differs). The table shows the average number of settled vertices (Nmb. settled, in thousands) and running times of sequential and parallel queries, using time limits $\tau = 100$ and $\tau = 500$. Best values (except Dijkstra wrt. space) are highlighted in bold.

Algorithm	Thr.	Custom			$\tau = 100$ min		$\tau = 500$ min	
		Time [s]	Space	[MiB]	Nmb. settled	Time [ms]	Nmb. settled	Time [ms]
isoDijkstra	1	–	646		460 k	68.32	7041 k	1184.06
isoCRP	1	**1.70**	900	(138)	101 k	15.44	354 k	60.67
isoGRASP	1	2.50	1856	(1094)	120 k	10.06	387 k	37.77
isoPHAST-CD	1	26.11	785		440 k	**6.09**	1501 k	31.63
isoPHAST-CP	1	1221.84	**781**		626 k	15.02	2029 k	31.00
isoPHAST-DT	1	1079.11	2935		597 k	9.96	1793 k	**24.80**
isoCRP	16	**1.70**	900	(138)	100 k	2.73	354 k	7.86
isoGRASP	16	2.50	1856	(1094)	120 k	2.35	387 k	5.93
isoPHAST-CD	16	38.07	769		918 k	**1.61**	4578 k	8.22
isoPHAST-CP	16	1432.39	**766**		944 k	4.47	5460 k	7.86
isoPHAST-DT	16	865.50	1066		914 k	1.74	2979 k	**3.80**

As expected, techniques based on multilevel overlays provide better customization times, while isoPHAST achieves the lowest query times (CD for medium-range and DT for long-range queries, respectively). Customization of isoCRP and isoGRASP is very practical (below three seconds). The lightweight preprocessing of isoPHAST-CD pays off as well, allowing customization in less than 30 s. The comparatively high preprocessing times of isoPHAST-CP and DT are mainly due to expensive core contraction. Still, metric-dependent preprocessing is far below half an hour, which is suitable for applications that do not require real-time metric updates. Compared to isoCRP, isoGRASP requires almost an order of magnitude of additional space for the downward graph (having about 110 million edges).

Executed sequentially, all approaches take well below 100 ms, which is significantly faster than isoDijkstra. The number of settled vertices is considerably larger for isoPHAST, however, data access is more cache-efficient. IsoPHAST provides faster queries than the multilevel overlay techniques for both limits, with the execption of isoPHAST-CP for small ranges (since the whole core graph is scanned). Again, the performance of isoPHAST-CD is quite notable, providing the fastest queries for (reasonable) medium-range limits and decent query times for the long-range limit. Finally, query times of isoPHAST-DT show best scaling behavior, with lowest running times for hardest queries.

The lower half of Table 1 reports parallel times for the same set of queries. Note that preprocessing times of isoPHAST change due to different parameter choices. Most approaches scale very well with the number of threads, providing a speedup of (roughly) 8 using 16 threads. Note that factors (according to the table) are much lower for isoPHAST, since we use tailored partitions for sequential queries. In fact, isoPHAST-DT scales best when run on the same preprocessed data (speedup of 11), since its sequential workflow (forward CH search, table scan) is very fast. Considering multilevel overlay techniques, isoGRASP scales worse than isoCRP (speedup of 6.5 compared to 7.7), probably because it is memory bandwidth bounded (while isoCRP comes with more computational overhead). Consequently, isoGRASP benefits greatly from storing a copy of the downward graph on each NUMA node. As one may expect, speedups are slightly lower for medium-range queries. The isoPHAST approaches yield best query times, below 2 ms for medium-range queries, and below 4 ms for the long-range limit. To summarize, all algorithms enable queries fast enough for practical applications, with speedups of more than two orders of magnitude compared to isoDijkstra.

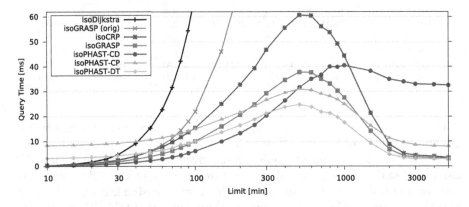

Fig. 1. Sequential query times for various time limits, ranging from 10 to (roughly) 4700 min (the diameter of our input graph).

Figure 1 shows how (sequential) query times scale with the time limit. For comparability, we also show (sequential) query times of original isoGRASP as described in [18] (computing distances to all in-range vertices, but no isochrone edges). Running times of all proposed algorithms (except isoDijkstra and original isoGRASP) follow a characteristic curve. Times first increase with the limit τ (the isochrone frontier is extended, intersecting more active cells), before dropping again once τ exceeds 500 min (the isochrone reaches the boundary of the network, decreasing the number of active cells). For $\tau > 4710$ min, all vertices are in range, making queries very fast, as there are no active cells. For small τ, the multilevel overlay techniques and isoPHAST-CD are fastest. IsoPHAST-CP is slowed down by the linear sweep over the core graph (taking about 6 ms, independent of τ), while isoPHAST-DT suffers from distance bounds not being tight. However, since Dijkstra's algorithm does not scale well, isoPHAST-CD becomes the slowest approach for large τ (while the other isoPHAST techniques benefit from good scaling behavior). Considering multilevel overlays, our isoGRASP is up to almost twice as fast as isoCRP, providing a decent trade-off between customization effort and query times. Note that while isoDijkstra is fast enough for some realistic time limits, it is not robust to user inputs. When executed in

Table 2. Impact of different outputs on the performance of isoCRP, isoGRASP, and a variant of isoPHAST (CP). We report sequential (seq.) and parallel (par.) query times as well as output size (# out., in thousands for vertices in range) when computing isochrone edges and vertices in range.

Algorithm	Limit [min]	Isochrone edges			Vertices in range		
		# out.	seq. [ms]	par. [ms]	# out.	seq. [ms]	par. [ms]
isoCRP	100	5937	15.44	2.73	460 k	15.83	2.77
	500	14718	60.67	7.86	7041 k	76.35	9.26
	5000	0	3.42	3.17	18010 k	46.64	6.64
isoGRASP	100	5937	10.06	2.35	460 k	11.07	2.50
	500	14718	37.77	5.93	7041 k	56.83	7.57
	5000	0	3.08	3.10	18010 k	46.09	6.44
isoPHAST	100	5937	15.02	4.47	460 k	16.40	4.70
	500	14718	31.00	7.86	7041 k	49.57	9.67
	5000	0	7.96	3.61	18010 k	50.86	7.03

parallel, query times follow the same characteristic curves (not reported in the figure). The linear sweep in the second phase of isoPHAST-CP becomes slightly faster, since the core is smaller (due to a different partition).

Alternative Outputs. Table 2 compares query times when computing different outputs (isochrone edges and vertices in range, respectively). For medium-range time limits ($\tau = 100$ min), both sequential and parallel query times increase by less than 10 %. When using long-range limits, where roughly half of the vertices are in range, sequential and parallel queries are slower by a factor of about 1.5, but still significantly faster than the original isoGRASP algorithm. Only when considering the graph diameter as time limit, sequential queries are significantly slower when computing all vertices in range, since variants reporting only isochrone edges can already terminate after the (very fast) upward phase.

Comparison with Related Work. Since we are not aware of any work solving our compact problem formulation, we cannot compare our algorithms directly to competitors. Hence, to validate the efficiency of our code, we compare our implementations of some basic building blocks to original publications. Table 3 reports running times for our implementations of Dijkstra's algorithm, GRASP, PHAST and RPHAST on one core

Table 3. Running times [ms] of basic one-to-all and one-to-many building blocks.

Algorithm	[our]	[9, 19]
Dij (1-to-all)	2653.18	–
PHAST	144.16	136.92
GRASP	171.11	169.00
Dij (1-to-many)	7.34	7.43
RPHAST (select)	1.29	1.80
RPHAST (query)	0.16	0.17

of a 4-core Intel Xeon E5-1630v3 clocked at 3.7 GHz, with 128 GiB of DDR4-2133 RAM, 10 MiB of L3 and 256 KiB of L2 cache (chosen as it most closely resembles the machines used in [9,19]). For comparison, we report running times (as-is) from [9,19]. For the one-to-many scenario, we adopt the methodology from [9], using a target and ball size of 2^{14}. Even when accounting for hardware differences, running times of our implementations are similar to the original publications.

8   Final Remarks

We proposed a compact definition of isochrones, and introduced a portfolio of speedup techniques for the resulting isochrone problem. While no single approach is best in all criteria (preprocessing effort, space consumption, query time, simplicity), the right choice depends on the application. If user-dependent metrics are needed, the fast and lightweight customization of isoCRP is favorable. Fast queries subject to frequent metric updates (e. g., due to real-time traffic) are enabled by our isoGRASP variant. If customization time below a minute is acceptable and time limits are low, isoPHAST-CD provides even faster query times. The other isoPHAST variants show best scaling behavior, making them suitable for long-range isochrones, or if customizability is not required.

Regarding future work, we are interested in integrating the computation of eccentricities into microcode [11], an optimization technique to accelerate customization of CRP. For isoPHAST, we want to separate metric-independent preprocessing and metric customization (exploiting, e. g., CCH [14]). We also explore approaches that do not (explicitly) require a partition of the road network. Another direction of research is the speedup of network Voronoi diagram computation [21,34], where multiple isochrones are grown simultaneously from a set of Voronoi generators. We are also interested in extending our speedup techniques to more involved scenarios, such as multimodal networks.

References

1. Abraham, I., Delling, D., Fiat, A., Goldberg, A.V., Werneck, R.F.: HLDB: location-based services in databases. In: Proceedings of the 20th ACM SIGSPATIAL International Symposium on Advances in Geographic Information Systems (GIS 2012), pp. 339–348. ACM Press, New York (2012)
2. Bast, H., Delling, D., Goldberg, A.V., Müller-Hannemann, M., Pajor, T., Sanders, P., Wagner, D., Werneck, R.F.: Route Planning in Transportation Networks. Technical report abs/1504.05140, ArXiv e-prints (2015)
3. Bauer, V., Gamper, J., Loperfido, R., Profanter, S., Putzer, S., Timko, I.: Computing isochrones in multi-modal, schedule-based transport networks. In: Proceedings of the 16th ACM SIGSPATIAL International Conference on Advances in Geographic Information Systems (GIS 2008), pp. 78:1–78:2. ACM Press, New York (2008)
4. Baum, M., Bläsius, T., Gemsa, A., Rutter, I., Wegner, F.: Scalable Isocontour Visualization in Road Networks via Minimum-Link Paths. Technical report abs/1602.01777, ArXiv e-prints (2016)
5. Delling, D., Goldberg, A.V., Nowatzyk, A., Werneck, R.F.: PHAST: hardware-accelerated shortest path trees. J. Parallel Distrib. Comput. **73**(7), 940–952 (2013)

6. Delling, D., Goldberg, A.V., Pajor, T., Werneck, R.F.: Customizable route planning. In: Pardalos, P.M., Rebennack, S. (eds.) SEA 2011. LNCS, vol. 6630, pp. 376–387. Springer, Heidelberg (2011)
7. Delling, D., Goldberg, A.V., Pajor, T., Werneck, R.F.: Customizable route planning in road networks. Transportation Science (2015)
8. Delling, D., Goldberg, A.V., Razenshteyn, I., Werneck, R.F.: Graph partitioning with natural cuts. In: Proceedings of the 25th International Parallel and Distributed Processing Symposium (IPDPS 2011), pp. 1135–1146. IEEE Computer Society (2011)
9. Delling, D., Goldberg, A.V., Werneck, R.F.: Faster batched shortest paths inroad networks. In: Proceedings of the 11th Workshop on Algorithmic Approachesfor Transportation Modeling, Optimization, and Systems (ATMOS 2011). OpenAccessSeries in Informatics, vol. 20, pp. 52–63. OASIcs (2011)
10. Delling, D., Holzer, M., Müller, K., Schulz, F., Wagner, D.: High-performance multi-level routing. In: The Shortest Path Problem: Ninth DIMACS Implementation Challenge, DIMACS Book, vol. 74, pp. 73–92. American Mathematical Society (2009)
11. Delling, D., Werneck, R.F.: Faster customization of road networks. In: Bonifaci, V., Demetrescu, C., Marchetti-Spaccamela, A. (eds.) SEA 2013. LNCS, vol. 7933, pp. 30–42. Springer, Heidelberg (2013)
12. Delling, D., Werneck, R.F.: Customizable point-of-interest queries in road networks. IEEE Trans. Knowl. Data Eng. **27**(3), 686–698 (2015)
13. Demetrescu, C., Goldberg, A.V., Johnson, D.S. (eds.): The Shortest Path Problem: Ninth DIMACS Implementation Challenge, DIMACS Book, vol. 74. American Mathematical Society (2009)
14. Dibbelt, J., Strasser, B., Wagner, D.: Customizable contraction hierarchies. In: Gudmundsson, J., Katajainen, J. (eds.) SEA 2014. LNCS, vol. 8504, pp. 271–282. Springer, Heidelberg (2014)
15. Dibbelt, J., Strasser, B., Wagner, D.: Customizable contraction hierarchies. ACM J. Exp. Algorithmics **21**(1), 108–122 (2016)
16. Dijkstra, E.W.: A note on two problems in connexion with graphs. Numer. Math. **1**(1), 269–271 (1959)
17. Efentakis, A., Grivas, N., Lamprianidis, G., Magenschab, G., Pfoser, D.: Isochrones, traffic and DEMOgraphics. In: Proceedings of the 21st ACM SIGSPATIAL International Conference on Advances in Geographic Information Systems (GIS 2013), pp. 548–551. ACM Press, New York (2013)
18. Efentakis, A., Pfoser, D.: GRASP. Extending graph separators for the single-source shortest-path problem. In: Schulz, A.S., Wagner, D. (eds.) ESA 2014. LNCS, vol. 8737, pp. 358–370. Springer, Heidelberg (2014)
19. Efentakis, A., Pfoser, D., Vassiliou, Y.: SALT. A unified framework for all shortest-path query variants on road networks. In: Bampis, E. (ed.) SEA 2015. LNCS, vol. 9125, pp. 298–311. Springer, Heidelberg (2015)
20. Efentakis, A., Theodorakis, D., Pfoser, D.: Crowdsourcing computing resources for shortest-path computation. In: Proceedings of the 20th ACM SIGSPATIAL International Symposium on Advances in Geographic Information Systems (GIS 2012), pp. 434–437. ACM Press, New York (2012)
21. Erwig, M.: The graph voronoi diagram with applications. Networks **36**(3), 156–163 (2000)
22. Foti, F., Waddell, P., Luxen, D.: A generalized computational framework for accessibility: from the pedestrian to the metropolitan scale. In: Proceedings of the

4th TRB Conference on Innovations in Travel Modeling. Transportation Research Board (2012)

23. Gamper, J., Böhlen, M., Cometti, W., Innerebner, M.: Defining isochrones in multi-modal spatial networks. In: Proceedings of the 20th ACM International Conference on Information and Knowledge Management (CIKM 2011), pp. 2381–2384. ACM Press, New York (2011)

24. Gamper, J., Böhlen, M., Innerebner, M.: Scalable computation of isochrones with network expiration. In: Ailamaki, A., Bowers, S. (eds.) SSDBM 2012. LNCS, vol. 7338, pp. 526–543. Springer, Heidelberg (2012)

25. Geisberger, R.: Advanced Route Planning in Transportation Networks. Ph.D. thesis, Karlsruhe Institute of Technology (2011)

26. Geisberger, R., Luxen, D., Sanders, P., Neubauer, S., Volker, L.: Fast detour computation for ride sharing. In: Proceedings of the 10th Workshop on Algorithmic Approaches for Transportation Modeling, Optimization, and Systems (ATMOS 2010). OpenAccess Series in Informatics, vol. 14, pp. 88–99. OASIcs (2010)

27. Geisberger, R., Sanders, P., Schultes, D., Vetter, C.: Exact routing in large road networks using contraction hierarchies. Transp. Sci. **46**(3), 388–404 (2012)

28. Grubwinkler, S., Brunner, T., Lienkamp, M.: Range prediction for EVs via crowd-sourcing. In: Proceedings of the 10th IEEE International Vehicle Power and Propulsion Conference (VPPC 2014), pp. 1–6. IEEE (2014)

29. Holzer, M., Schulz, F., Wagner, D.: Engineering multilevel overlay graphs for shortest-path queries. ACM J. Exp. Algorithmics **13**, 1–26 (2008)

30. Innerebner, M., Böhlen, M., Gamper, J.: ISOGA: a system for geographical reachability analysis. In: Liang, S.H.L., Wang, X., Claramunt, C. (eds.) W2GIS 2013. LNCS, vol. 7820, pp. 180–189. Springer, Heidelberg (2013)

31. Jung, S., Pramanik, S.: An efficient path computation model for hierarchically structured topographical road maps. IEEE Trans. Knowl. Data Eng. **14**(5), 1029–1046 (2002)

32. Knopp, S., Sanders, P., Schultes, D., Schulz, F., Wagner, D.: Computing many-to-many shortest paths using highway hierarchies. In: Proceedings of the 9th Workshop on Algorithm Engineering and Experiments (ALENEX 2007), pp. 36–45. SIAM (2007)

33. Marciuska, S., Gamper, J.: Determining objects within isochrones in spatial network databases. In: Catania, B., Ivanović, M., Thalheim, B. (eds.) ADBIS 2010. LNCS, vol. 6295, pp. 392–405. Springer, Heidelberg (2010)

34. Okabe, A., Satoh, T., Furuta, T., Suzuki, A., Okano, K.: Generalized network voronoi diagrams: concepts, computational methods, and applications. Int. J. Geogr. Inf. Sci. **22**(9), 965–994 (2008)

35. O'Sullivan, D., Morrison, A., Shearer, J.: Using desktop GIS for the investigation of accessibility by public transport: an isochrone approach. Int. J. Geogr. Inf. Sci. **14**(1), 85–104 (2000)

36. Pothen, A., Simon, H.D., Liou, K.P.: Partitioning sparse matrices with eigenvectors of graphs. SIAM J. Matrix Anal. Appl. **11**, 430–452 (1990)

37. Sanders, P., Schulz, C.: Distributed evolutionary graph partitioning. In: Proceedings of the 14th Meeting on Algorithm Engineering and Experiments (ALENEX 2012), pp. 16–29. SIAM (2012)

38. Schulz, C.: High Quality Graph Partitioning. Ph.D. thesis, Karlsruhe Institute of Technology (2013)

Dynamic Time-Dependent Route Planning in Road Networks with User Preferences

Moritz Baum[1], Julian Dibbelt[1(✉)], Thomas Pajor[2], and Dorothea Wagner[1]

[1] Karlsruhe Institute of Technology, Karlsruhe, Germany
{moritz.baum,julian.dibbelt,dorothea.wagner}@kit.edu
[2] Cupertino, CA, USA

Abstract. Algorithms for computing driving directions on road networks often presume constant costs on each arc. In practice, the current traffic situation significantly influences the travel time. One can distinguish traffic congestion that can be predicted using historical traffic data, and congestion due to unpredictable events, e. g., accidents. We study the *dynamic and time-dependent* route planning problem, which takes both live traffic and long-term prediction into account. We propose a practical algorithm that, while robust to user preferences, is able to integrate global changes of the time-dependent metric faster than previous approaches and allows queries in the order of milliseconds.

1 Introduction

To enable responsive route planning applications on large-scale road networks, *speedup techniques* have been proposed [1], employing *preprocessing* to accelerate Dijkstra's shortest-path algorithm [18]. A successful approach [4,9,16,21,28,30] exploits that road networks have small separators [10,22,27,40,41], computing coarsened overlays that maintain shortest path distance. An important aspect [14] in practice is the consideration of traffic patterns and incidents. In *dynamic, time-dependent* route planning, costs vary as a function of time [6,19]. These functions are derived from historic knowledge of traffic patterns [39], but have to be updated to respect traffic incidents or short-term predictions [15]. In this work, we investigate the challenges that arise when extending a separator-based overlay approach to the dynamic, time-dependent route planning scenario.

Related Work. In time-dependent route planning, there are two major query variants: (1) Given the departure time at a source, compute the *earliest arrival time (EA)* at the target; (2) compute earliest arrival times for all departure times of a day (*profile search*). Dijkstra's algorithm [18] can be extended to solve these problems for cost functions with reasonable properties [6,19,38]. However, functional representations of profiles (typically by piecewise-linear functions) are quite complex on realistic instances [13]. Many speedup techniques have been

---

Partially supported by EU grants 288094 (eCOMPASS) and 609026 (MOVE-SMART).

A.V. Goldberg and A.S. Kulikov (Eds.): SEA 2016, LNCS 9685, pp. 33–49, 2016.
DOI: 10.1007/978-3-319-38851-9_3

adapted to time-dependency. Some use (scalar) lower bounds on the travel time functions to guide the graph search [11,12,37]. TD-CALT [11] yields reasonable EA query times for approximate solutions, allowing dynamic traffic updates, but no profile search. TD-SHARC [8] offers profile search on a country-scale network. Time-dependent Contraction Hierarchies (TCH) [2] enable fast EA and profile searches on continental networks. During preprocessing, TCH computes overlays by iteratively inserting shortcuts [25] obtained from profile searches. Piecewise-linear function approximation [29] is used to reduce shortcut complexity, dropping optimality. A multi-phase extension (ATCH) restores exact results [2]. Time-dependent shortest path oracles described in [33–35] approximate distances in sublinear query time after subquadratic preprocessing. In practical experiments, however, preprocessing effort is still substantial [31,32].

TCH has been generalized to combined optimization of functional travel time and scalar, other costs [3], which poses an NP-hard problem. While this hardness result would of course impact any approach, interestingly, the experiments in [3] suggest that TCH on its own is not particularly robust against user preferences: In a scenario that amounts to the avoidance of highways, preprocessing effort doubles and query performance decreases by an order of magnitude. (Our experiments will confirm this on a non NP-hard formulation of highway avoidance.)

Other works focus on unforeseen dynamic changes (e. g., congestion due to an accident), often by enabling partial updates of preprocessed data [12,20]. Customizable Route Planning (CRP) [9] offloads most preprocessing effort to a metric-independent, separator-based phase. Preprocessed data is then *customized* to a given routing metric for the whole network within seconds or below. This also enables robust integration of user preferences. Customizable Contraction Hierarchies (CCH) [16] follows a similar approach. However, CRP and CCH handle only scalar metrics. To the best of our knowledge, non-scalar metrics for separator-based approaches have only been investigated in the context of electric vehicles (EVCRP) [5], where energy consumption depends on state-of-charge, but functional complexity is very low. On the other hand, the use of scalar approaches for handling live traffic information yields inaccurate results for medium and long distances: Such methods wrongly consider current traffic even at far away destinations—although it will have dispersed once reaching the destination. For realistic results, a combination of dynamic and time-dependent (non-scalar, functional) route planning accounts for current traffic, short-term predictions, and historic knowledge about recurring traffic patterns.

Our Contribution. We carefully extend CRP [9] to time-dependent functions. As such, we are the first to evaluate partition-based overlays on a challenging non-scalar metric. To this end, we integrate profile search into CRP's customization phase and compute time-dependent overlays. Unlike EVCRP and TCH, a naïve implementation fails: Shortcuts on higher-level overlays are too expensive to be kept in memory (and too expensive to evaluate during queries). To reduce functional complexity, we approximate overlay arcs. In fact, approximation subject to a very small error suffices to make our approach practical, in accordance to theory [23]. The resulting algorithmic framework enables interactive queries with low average and maximum error in a very realistic scenario consisting of live traffic, short-term traffic predictions, and historic traffic patterns. Moreover, it supports user preferences such as lower maximum driving speeds or the avoidance of highways. In an extensive experimental setup, we demonstrate that our approach enables integration of custom updates much faster than previous approaches, while allowing fast queries that enable interactive applications. It is also robust to changes in the metric that turn out to be much harder for previous techniques.

2 Preliminaries

A road network is modeled as a directed *graph* $G = (V, A)$ with $n = |V|$ *vertices* and $m = |A|$ *arcs*, where vertices $v \in V$ correspond to intersections and arcs $(u, v) \in A$ to road segments. An *s–t-path* P (in G) is a sequence $P_{s,t} = [v_1 = s, v_2, \ldots, v_k = t]$ of vertices such that $(v_i, v_{i+1}) \in A$. If s and t coincide, we call P a *cycle*. Every arc a has assigned a *periodic travel-time function* $f_a : \Pi \to \mathbb{R}^+$, mapping departure time within period $\Pi = [0, \pi]$ to travel time. Given a departure time τ at s, the (time-dependent) travel time $\tau_{[s,\ldots,t]}$ of an s–t-path is obtained by consecutive function evaluation, i. e., $\tau_{[s,\ldots,v_i]} = f_{(v_{i-1},v_i)}(\tau_{[s,\ldots,v_{i-1}]})$. We assume that functions are piecewise linear and represented by *breakpoints*. We denote by $|f|$ the number of breakpoints of a function f. Moreover, we define $f^{\max}$ as the maximum value of f, i.e., $f^{\max} = \max_{\tau \in \Pi} f(\tau)$. Analogously, $f^{\min}$ is the minimum value of f. A function f is *constant* if $f \equiv c$ for some $c \in \Pi$. We presume that functions fulfill the *FIFO property*, i. e., for arbitrary $\sigma \leq \tau \in \Pi$, the condition $\sigma + f(\sigma) \leq \tau + f(\tau)$ holds (waiting at a vertex never pays off). Unless waiting is allowed at vertices, the shortest-path problem becomes $\mathcal{NP}$-hard if this condition is not satisfied for all arcs [7,42]. Given two functions f, g, the *link* operation is defined as $\mathrm{link}(f, g) := f + g \circ (\mathrm{id} + f)$, where id is the identity function and $\circ$ is function composition. The result $\mathrm{link}(f, g)$ is piecewise linear again, with at most $|f| + |g|$ breakpoints (namely, at departure times of breakpoints of f and backward projections of departure times of points of g). We also define *merging* of f and g by $\mathrm{merge}(f, g) := \min(f, g)$. The result of merging piecewise linear functions is piecewise linear, and the number of breakpoints is in $\mathcal{O}(|f| + |g|)$ (containing breakpoints of the two original functions and at most one intersection per linear segment). Linking and merging are implemented by coordinated linear sweeps over the breakpoints of the corresponding functions.

The *(travel-time) profile* of a path $P = [v_1, \ldots, v_k]$ is the function $f_P : \Pi \to \mathbb{R}^+$ that maps departure time τ at v_1 to travel time on P. Starting at $f_{[v_1,v_2]} = f_{(v_1,v_2)}$, we obtain the desired profile by consecutively applying the link operation, i. e., $f_{[v_1,\ldots,v_i]} = \mathrm{link}(f_{[v_1,\ldots,v_{i-1}]}, f_{(v_{i-1},v_i)})$. Given a set $\mathcal{P}$ of s–t-paths, the corresponding *s–t-profile* is $f_{\mathcal{P}}(\tau) = \min_{P \in \mathcal{P}} f_P(\tau)$ for $\tau \in \Pi$, i. e., the *minimum profile* over all paths in $\mathcal{P}$. The s–t-profile maps departure time to minimum travel time for the given paths. It is obtained by (iteratively) merging the respective paths.

A *partition* of V is a set $\mathcal{C} = \{C_1, \ldots, C_k\}$ of disjoint vertex sets such that $\bigcup_{i=1}^{k} C_i = V$. More generally, a *nested multi-level partition* consists of sets $\{\mathcal{C}^1, \ldots, \mathcal{C}^L\}$ such that $\mathcal{C}^\ell$ is a partition of V for all $\ell \in \{1, \ldots, L\}$, and additionally for each cell C_i in $\mathcal{C}^\ell$, $\ell < L$, there is a partition $\mathcal{C}^{\ell+1}$ at level $\ell + 1$ containing a cell C_j with $C_i \subseteq C_j$. We call C_j the *supercell* of C_i. For consistency, we define $\mathcal{C}^0 = \{\{v\} \mid v \in V\}$ and $\mathcal{C}^{L+1} = \{V\}$. Vertices u and v are *boundary vertices* on level ℓ if they are in different cells of $\mathcal{C}^\ell$. Accordingly, the arc $(u, v) \in A$ is a *boundary arc* on level ℓ.

Query Variants and Algorithms. Given a departure time τ and vertices s and t, an *earliest-arrival (EA)* query asks for the minimum travel time from s to t when departing at time τ. Similarly, a *latest-departure (LD)* query asks for the minimum travel time of an s–t-path arriving at time τ. A *profile query* for given source s and target t asks for the minimum travel time at every possible departure time τ, i.e., a profile $f_{s,t}$ from s to t (over all s–t-paths in G). EA queries can be handled by a time-dependent variant of Dijkstra's algorithm [19], which we refer to as *TD-Dijkstra*. It maintains (scalar) *arrival time labels* $d(\cdot)$ for each vertex, initially set to τ for the source s (∞ for all other vertices). In each step, a vertex u with minimum $d(u)$ is extracted from a priority queue (initialized with s). Then, the algorithm *relaxes* all outgoing arcs (u, v): if $d(u) + f_{(u,v)}(d(u))$ improves $d(v)$, it updates $d(v)$ accordingly and adds v to the priority queue (unless it is already contained). LD queries are handled analogously by running the algorithm from t, relaxing incoming instead of outgoing arcs, and maintaining departure time labels.

Profile queries can be solved by *Profile-Dijkstra* [13], which is based on linking and merging. It generalizes Dijkstra's algorithm, maintaining s–v profiles f_v at each vertex $v \in V$. Initially, it sets $f_s \equiv 0$, and $f_v \equiv \infty$ for all other vertices. The algorithm continues along the lines of TD-Dijkstra, using a priority queue with scalar keys $f_v^{\min}$. For extracted vertices u, arc relaxations propagate profiles rather than travel times, computing $g := \mathrm{link}(f_u, f_{(u,v)})$ and $f_v := \mathrm{merge}(f_v, g)$ for outgoing arcs (u, v). As shown by Foschini et al. [23], the number of breakpoints of the profile of an s–v-paths can be superpolynomial, and hence, so is space consumption *per vertex label* and the running time of Profile-Dijkstra in the worst case. Accordingly, it is not feasible for large-scale instances, even in practice [13].

3   Our Approach

We propose *Time-Dependent CRP (TDCRP)*, a speedup technique for time-dependent route planning allowing fast integration of user-dependent metric changes. Additionally, we enable current and/or predicted traffic updates with limited departure time horizon (accounting for the fact that underlying traffic situations resolve over time). To take historic knowledge of traffic patterns into account, we use functions of departure time at arcs. This conceptual change has important consequences: For plain CRP, the topology data structures is fixed after preprocessing, enabling several micro-optimizations with significant impact on customization and query [9]. In our case, functional complexity is metric-dependent (influenced by, e. g., user preferences) and has to be handled dynamically during customization. Hence, for adaptation to dynamic time-dependent scenarios, we require new data structures and algorithmic changes during customization. Below, we recap the three-phase workflow of CRP [9] that allows fast integration of user-dependent routing preferences, describing its extension to TDCRP along the way. In particular, we incorporate *profile queries* into the customization phase to obtain *time-dependent* shortcuts. Moreover, we adapt the query phase to efficiently compute time-dependent shortest routes.

3.1 Preprocessing

The (metric-independent) preprocessing step of CRP computes a multi-level partition of the vertices, with given number L of levels. Several graph partition algorithms tailored to road networks exist, providing partitions with balanced cell sizes and small cuts [10,27,40,41]. For each level $\ell \in \{1, \ldots, L\}$, the respective partition $\mathcal{C}^\ell$ induces an *overlay graph* H^ℓ, containing all boundary vertices and boundary arcs in $\mathcal{C}^\ell$ and *shortcut* arcs between boundary vertices within each cell $C_i^\ell \in \mathcal{C}^\ell$. We define $\mathcal{C}^0 = \{\{v\} \mid v \in V\}$ and $H^0 := G$ for consistency. Building the overlay, we use the clique matrix representation, storing cliques of boundary vertices in matrices of contiguous memory [9]. Matrix entries represent *pointers* to functions (whose complexity is not known until customization). This dynamic data structure rules out some optimizations for plain CRP, such as microcode instructions, that require preallocated ranges of memory for the metric [9]. To improve locality, all functions are stored in a single array, such that profiles corresponding to outgoing arcs of a boundary vertex are in contiguous memory.

3.2 Customization

In the customization phase, costs of all shortcuts (added to the overlay graphs during preprocessing) are computed. We run profile searches to obtain these time-dependent costs. In particular, we require, for each boundary vertex u (in some cell C_i at level $\ell \geq 1$), the time-dependent distances for all $\tau \in \Pi$ to all boundary vertices $v \in C_i$. To this end, we run a profile query on the overlay $H^{\ell-1}$. By design, this query is restricted to *subcells* of C_i, i.e., cells C_j on level $\ell-1$ for which $C_j \subseteq C_i$ holds. This yields profiles for all outgoing (shortcut) arcs (u, v) in C_i from u. On higher levels, previously computed overlays are used for faster computation of shortcuts. Unfortunately, profile queries are expensive in terms of both running time and space consumption. Below, we describe improvements to remedy these effects, mostly by tuning the profile searches.

Improvements. The main bottleneck of profile search is performing link and merge operations, which require linear time in the function size (cf. Sect. 2). To avoid unnecessary operations, we explicitly compute and store the minimum $f^{\min}$ and the maximum $f^{\max}$ of a profile f in its corresponding label and in shortcuts of overlays. These values are used for early pruning, avoiding costly link and merge operations: Before relaxing an arc (u, v), we check whether $f_u^{\min} + f_{(u,v)}^{\min} > f_v^{\max}$, i. e., the minimum of the linked profile exceeds the maximum of the label at v. If this is the case, the arc (u, v) does not need to be relaxed. Otherwise, the functions are linked. We distinguish four cases, depending on whether the first or second function are constant, respectively. If both are constant, linking becomes trivial (summing up two integers). If one of them is constant, simple shift operations suffice (we need to distinguish two cases, depending on which of the two functions is constant). Only if no function is constant, we apply the link operation.

After linking $f_{(u,v)}$ to f_u, we obtain a tentative label $\tilde{f}_v$ together with its minimum $\tilde{f}_v^{\min}$ and maximum $\tilde{f}_v^{\max}$. Before merging f_v and $\tilde{f}_v$, we run additional checks to avoid unnecessary merge operations. First, we perform bound checks: If $\tilde{f}_v^{\min} > f_v^{\max}$, the function f_v remains unchanged (no merge necessary). Note that this may occur although we checked bounds before linking. Conversely, if $\tilde{f}_v^{\max} < f_v^{\min}$, we simply replace f_v by $\tilde{f}_v$. If the checks fail, and one of the two functions is constant, we must merge. But if f_v and $\tilde{f}_v$ are both nonconstant, one function might still dominate the other. To test this, we do a coordinated linear-time sweep over the breakpoints of each function, evaluating the current line segment at the next breakpoint of the other function. If during this test $\tilde{f}_v(\tau) < f_v(\tau)$ for any point $(\tau, \cdot)$, we must merge. Otherwise we can avoid the merge operation and its numerically unstable line segment intersections.

Additionally, we use *clique flags*: For a vertex v, define its *parents* as all direct predecessors on paths contributing to the profile at the current label of v. For each vertex v of an overlay H^ℓ, we add a flag to its label that is true if *all* parents of v belong to the same cell at level ℓ. This flag is set to true whenever the corresponding label f_v is replaced by the tentative function $\tilde{f}_v$ after relaxing a clique arc (u, v), i. e., the label is set for the first time or the label f_v is dominated by the tentative function $\tilde{f}_v$. It is set to false if the vertex label is partially improved after relaxing a boundary arc. For flagged vertices, we do not relax outgoing clique arcs, as this cannot possibly improve labels within the same cell (due to the triangle inequality and the fact that we use full cliques).

Parallelization. Cells on a given level are processed independently, so customization can be parallelized naturally, assigning cells to different threads [9]. In our scenario, however, workload is strongly correlated with the number of time-dependent arcs in the search graph. It may differ significantly between cells: In realistic data sets, the distribution of time-dependent arcs is clearly not uniform, as it depends on the road type (highways vs. side roads) and the area (rural vs. urban). To balance load, we parallelize per boundary vertex (and not per cell).

Shortcut profiles are written to dynamic containers, as the number of breakpoints is not known in advance. Thus, we must prohibit parallel (writing) access to these data structure. One way to solve this is to make use of locks. However, this is expensive if many threads try to write profiles at the same time. Instead, we use thread-local profile containers, i. e., each thread uses its own container to store profiles. After customization of each level, we synchronize data by copying profiles to the global container sequentially. To improve spatial locality during queries, we maintain the relative order of profiles wrt. the matrix layout (so profiles of adjacent vertices are likely to be contiguous in memory). Since relative order within each thread-local containers is maintained easily (by running queries accordingly), we can use merge sort when writing profiles to the global container.

Approximation. On higher levels of the partition, shortcuts represent larger parts of the graph. Accordingly, they contain more breakpoints and consume more space. This makes profile searches fail on large graphs due to insufficient memory, even on modern hardware. Moreover, running time is strongly correlated to the complexity of profiles. To save space and time, we *simplify* functions during customization. To this end, we use the algorithm of Imai and Iri [29]. For a maximum (relative or absolute) error bound ε, it computes an approximation of a given piecewise linear function with minimum number of breakpoints. In TCH [2], this technique is applied after preprocessing to reduce space consumption. Instead, we use the algorithm to simplify profiles after computing all shortcuts of a certain level. Therefore, searches on higher levels use approximated functions from lower levels, leading to slightly less accurate profiles but faster customization; see Sect. 4. The bound ε is a tuning parameter: Larger values allow faster customization, but decrease quality. Also, approximation is not necessarily applied on all levels, but can be restricted to the higher ones. Note that after approximating shortcuts, the triangle inequality may no longer hold for the corresponding overlay. This is relevant when using clique flags: They yield faster profile searches, but slightly decrease quality (additional arc relaxations may improve shortcut bounds).

3.3   Live Traffic and Short-Term Traffic Predictions

Updates due to, e. g., live traffic, require that we rerun parts of the customization. Clearly, we only have to run customization for *affected* cells, i. e., cells containing arcs for which an update is made. We can do even better if we exploit that live traffic and short-term updates only affect a limited *time horizon*. Thus, we do not propagate updates to boundary vertices that cannot reach an affected arc before the end of its time horizon.

We assume that short-term updates are *partial functions* $f \colon [\pi', \pi''] \to \mathbb{R}^+$, where $\pi' \in \Pi$ and $\pi'' \in \Pi$ are the *beginning* and *end* of the time horizon, respectively. Let $a_1 = (u_1, v_1), \ldots, a_k = (u_k, v_k)$ denote the *updated* arcs inside some cell C_i at level ℓ, and let $f_1, \ldots, f_k$ be the corresponding partial functions representing time horizons. Moreover, let τ be the current point in time. To update C_i we run, on its induced subgraph, a backward *multi-target* latest departure (LD) query from the tails of all updated arcs. In other words, we initially insert the vertices $u_1, \ldots, u_k$ into the priority queue. For each $i \in \{1, \ldots, k\}$, the label of u_i is set to π_i'', i. e., the end of the time horizon $[\pi_i', \pi_i'']$ of the partial function f_i. Consequently, the LD query computes, for each vertex of the cell C_i, the latest possible departure time such that some affected arc is reached before the end of its time horizon. Whenever the search reaches a boundary vertex of the cell, it is marked as *affected* by the update. We stop the search as soon as the departure time label of the current vertex is below τ. (Recall that LD visits vertices in decreasing order of departure time.) Thereby, we ensure that only such boundary vertices are marked from which an updated arc can be reached in time.

Afterwards, we run profile searches for C_i as in regular customization, but only from affected vertices. For profiles obtained during the searches, we test whether they improve the corresponding stored shortcut profile. If so, we add the *affected* interval of the profile for which a change occurs to the set of time horizons of the next level. If shortcuts are approximations, we test whether the change is *significant*, i.e., the maximum difference between the profiles exceeds some bound. We continue the update process on the next level accordingly.

3.4   Queries

The query algorithm makes use of shortcuts computed during customization to reduce the search space. Given a source s and a target t, the search graph consists of the overlay graph induced by the top-level partition $\mathcal{C}^L$, all overlays of cells of lower levels containing s or t, and the level-0 cells in the input graph G that contain s or t. Note that the search graph does not have to be constructed explicitly, but can be obtained on-the-fly [9]: At each vertex v, one computes the highest levels $\ell_{s,v}$ and $\ell_{v,t}$ of the partition such that v is not in the same cell of the partition as s or t, respectively (or 0, if v is in the same level-1 cell as s or t). Then, one relaxes outgoing arcs of v only at level $\min\{\ell_{s,v}, \ell_{v,t}\}$ (recall that $H^0 = G$).

To answer EA queries, we run TD-Dijkstra on this search graph. For faster queries, we make use of the minimum values $f_{(u,v)}^{\min}$ stored at arcs: We do not relax an arc (u,v) if $d(u) + f_{(u,v)}^{\min}$ does not improve $d(v)$. Thereby, we avoid costly function evaluation. Note that we do not use clique flags for EA queries, since we have observed rare but high maximum errors in our implementation when combined with approximated clique profiles.

To answer profile queries, Profile-Dijkstra can be run on the CRP search graph, using the same optimizations as described in Sect. 3.2.

4   Experiments

We implemented all algorithms in C++ using g++ 4.8 (flag -O3) as compiler. Experiments were conducted on a dual 8-core Intel Xeon E5-2670 clocked at 2.6 GHz, with 64 GiB of DDR3-1600 RAM, 20 MiB of L3 and 256 KiB of L2 cache. We ran customization in parallel (using all 16 threads) and queries sequentially.

Input Data and Methodology. Our main test instance is the road network of Western Europe ($|V| = 18$ million, $|A| = 42.2$ million), kindly provided by PTV AG. For this well-established benchmark instance [1], travel time functions were generated synthetically [37]. We also evaluate the subnetwork of

Table 1. Customization performance on Europe for varying approximation parameters (ε). We report, per level, the number of breakpoints (bps, in millions) in the resulting overlay, the percentage of clique arcs that are time-dependent (td.clq.arcs), average complexity of time-dependent arcs (td.arc.cplx), as well as customization time. Without approximation, Levels 5 and 6 cannot be computed as they do not fit into main memory.

ε		Lvl1	Lvl2	Lvl3	Lvl4	Lvl5	Lvl6	Total
—	bps [10^6]	99.1	398.4	816.4	1363.4	—	—	2677.4
	td.clq.arcs [%]	17.0	52.6	76.0	84.2	—	—	—
	td.arc.cplx	21.0	68.9	189.0	509.3	—	—	—
	time [s]	11.4	52.0	152.9	206.2	—	—	375.7
0.01 %	bps [10^6]	75.7	182.7	244.6	240.8	149.3	59.2	952.2
	td.clq.arcs [%]	17.0	52.6	76.0	84.2	85.2	82.5	—
	td.arc.cplx	16.0	31.6	56.6	90.0	108.6	108.0	—
	time [s]	4.5	18.0	32.7	82.1	150.3	151.5	439.1
0.1 %	bps [10^6]	60.7	107.5	111.5	87.9	47.9	17.6	432.9
	td.clq.arcs [%]	17.0	52.7	76.0	84.2	85.2	82.5	—
	td.arc.cplx	12.9	18.6	25.8	32.8	34.8	32.1	—
	time [s]	4.2	16.0	21.4	40.7	62.4	55.0	199.7
1.0 %	bps [10^6]	45.7	58.0	45.6	29.2	14.7	5.4	198.5
	td.clq.arcs [%]	17.0	52.7	76.0	84.2	85.2	82.5	—
	td.arc.cplx	9.7	10.0	10.6	10.9	10.7	9.8	—
	time [s]	4.1	14.1	14.8	22.7	29.6	24.1	109.2

Germany ($|V| = 4.7$ million, $|A| = 10.8$ million), where time-dependent data from historical traffic is available (we extract the 24 h profile of a Tuesday).[1] For partitioning, we use PUNCH [10], which is explicitly developed for road networks and aims at minimizing the number of boundary arcs. For Europe, we consider a 6-level partition, with maximum cell sizes $2^{[4:8:11:14:17:20]}$. For Germany, we use a 5-level partition, with cell sizes of $2^{[4:8:12:15:18]}$. Compared to plain CRP, we use partitions with more levels, to allow fine-grained approximation. Computing the partition took 5 min for Germany, and 23 min for Europe. Given that road topology changes rarely, this is sufficiently fast in practice.

Evaluating Customization. Table 1 details customization for different approximation parameters ε on the Europe instance. We report, for several choices of ε and for each level of the partition, figures on the complexity of shortcuts in the overlays and the parallelized customization time. The first block shows figures for exact profile computation. Customization had to be aborted after the fourth level, because the 64 GiB of main memory were not sufficient to store

---

[1] The Germany and Europe instances can be obtained easily for scientific purposes, see http://i11www.iti.uni-karlsruhe.de/resources/roadgraphs.php.

Table 2. Query performance on Europe as a trade-off between customization effort and approximation. For customization, we set different approximation parameters (ε) and disable (o) or enable (•) clique flags (Cl.). For the different settings, we report query performance in terms of number of vertices extracted from the queue, scanned arcs, evaluated function breakpoints (# Bps), running time, and average and maximum error, each averaged over 100 000 random queries. As we employ approximation per level, resulting query errors can be higher than the input parameter.

Customization			Query					
Approx. ε	Cl.	Time [s]	# Vertices	# Arcs	# Bps	Time [ms]	Err. [%]	
							avg.	max.
0.01 %	o	1 155.1	3 499	541 091	433 698	14.69	<0.01	0.03
0.01 %	•	439.1	3 499	541 090	434 704	14.53	<0.01	0.03
0.10 %	o	533.0	3 499	541 088	96 206	7.63	0.04	0.28
0.10 %	•	199.7	3 499	541 088	99 345	6.47	0.04	0.29
1.00 %	o	284.4	3 499	541 080	67 084	5.66	0.51	3.15
1.00 %	•	109.2	3 499	541 058	70 202	5.75	0.54	3.21

the profiles of all vertex labels. For remaining levels, we clearly see the strong increase in the total number of breakpoints per level. Also, the relative amount of time-dependent arcs rises with each level, since shortcuts become longer. Customization time clearly correlates with profile complexity, from 10 s on the lowest level, to more then three minutes on the fourth. When approximating, we see that customization becomes faster for larger values of ε. We apply approximation to all levels of the partition (using it only on the topmost levels did not provide significant benefits in preliminary experiments). Recall that higher levels work on approximated shortcuts of previous levels, so ε does not provide a bound on the error of the shortcuts. We see that even a very small value (0.01 %) yields a massive drop of profile complexity (more than a factor 5 at Level 4), and immediately allows full customization. For reasonably small values ($\varepsilon = 0.1\,\%, \varepsilon = 1.0\,\%$), we see that customization becomes much faster (less than two minutes for $\varepsilon = 1.0\,\%$). In particular, this is fast enough for traffic updates. Even for larger values of ε, the higher levels are far more expensive: This is due to the increasing amount of time-dependent arcs, slowing down profile search.

Evaluating Customization and Queries. In Table 2, we show query performance for different values of the approximation parameter ε on the Europe instance. We also show the effect of using clique flags during customization: they improve customization performance by about a factor of 2.6, while having a negligible influence on query results. For each value of ε, we report timings as well as average and maximum error for 100 000 point-to-point queries. For each query, the source and target vertex and the departure time were picked uniformly at random. Similar to customization, the data shows that query times decrease with higher approximation ratio. Again, this is due to the smaller number of break-

Table 3. Robustness comparison for TCH [2] and TDCRP. For different input instances, we report timing of metric-dependent preprocessing (always run on 16 cores) and sequential queries. Query times are averaged over the same 100 000 random queries as in Table 2.

Network	TCH		TDCRP	
	Prepro. [s]	Query [ms]	Custom. [s]	Query [ms]
Europe	1 479	1.37	109	5.75
Europe, bad traffic	7 772	5.87	208	8.01
Europe, avoid highways	8 956	19.54	127	8.29

points in profiles (observe that the number of visited vertices and arcs is almost identical in all cases). As expected, both average and maximum error clearly correlate with (but are larger than) ε. There are two reasons for this: As shown in [24, 32, 35], query errors not only depend on ε but also on the maximum slope of any approximated function. Moreover, since we apply approximation per level, the error bound in [24] applies recursively, leading to a higher theoretical bound. Still, we observe that even for the parameter choice $\varepsilon = 1.0\,\%$, the maximum error is very low (about 3 %). Moreover, query times are quite practical for all values of ε, ranging from 5 ms to 15 ms. In summary, our approach allows query times that are fast enough for interactive applications, if a reasonable, small error is allowed. Given that input functions are based on statistical input with inherent inaccuracy, the error of TDCRP is more than acceptable for realistic applications.

Evaluating Robustness. We also evaluate robustness of our approach against dynamic updates and user-dependent custom metrics. The first scenario (bad traffic) simulates a highly congested graph: for every time-dependent arc in the Western Europe instance with associated travel-time function f, we replace f by f' defined as $f'(\tau) := 2(f(\tau) - f^{\min}(\tau)) + f^{\min}(\tau)$, while maintaining the FIFO property on f'. In the second scenario, we consider user restrictions (avoid highways). For each scenario, customization and the same set of 100 000 random queries as before are run on the respective modified instance. (Hence, we do not remove highways for the second scenario, setting very high costs instead.) Table 3 compares results of the original instance (Europe) to the modified ones.

Besides our approach, which is run using parameter $\varepsilon = 1.0$ for customization, we also evaluate TCH [2], the fastest known approach for time-dependent route planning. All measurements for TCH are based on this freely available implementation: https://github.com/GVeitBatz/KaTCH. While TCH allows faster queries on the original instance, we see that running times increase significantly for the modified ones. Preprocessing time also increases to several hours in both cases. In the first scenario (bad traffic), this can be explained by a larger number of paths that are relevant at different points in time (more congested roads need to be bypassed). Consequently, customization time of TDCRP rises as well but by a much smaller factor. In the second scenario (avoid highways), the TCH hierarchy clearly deteriorates. While TDCRP is quite robust to this change (both customization and query times increase by less than 50 %), TCH queries slow down by more than an order of magnitude.

While possibly subject to implementation, our experiment indicates that underlying vertex orderings of TCH are not robust against less well-behaved metrics. Similar effects can be shown for scalar Contraction Hierarchies (CH) on metrics reflecting, e. g., travel distance [9,25]. In summary, TDCRP is much more robust in both scenarios.

Comparison with Related Work. Finally, Table 4 provides an overview comparing our results to the most relevant existing approaches for time-dependent route planning. For the related work, we show measurements in the fastest reported variant (e. g., if parallelized) but we scale all timings to our hardware as detailed in Table 5 using a benchmark tool [1] available at http://tpajor.com/projects/.

For TCH and ATCH [2], preprocessing can be further split into *node order computation* and *contraction*. Since it has been shown in [2] that node orders can be re-used for certain other metrics (e. g., other week days), we report running times of the contraction as rudimentary customization times. Recall, however, that our robustness tests in Table 3 suggest that there is a limit to the applicability of such a customization approach based on current TCH orders.

We evaluated our approach on both benchmark instances (Germany, and Europe) for the two fastest variants ($\varepsilon = 0.1$ and $\varepsilon = 1.0$) and we see that it competes very well with the previous techniques: While providing query times similar to the fastest existing approaches, TDCRP has by far the lowest metric-dependent preprocessing time (i. e., customization time) and a good parallel speedup (factor 13.9 to 14.2 on Europe for 16 threads). At the same time, resulting average and maximum errors (due to approximating profiles during customization) are similar to previous results and low enough for practical purposes. When parallelized, customization of the whole network is fast enough for regular live-traffic updates: 8 to 16 s on Germany, and 2 to 3 min on Europe. Note, however, that other approaches are also able to handle live traffic by providing *partial updates* of the preprocessed data: For example, by exploiting the fact that effects of live traffic are locally and temporally limited, FLAT [32]) and TDCALT [11] achieve partial update times in well below a minute (for 1,000 traffic-affected arcs).

Interestingly, TDCALT's preprocessing is also quite fast. This could make it an interesting alternative candidate for our scenario (metric customization); since it is mostly based on lower bounds and only light contraction, it might also be fairly robust to sensible, user-defined metrics (unlike TCH, cf. Table 3). Note, however, that TDCALT on Europe requires a significantly higher approximation to achieve a similar level of query performance (even scaled), yielding a high maximum error. Furthermore, in the evaluated variant, landmarks are chosen after the graph contraction routine, making it hard to parallelize the preprocessing (which also has not been attempted). Additionally, TDCALT allows no practical profile search on large instances [8,11], making it a less versatile approach.

Table 4. Comparison of time-dependent speedup techniques on instances of Germany, and Europe. We present figures for variants of TDCALT [11], SHARC [8], TCH and ATCH [2], FLAT [32], and TDCRP. For better comparability across different hardware, we scale all sequential (Seq.) and parallel (Par.) timings to our machine; see Table 5 for factors. For preprocessing, customization, and live traffic updates, we show the number of threads used (Thr.). For EA queries, we present average numbers on queue extractions (# Vert.), scanned arcs, sequential running time in milliseconds, and average and maximum relative error.

Algorithm	Inst.	Thr.	Preprocessing Par. [h:m:s]	Preprocessing Space [B/n]	Customization Par. [m:s]	Customization Seq. [m:s]	Customization Space [B/n]	Traffic Par. [m:s]	EA Queries # Vert.	EA Queries # Arcs	EA Queries Seq. [ms]	EA Queries Err. [%] avg.	EA Queries Err. [%] max.
TDCALT	Germany	1	3:14	50	—	—	—	n/a	3 190	12 255	1.93	—	—
TDCALT-K1.15	Germany	1	3:14	50	—	—	—	n/a	1 593	5 339	0.67	0.05	13.84
eco L-SHARC	Germany	1	28:03	219	—	—	—	—	2 776	19 005	2.27	—	—
heu SHARC	Germany	1	1:14:06	137	—	—	—	—	818	1 611	0.25	n/a	0.61
ATCH (1.0)	Germany	8	5:50	239	1:09	6:59	239	—	588	7 993	1.15	—	—
inex. TCH (0.1)	Germany	8	5:50	286	1:09	6:59	286	—	642	7 138	0.65	0.02	0.10
inex. TCH (2.5)	Germany	8	5:50	172	1:09	6:59	172	—	668	7 429	0.67	0.79	2.44
FLAT/FCA	Germany	6	>1 day	>10 000	—	—	—	0:44	1 122	n/a	1.51	n/a	1.53
TDCRP (0.1)	Germany	16	4:33	29	0:16	3:30	166	0:16	2 152	167 263	1.92	0.05	0.25
TDCRP (1.0)	Germany	16	4:33	29	0:08	1:43	77	0:08	2 152	167 305	1.66	0.68	2.85
TDCALT	Europe	1	21:35	61	—	—	—	0:22	60 961	356 527	43.67	—	—
TDCALT-K1.05	Europe	1	21:35	61	—	—	—	0:22	32 405	n/a	22.48	0.01	3.94
TDCALT-K1.15	Europe	1	21:35	61	—	—	—	0:22	6 365	32 719	3.31	0.26	8.69
eco L-SHARC	Europe	2	2:27:07	198	—	—	—	—	18 289	165 382	13.77	—	—
heu SHARC	Europe	17	7:59:08	127	—	—	—	—	5 031	8 411	1.06	n/a	1.60
ATCH (1.0)	Europe	8	42:21	208	7:26	48:07	208	—	1 223	20 336	2.68	—	—
inex. TCH (0.1)	Europe	8	42:21	239	7:26	48:07	239	—	1 722	24 389	2.50	0.02	0.15
inex. TCH (2.5)	Europe	8	42:21	175	7:26	48:07	175	—	1 875	26 948	2.72	0.48	3.37
TDCRP (0.1)	Europe	16	22:33	32	3:20	47:10	237	3:20	3 499	541 088	6.47	0.04	0.29
TDCRP (1.0)	Europe	16	22:33	32	1:49	25:16	133	1:49	3 499	541 058	5.75	0.54	3.21

Table 5. Scaling factors for different machines, used in Table 4. Scores were determined by a shared Dijkstra implementation [1] on the same graph. These factors have to be taken with a grain of salt, since Dijkstra's algorithm is not a good indicator of cache performance. When scaling on TDCRP performance, instead, we observe a factor of 2.06–2.18 for the Opteron 2218 (which we have access to), depending on the instance.

Machine	Used by	Score [ms]	Factor
2× 8-core Intel Xeon E5-2670, 2.6 GHz	TDCRP	36 582	—
AMD Opteron 2218, 2.6 GHz	TDCALT [11], SHARC [8]	101 552	2.78
2× 4-core Intel Xeon X5550, 2.66 GHz	TCH, ATCH [2]	39 684	1.08
6-core Intel Xeon E5-2643v3, 3.4 Ghz	FLAT/FCA [32]	30 901	0.84

To summarize, we see that TDCRP clearly broadens the state-of-the-art of time-dependent route planning, handling a wider range of practical requirements (e.g., fast metric-dependent preprocessing, robustness to user preferences, live traffic) with a query performance close to the fastest known approaches.

5   Conclusion

In this work, we introduced TDCRP, a separator-based overlay approach for dynamic, time-dependent route planning. We showed that, unlike its closest competitor (A)TCH, it is robust against user-dependent metric changes, very much like CRP is more robust than CH. Most importantly, unlike scalar CRP, we have to deal with time-dependent shortcuts, and a strong increase in functional complexity on higher levels; To reduce memory consumption, we approximate the overlay arcs at each level, accelerating customization and query times. As a result, we obtain an approach that enables fast near-optimal, time-dependent queries, with quick integration of user preferences, live traffic, and traffic predictions.

There are several aspects of future work. Naturally, we are interested in alternative customization approaches that avoid label-correcting profile searches. This could be achieved, e.g., by using kinetic data structures [23], or balanced contraction [2] within cells. It would be interesting to re-evaluate (A)TCH in light of Customizable CH [16,17]. Also, while we customized time-dependent overlay arcs with both historic travel time functions (changes seldom) and user preferences (changes often) at once, in practice, it might pay off to separate this into two further phases (yielding a 4-phase approach). Furthermore, one could aim at exact queries based on approximated shortcuts as in ATCH.

While our approach is customizable, it requires arc cost functions that map time to time. This allows to model avoidance of highways or driving slower than the speed limit, but it cannot handle combined linear optimization of (time-dependent) travel time and, e.g., toll costs. For that, one should investigate the application of generalized time-dependent objective functions as proposed in [3].

Finally, functional complexity growth of time-dependent shortcuts is problematic, and from what we have seen, it is much stronger than the increase in the number of corresponding paths. It seems wasteful to apply the heavy machinery of linking and merging during preprocessing, when time-dependent evaluation of just a few paths (more than one is generally needed) would give the same results. This might explain why TDCALT, which is mostly based just on scalar lower bounds, is surprisingly competitive. So re-evaluation seems fruitful, possibly exploiting insights from [20]. Revisiting hierarchical preprocessing techniques that are not based on shortcuts [26,36] could also be interesting.

Acknowledgements. We thank Gernot Veit Batz, Daniel Delling, Moritz Kobitzsch, Felix König, Spyros Kontogiannis, and Ben Strasser for interesting conversations.

References

1. Bast, H., Delling, D., Goldberg, A.V., Müller-Hannemann, M., Pajor, T., Sanders, P., Wagner, D., Werneck, R.F.: Route Planning in Transportation Networks. CoRR abs/1504.05140 (2015)
2. Batz, G.V., Geisberger, R., Sanders, P., Vetter, C.: Minimum time-dependent travel times with contraction hierarchies. ACM J. Exp. Algorithmics **18**(1.4), 1–43 (2013)
3. Batz, G.V., Sanders, P.: Time-dependent route planning with generalized objective functions. In: Epstein, L., Ferragina, P. (eds.) ESA 2012. LNCS, vol. 7501, pp. 169–180. Springer, Heidelberg (2012)
4. Bauer, R., Columbus, T., Rutter, I., Wagner, D.: Search-space size in contraction hierarchies. In: Fomin, F.V., Freivalds, R., Kwiatkowska, M., Peleg, D. (eds.) ICALP 2013, Part I. LNCS, vol. 7965, pp. 93–104. Springer, Heidelberg (2013)
5. Baum, M., Dibbelt, J., Pajor, T., Wagner, D.: Energy-optimal routes for electric vehicles. In: SIGSPATIAL 2013, pp. 54–63. ACM Press (2013)
6. Cooke, K., Halsey, E.: The shortest route through a network with time-dependent internodal transit times. J. Math. Anal. Appl. **14**(3), 493–498 (1966)
7. Dean, B.C.: Algorithms for minimum-cost paths in time-dependent networks with waiting policies. Networks **44**(1), 41–46 (2004)
8. Delling, D.: Time-dependent SHARC-routing. Algorithmica **60**(1), 60–94 (2011)
9. Delling, D., Goldberg, A.V., Pajor, T., Werneck, R.F.: Customizable route planning in road networks. Transport. Sci. (2015)
10. Delling, D., Goldberg, A.V., Razenshteyn, I., Werneck, R.F.: Graph partitioning with natural cuts. In: IPDPS 2011, pp. 1135–1146. IEEE Computer Society (2011)
11. Delling, D., Nannicini, G.: Core routing on dynamic time-dependent road networks. Informs J. Comput. **24**(2), 187–201 (2012)
12. Delling, D., Wagner, D.: Landmark-based routing in dynamic graphs. In: Demetrescu, C. (ed.) WEA 2007. LNCS, vol. 4525, pp. 52–65. Springer, Heidelberg (2007)
13. Delling, D., Wagner, D.: Time-dependent route planning. In: Ahuja, R.K., Möhring, R.H., Zaroliagis, C.D. (eds.) Robust and Online Large-Scale Optimization. LNCS, vol. 5868, pp. 207–230. Springer, Heidelberg (2009)
14. Demiryurek, U., Banaei-Kashani, F., Shahabi, C.: A case for time-dependent shortest path computation in spatial networks. In: SIGSPATIAL 2010, pp. 474–477. ACM Press (2010)

15. Diamantopoulos, T., Kehagias, D., König, F., Tzovaras, D.: Investigating the effect of global metrics in travel time forecasting. In: ITSC 2013, pp. 412–417. IEEE (2013)
16. Dibbelt, J., Strasser, B., Wagner, D.: Customizable contraction hierarchies. In: Gudmundsson, J., Katajainen, J. (eds.) SEA 2014. LNCS, vol. 8504, pp. 271–282. Springer, Heidelberg (2014)
17. Dibbelt, J., Strasser, B., Wagner, D.: Customizable contraction hierarchies. J. Exp. Algorithmics. **21**(1), 1.5:1–1.5:49 (2016). doi:10.1145/2886843
18. Dijkstra, E.W.: A note on two problems in connexion with graphs. Numer. Math. **1**(1), 269–271 (1959)
19. Dreyfus, S.E.: An appraisal of some shortest-path algorithms. Oper. Res. **17**(3), 395–412 (1969)
20. Efentakis, A., Pfoser, D.: Optimizing landmark-based routing and preprocessing. In: IWCTS 2013, pp. 25:25–25:30. ACM Press (2013)
21. Efentakis, A., Pfoser, D., Vassiliou, Y.: SALT. a unified framework for all shortest-path query variants on road networks. In: Bampis, E. (ed.) SEA 2015. LNCS, vol. 9125, pp. 298–311. Springer, Heidelberg (2015)
22. Eppstein, D., Goodrich, M.T.: Studying (non-planar) road networks through an algorithmic lens. In: SIGSPATIAL 2008, pp. 16:1–16:10. ACM Press (2008)
23. Foschini, L., Hershberger, J., Suri, S.: On the complexity of time-dependent shortest paths. Algorithmica **68**(4), 1075–1097 (2014)
24. Geisberger, R., Sanders, P.: Engineering time-dependent many-to-many shortest paths computation. In: ATMOS 2010, pp. 74–87. OASIcs (2010)
25. Geisberger, R., Sanders, P., Schultes, D., Vetter, C.: Exact routing in large road networks using contraction hierarchies. Transp. Sci. **46**(3), 388–404 (2012)
26. Gutman, R.J.: Reach-based routing: a new approach to shortest path algorithms optimized for road networks. In: ALENEX 2004, pp. 100–111. SIAM (2004)
27. Hamann, M., Strasser, B.: Graph bisection with pareto-optimization. In: ALENEX 2016, pp. 90–102. SIAM (2016)
28. Holzer, M., Schulz, F., Wagner, D.: Engineering multilevel overlay graphs for shortest-path queries. ACM J. Exp. Algorithmics **13**(2.5), 1–26 (2008)
29. Imai, H., Iri, M.: An optimal algorithm for approximating a piecewise linear function. J. Inf. Process. **9**(3), 159–162 (1986)
30. Jung, S., Pramanik, S.: An efficient path computation model for hierarchically structured topographical road maps. IEEE Trans. Knowl. Data Eng. **14**(5), 1029–1046 (2002)
31. Kontogiannis, S., Michalopoulos, G., Papastavrou, G., Paraskevopoulos, A., Wagner, D., Zaroliagis, C.: Analysis and experimental evaluation of time-dependent distance oracles. In: ALENEX 2015, pp. 147–158. SIAM (2015)
32. Kontogiannis, S., Michalopoulos, G., Papastavrou, G., Paraskevopoulos, A., Wagner, D., Zaroliagis, C.: Engineering oracles for time-dependent road networks. In: ALENEX 2016, pp. 1–14. SIAM (2016)
33. Kontogiannis, S., Wagner, D., Zaroliagis, C.: Hierarchical Oracles for Time-Dependent Networks. CoRR abs/1502.05222 (2015)
34. Kontogiannis, S., Zaroliagis, C.: Distance oracles for time-dependent networks. In: Esparza, J., Fraigniaud, P., Husfeldt, T., Koutsoupias, E. (eds.) ICALP 2014. LNCS, vol. 8572, pp. 713–725. Springer, Heidelberg (2014)
35. Kontogiannis, S., Zaroliagis, C.: Distance oracles for time-dependent networks. Algorithmica **74**(4), 1404–1434 (2015)
36. Maervoet, J., Causmaecker, P.D., Berghe, G.V.: Fast approximation of reach hierarchies in networks. In: SIGSPATIAL 2014, pp. 441–444. ACM Press (2014)

37. Nannicini, G., Delling, D., Liberti, L., Schultes, D.: Bidirectional A* search on time-dependent road networks. Networks **59**, 240–251 (2012)
38. Orda, A., Rom, R.: Shortest-path and minimum delay algorithms in networks with time-dependent edge-length. J. ACM **37**(3), 607–625 (1990)
39. Pfoser, D., Brakatsoulas, S., Brosch, P., Umlauft, M., Tryfona, N., Tsironis, G.: Dynamic travel time provision for road networks. In: SIGSPATIAL 2008, pp. 68:1–68:4. ACM Press (2008)
40. Sanders, P., Schulz, C.: Distributed evolutionary graph partitioning. In: ALENEX 2012, pp. 16–29. SIAM (2012)
41. Schild, A., Sommer, C.: On balanced separators in road networks. In: Bampis, E. (ed.) SEA 2015. LNCS, vol. 9125, pp. 286–297. Springer, Heidelberg (2015)
42. Sherali, H.D., Ozbay, K., Subramanian, S.: The time-dependent shortest pair of disjoint paths problem: complexity, models, and algorithms. Networks **31**(4), 259–272 (1998)

UKP5: A New Algorithm for the Unbounded Knapsack Problem

Henrique Becker[(⊠)] and Luciana S. Buriol

Federal University of Rio Grande do Sul (UFRGS), Porto Alegre, Brazil
{hbecker,buriol}@inf.ufrgs.br
http://ppgc.inf.ufrgs.br/

Abstract. In this paper we present UKP5, a novel algorithm for solving the unbounded knapsack problem. UKP5 is based on dynamic programming, but implemented in a non traditional way: instead of looking backward for stored values of subproblems, it stores incremental lower bounds forward. UKP5 uses sparsity, periodicity, and dominance for speeding up computation. UKP5 is considerably simpler than EDUK2, the state-of-the-art algorithm for solving the problem. Moreover, it can be naturally implemented using the imperative paradigm, differently from EDUK2. We run UKP5 and EDUK2 on a benchmark of hard instances proposed by the authors of EDUK2. The benchmark is composed by 4540 instances, divided into five classes, with instances ranging from small to large inside each class. Speedups were calculated for each class, and the overall speedup was calculated as the classes speedups average. The experimental results reveal that UKP5 outperforms EDUK2, being 47 times faster on the overall average.

Keywords: Unbounded knapsack problem · Dynamic programming · Combinatorial optimization

1 Introduction

The unbounded knapsack problem (UKP) is a simpler variation of the well-known bounded knapsack problem (BKP). UKP allows the allocation of an unbounded quantity of each item type. The UKP is NP-Hard, and thus has no known polynomial-time algorithm for solving it. However, it can be solved by a pseudo-polynomial dynamic programming algorithm. UKP arises in real world problems mainly as a subproblem of the Bin Packing Problem (BPP) and Cutting Stock Problem (CSP). Both BPP and CSP are of great importance for the industry [3], [5,6]. The currently fastest known solver for BPP/CSP [2,3] uses a column generation technique (introduced in [5]) that needs to solve an UKP instance as the pricing problem at each iteration of a column generation approach. The need for efficient algorithms for solving the UKP is fundamental for the overall performance of the column generation.

Two techniques are often used for solving UKP: dynamic programming (DP) [1], [4, p. 214], [7, p. 311] and branch and bound (B&B) [10]. The DP approach

© Springer International Publishing Switzerland 2016
A.V. Goldberg and A.S. Kulikov (Eds.): SEA 2016, LNCS 9685, pp. 50–62, 2016.
DOI: 10.1007/978-3-319-38851-9_4

has a stable pseudo-polynomial time algorithm linear on the capacity and number of items. The B&B approach can be less stable. It can be faster than DP on instances with some characteristics, such as when the remainder of the division between the weight of the best item by the capacity is small; or the items have a big efficiency variance. Nonetheless, B&B has always the risk of an exponential time worst case.

The state-of-the-art solver for the UKP, introduced by [12], is a hybrid solver that combines DP and B&B. It tries to solve the problem by B&B, and if this fails to solve the problem quickly, it switches to DP using some data gathered by the B&B to speed up the process. The solver's name is PYAsUKP, and it is an implementation of the EDUK2 algorithm.

1.1   UKP Formal Notation

The following notation of the UKP will be used for the remainder of the paper. An UKP instance is composed by a capacity c, and a list of n items. Each item can be referenced by its index in the item list $i \in \{1 \ldots n\}$. Each item i has a weight value w_i, and a profit value p_i. A solution is an item multiset, i.e., a set that allows multiple copies of the same element. The sum of the items weight, or profit, of a solution s is denoted by w_s, or p_s. A valid solution s has $w_s \leq c$. An optimal solution s^* is a valid solution with the greatest profit among all valid solutions. The UKP objective is to find an optimal solution for the given UKP instance. The mathematical formulation of UKP is:

$$maximize \sum_{i=1}^{n} p_i x_i \tag{1}$$

$$subject\ to \sum_{i=1}^{n} w_i x_i \leq c \tag{2}$$

$$x_i \in \mathbb{N}_0 \tag{3}$$

The quantities of each item i in an optimal solution are denoted by x_i, and are restricted to the non-negative integers, as (3) indicates. We assume that the capacity c, the quantity of items n and the weights of the items w_i are positive integers. The profits of the items p_i are positive real numbers.

The efficiency of an item i is the ratio $\frac{p_i}{w_i}$, and is denoted by e_i. We use w_{min} and w_{max} to denote the smallest item weight, and the biggest item weight, respectively. Also, we refer to the item with the lowest weight among the ones tied with the greatest efficiency as the *best item*, and the item with the lowest weight among all items as the *smallest item*. If two or more items have the same weight we consider only the one with the best profit (the others can be discarded without loss to the optimal solution value); if they have the same weight and profit we consider them the same item.

1.2   Dominance

Dominance, in the UKP context, is a technique for discarding items without affecting the optimal solution value. By this definition, every item that isn't used in an optimal solution could be discarded, but this would need the knowledge of the solution beforehand. Some dominances can be verified in polynomial time over n, and can speed up the resolution of an NP-Hard problem by reducing the instance input size. Instances where many items can be excluded by the two simplest dominances (simple dominance and multiple dominance) are known as "easy" instances. Research on these two dominances was done to a large extent, leading to the following statement by Pisinger in 1995 "[...] perhaps too much effort has previously been used on the solution of easy data instances." [11, p. 20].

Other two important dominances are collective dominance and threshold dominance [12]. These two dominances are too time demanding to be applied at a preprocessing phase, differently from simple and multiple dominances. They are often integrated in the UKP algorithm, and remove items while the algorithm executes. The collective dominance needs to know the $opt(y)$ to exclude an item i with $w_i = y$, where $opt(y)$ is the optimal solution value for a capacity y. The threshold dominance needs to know the $opt(\alpha \times w_i)$ to exclude the item i from capacity $y = \alpha \times w_i$ onwards, where α is any positive integer.

1.3   Periodicity

A periodicity bound y is an upper capacity bound for the existence of optimal solutions without the best item. In another words, it's a guarantee that any optimal solution for an instance where $c \geq y$ has at least one copy of the best item. The periodicity bound is specially useful because it can be applied repeatedly. For example, let $c = 1000$, $y = 800$ and $w_b = 25$ where b is the best item; because of $c \geq y$ we know that any optimal solution has a copy of b, so we can add one b to the solution and combine with an optimal solution for $c = 975$; but 975 is yet bigger than 800, so we can repeat the process until $c = 775$. This way, for any UKP instance where $c \leq y$ we can reduce the instance capacity by $max(1, \lceil (c - y^*)/w_b \rceil) \times w_b$. After solving this instance with reduced capacity we can add $max(1, \lceil (c - y^*)/w_b \rceil)$ copies of b to the optimal solution to obtain an optimal solution for the original instance.

There exist many proposed periodicity bounds, but some are time consuming (as $O(n^2)$ [8]), others depend on specific instance characteristics (as [9][12]). We used only a UKP5-specific periodicity bound described later and the y^* bound described in [4, p. 223]. The y^* is $O(1)$ on an item list ordered by non-increasing efficiency, and it is generic, being successfully applied on instances of most classes. Assuming i is the best item, and j is the second most efficient item, then $y^* = p_i/(e_i - e_j)$.

1.4   Sparsity

For some UKP instances, not every non-zero capacity value can be obtained by a linear combination of the items weight. If w_{min} is small, for example $w_{min} = 1$, we have the guarantee that every non-zero capacity has at least one solution with weight equal to the capacity value. But if w_{min} is big, for example $w_{min} = 10^4$, there can be a large number of capacities with no solution comprising weight equal to the capacity. These capacities have an optimal solution that don't fill the capacity completely. The UKP5 exploits sparsity in the sense that it avoids computing the optimal solution value for those unfulfilled capacities. The array that stores the optimal solutions value is, therefore, sparse.

2   UKP5: The Proposed Algorithm

UKP5 is inspired by the DP algorithm described by Garfinkel [4, p. 221]. The name "UKP5" is due to five improvements applied over that algorithm:

1. **Symmetry pruning:** symmetric solutions are pruned in a more efficient fashion than in [4];
2. **Sparsity:** not every position of the optimal solutions value array has to be computed;
3. **Dominated solutions pruning:** dominated solutions are pruned;
4. **Time/memory tradeoff:** the test $w_i \leq y$ from the algorithm in [4] was removed in cost of more $O(w_{max})$ memory;
5. **Periodicity:** the periodicity check suggested in [4] (but not implemented there) was adapted and implemented.

A pseudocode of our algorithm is presented in Algorithm 1. We have two main data structures, the arrays g and d, both with dimension $c + w_{max}$. The g is a sparse array where we store solutions profit. If $g[y] > 0$ then there exists a non-empty solution s with $w_s = y$ and $p_s = g[y]$. The d array stores the index of the last item used on a solution. If $g[y] > 0 \wedge d[y] = i$ then the solution s with $w_s = y$ and $p_s = g[y]$ has at least one copy of item i. This array makes it trivial to recover the optimal solution, but its main use is to prune solution symmetry.

Our first loop (lines 4 to 9) simply stores all solutions comprised of a single item in the arrays g and d. For a moment, let's ignore lines 12 to 14, and replace $d[y]$ (at line 16) by n. With these changes, the second loop (between lines 11 and 22) iterates g and when it finds a stored solution ($g[y] > 0$) it tests n new solutions (the combinations of the current solution with every item). The new solutions are stored at g and d, replacing solutions already stored if the new solution has the same weight but a greater profit value.

When we add the lines 12 to 14 to the algorithm, it stops creating new solutions from dominated solutions. If a solution s with a smaller weight ($w_s < y$) has a bigger profit ($p_s = opt > p_t$, where $w_t = y \wedge p_t = g[y]$), then s dominates t. If a solution s dominates t then, for any item i, the $s \cap \{i\}$ solution will dominate the $t \cap \{i\}$ solution. This way, new solutions created from t are guaranteed to be dominated by the solutions created from s. A whole superset of t can be discarded without loss to solution optimality.

The change from n to $d[y]$ is based on the algorithm from [4] and it prunes symmetric solutions. In a naive DP algorithm, if the item multiset $\{5, 3, 3\}$ is a

Algorithm 1. UKP5 – Computation of opt

1: **procedure** UKP5($n, c, w, p, w_{min}, w_{max}$)
2:     $g \leftarrow$ array of $c + w_{max}$ positions each one initialized with 0
3:     $d \leftarrow$ array of $c + w_{max}$ positions each one initialized with n
4:     **for** $i \leftarrow 1, n$ **do**   ▷ Stores one-item solutions
5:         **if** $g[w_i] < p_i$ **then**
6:             $g[w_i] \leftarrow p_i$
7:             $d[w_i] \leftarrow i$
8:         **end if**
9:     **end for**
10:     $opt \leftarrow 0$
11:     **for** $y \leftarrow w_{min}, c$ **do**   ▷ Can end early because of periodicity check
12:         **if** $g[y] \leq opt$ **then** ▷ Handles sparsity and pruning of dominated solutions
13:             **continue**   ▷ Ends current iteration and begins the next
14:         **end if**
15:         $opt \leftarrow g[y]$
16:         **for** $i = 1, d[y]$ **do**   ▷ Creates new solutions (never symmetric)
17:             **if** $g[y + w_i] < g[y] + p_i$ **then**
18:                 $g[y + w_i] \leftarrow g[y] + p_i$
19:                 $d[y + w_i] \leftarrow i$
20:             **end if**
21:         **end for**
22:     **end for**
23:     **return** opt
24: **end procedure**

---

valid solution, then every permutation of it is reached in different ways, wasting processing time. To avoid computing symmetric solutions, we enforce non-increasing order of the items index. Any item inserted on a solution s has an index that is equal to or lower than the index of the last item inserted on s. This way, solution $\{10, 3, 5, 3\}$ cannot be reached. However, this is not a problem because this solution is equal to $\{10, 5, 3, 3\}$, and this solution can be reached.

When the two changes are combined, and the items are sorted by non-increasing efficiency, UKP5 gains in performance. The UKP5 iterates by the item list only when it finds a non-dominated solution, i.e., $g[y] \geq 0$ (line 12). Non-dominated solutions are more efficient (larger ratio of profit by weight) than the skipped dominated solutions. Therefore, the UKP5 inner loop (lines 16 to 21) often iterates up to a low $d[y]$ value. Experimental results show that, after some threshold capacity, the UKP5 inner loop consistently iterates only for a small fraction of the item list.

The algorithm ends with the optimal solution stored at opt. The solution assemble phase isn't described in Algorithm 1, but it's similar to the one used by the DP method described in [4, p. 221, Steps 6–8]. Let y_{opt} be a capacity where $g[y_{opt}] = opt$. We add a copy of item $i = d[y_{opt}]$ to the solution, then we add a copy of item $j = d[y_{opt} - w_i]$, and so on, until $d[0]$ is reached. This phase has a $O(c)$ time complexity, as the solution can be composed of c copies of an item i with $w_i = 1$.

A Note About UKP5 Performance. In the computational results section we will show that UKP5 outperforms PYAsUKP in about two orders of magnitude. We grant the majority of the algorithm performance to the ability of applying sparsity, solution dominance and symmetry pruning with almost no overhead. At each iteration of capacity y sparsity and solution dominance are integrated in a single constant time test (line 12). This test, when combined with an item list sorted by non-increasing efficiency, also helps to avoid propagating big index values for the next positions of d, benefiting the performance of the solution generation with symmetry pruning (the use of $d[y]$ on line 16).

2.1 Solution Dominance

In this section we will give a more detailed explanation of the workings of the previously cited solution dominance. We use the $min_{ix}(s)$ notation to refer to the lowest index between the items that compose the solution s. The $max_{ix}(s)$ notation has analogue meaning.

When a solution t is pruned because s dominates t (lines 12 to 14), some solutions u, where $u \supsetneq t$, are not generated. If s dominates t and $u \supsetneq t$, and $max_{ix}(u \backslash t) \leq min_{ix}(t)$, then u is not generated by UKP5. In other words, if $\{3, 2\}$ is dominated, then $\{3, 2, 2\}$ and $\{3, 2, 1\}$ are not generated by UKP5, but $\{3, 2, 3\}$ or $\{3, 2, 5\}$ could yet be generated. Ideally, any u where $u \supsetneq t$ should not be generated as it will be dominated by a solution u' where $u' \supsetneq s$ anyway. It's interesting to note that this happens eventually, as any $t \cap \{i\}$ where $i > min_{ix}(t)$ will be dominated by $s \cap \{i\}$ (or by a solution that dominates $s \cap \{i\}$), and at some point no solution that is a superset of t is generated anymore.

2.2 Implementation Details

With the purpose of making the initial explanation simpler, we have omitted some steps that are relevant to the algorithm performance, but not essential for assessing its correctness. A complete overview of the omitted steps is presented at this section.

All the items are sorted by non-increasing efficiency and, between items with the same efficiency, by increasing weight. This speed ups the algorithm, but does not affect its correctness.

The y^* periodicity bound is computed as in [4, p. 223], and used to reduce the c value. We further proposed an UKP5-specific periodicity check that was successfully applied. This periodicity check isn't used to reduce the c capacity before starting UKP5, as y^*. The periodicity check is a stopping condition inside UKP5 main loop (11 and 22). Let y be the value of the variable y at line 11, and let y' be the biggest capacity where $g[y'] \neq 0 \wedge d[y'] > 1$. If at some moment $y > y'$ then we can stop the computation and fill the remaining capacity with copies of the first item (item of index 1). This periodicity check works only if the first item is the best item. If this assumption is false, then the described condition will never happen, and the algorithm will iterate until $y = c$ as usual. The algorithm correctness isn't affected.

There's an *else if* test at line 20. If $g[y + w_i] = g[y] + p_i \wedge i < d[y + w_i]$ then $d[y] \leftarrow i$. This may seem unnecessary, as appears to be an optimization of a rare case, where two solutions comprised from different item multisets have the same weight and profit. Nonetheless, without this test, the UKP5 was about 1800 (one thousand and eight hundreds) times slower on some subset-sum instance datasets.

We iterate only until $c - w_{min}$ (instead of c, in line 11), as it is the last y value that can affect $g[c]$). After this we search for a value greater than *opt* in the range $g[c - w_{min} + 1]$ to $g[c]$ and update *opt*.

3   Computational Results

In this section we describe the experiments environment, instance sets and results. We compare our UKP5 implementation, and the EDUK2 implementation provided by [12] (called PYAsUKP). The used source codes can be found at https://github.com/henriquebecker91/masters/tree/v0.1[1]. The times reported were given by the tools themselves and do not count the instance loading time. The runs external time[2] were also captured and no significant discrepancy was observed. Therefore, we have chosen to use the times reported by PYAsUKP and UKP5 (as is the common practice). For all instances, the weight, profit and capacity are integral.

We use the following notation: $rand(x, y)$ means a random integer between x and y (both inclusive); $x\overline{n}$ means x as a string concatenated with the value of variable n as a string. For example: if $n = 5000$ then $10\overline{n} = 105000$.

3.1   Environment

The computer used on the experiments was an ASUS R552JK-CN159H. The CPU has four physical cores (Intel Core i7-4700HQ Processor, 6M Cache, 3.40 GHz). The operating system used was Linux 4.3.3-2-ARCH x86_64 GNU/Linux (i.e. Arch linux). Three of the four cores were isolated using the *isolcpus* kernel flag. The *taskset* utility was used to execute UKP5 and PYAsUKP in parallel

---

[1] The UKP5 implementation is at **codes/cpp/** and two versions of PYAsUKP are at **codes/ocaml/**. The *pyasukp_site.tgz* is the version used to generate the instances, and was also available at http://download.gna.org/pyasukp/pyasukpsrc.html. A more stable version was provided by the authors. This version is in *pyasukp_mail.tgz* and it was used to solve the instances the results presented in Table 1. The *create_*_instances.sh* scripts inside **codes/sh/** were used to generate the instance datasets.

[2] Given by the *time* application, available at https://www.archlinux.org/packages/ extra/x86_64/time/. The bash internal command was *not* used.

on the isolated cores. The computer memory was never completely used (so no swapping was done). The UKP5 code was compiled with gcc (g++) version 5.3.0 (the *-O3 -std=c++11* flags were enabled).

3.2   Instance Sets

The instance sets aim to reproduce the ones described in [12]. The same tool was used to generate the datasets (PYAsUKP), and the same parameters were used, otherwise noted the contrary. In Subsect. 5.1.1 *Known "hard" instances* of [12] some sets of easy instances are used to allow comparison with MTU2. However, the authors reported integer overflow problems with MTU2 on harder instances. With exception of the subset-sum dataset, all datasets have a similar harder set (Subsect. 5.2.1 *New hard UKP instances* [12]). Thus, we considered in the runs only the harder ones. Each instance has a random capacity value within intervals shown in Table 1. The PYAsUKP parameters *-wmin w_{min} -cap c -n n* were used in all instances generation. We found some small discrepancies between the formulas presented in [12] and the ones used in PYAsUKP code. We opted for using the ones from PYAsUKP code, and they are presented below.

Subset-Sum. Instances generated with $p_i = w_i = rand(w_{min}, w_{max})$. The majority of the subset-sum instances used in [12] were solved on less than a centisecond in our experiments. This makes it easy to have imprecise measuring. Because of this, in this paper, we use a similar dataset, but with each parameter multiplied by ten. Therefore, we generated 10 instances for each possible combination of: $w_{min} \in \{10^3, 5 \times 10^3, 10^4, 5 \times 10^4, 10^5\}$; $w_{max} \in \{5 \times 10^5, 10^6\}$ and $n \in \{10^3, 2 \times 10^3, 5 \times 10^3, 10^4\}$, totaling 400 instances. We do not discriminate each combination in Table 1 for brevity. The PYAsUKP *-form ss -wmax w_{max}* parameters were used.

Strong Correlation. Instances generated using the following formula: $w_i = w_{min} + i - 1$ and $p_i = w_i + \alpha$, for a given w_{min} and α. Note that, except by the random capacity, all instances with the same α, **n**, and w_{min} combination are equal. The formula doesn't rely on random numbers. The PYAsUKP *-form chung -step α* parameters were used.

Postponed Periodicity. This family of instances is generated by the following method: **n** distinct weights are generated with $rand(w_{min}, w_{max})$ and then sorted by increasing order; $p_1 = w_1 + rand(1, 500)$; and $\forall i \in [2, n]$. $p_i = p_{i-1} + rand(1, 125)$. The w_{max} is computed as $10\bar{n}$. The PYAsUKP *-form nsds2 -step 500 -wmax w_{max}* parameters were used.

No Collective Dominance. This family of instances is generated by the following method: **n** distinct weights are generated with $rand(w_{min}, w_{max})$ and then sorted by increasing order; $p_1 = p_{min} + rand(0, 49)$; and $\forall i \in [2, n]$. $p_i =$

$\lfloor w_i \times ((p_{i-1}/w_{i-1})+0.01) \rfloor + rand(1,10)$. The given values are: $w_{min} = p_{min} = \mathbf{n}$ and $w_{max} = 10\overline{n}$. The PYAsUKP *-form hi -pmin p_{min} -wmax w_{max}* parameters were used.

SAW. This family of instances is generated by the following method: generate $\mathbf{n}$ random weights between w_{min} and $w_{max} = 1\overline{n}$ with the following property: $\forall i \in [2,n]$. $w_i \bmod w_1 > 0$ (w_1 is the smallest weight); sort by increasing order; then $p_1 = w_1 + \alpha$ where $\alpha = rand(1,5)$, and $\forall i \in [2,n]$. $p_i = rand(l_i, u_i)$ where $l_i = max(p_{i-1}, q_i)$, $u_i = q_i + m_i$, $q_i = p_1 \times \lfloor w_i/w_1 \rfloor$, and $m_i = w_i \bmod w_1$. The PYAsUKP *-form saw -step α -wmax w_{max}* parameters were used.

3.3   Results and Analysis

Table 1 presents the times used by UKP5 and PYAsUKP to solve the instance classes previously described. No time limit was defined. Figure 1 presents the same data, in logarithmic scale.

Based on Table 1, except by one instance set that we will talk about later, we can make two statements: (1) the average time, standard deviation, and maximal time of UKP5 are always smaller than the PYAsUKP ones; (2) the minimal PYAsUKP time is always smaller than the UKP5 one.

Let's begin with the second statement. As EDUK2 uses a branch-and-bound (B&B) algorithm before resorting to dynamic programming (DP), this is an expected result. Instances with big capacities and solutions that are composed by a large quantity of the best item, and a few non-best most efficient items, can be quickly solved by B&B. Our exception dataset (Strong Correlation, $\alpha = 5$, $n = 10$ and $w_{min} = 10$) is exactly this case. As said before, the strong correlation formula does not make use of random numbers, so all twenty instances of that dataset have the same items. The only thing that changes is the capacity. All solutions of this dataset are composed by hundreds of copies of the best item (that is also the smallest item, making the dataset even easier) and exactly one non-best item for making better use of the residual capacity ($c \bmod w_1$). All other datasets have instances that present the same characteristics, and because of that, the PYAsUKP minimal time is always close to zero. In Fig. 1 it is possible to observe that there are many instances solved in less than 10 s by PYAsUKP which took longer for UKP5 to solve. The number of instances where PYAsUKP was faster than UKP5 by instance class are: Subset-sum: 264 ($\approx 65\%$); Strong correlation: 60 (25 %); Postponed periodicity: 105 ($\approx 13\%$); No collective dominance: 259 ($\approx 13\%$); SAW: 219 ($\approx 20\%$). This from a total of 4540 instances.

For the instances that are solved by B&B in short time, the DP is not competitive against B&B. The UKP5 can't compete with PYAsUKP on easy datasets, as only the time for initializing an array of size c is already greater than the B&B's time. Nonetheless, for hard instances of combinatorial problems, B&B is known to show a bad worst case performance (exponential time). As EDUK2 combines B&B and DP with the intent of getting the strengths of both, and

Table 1. Columns **n** and w_{min} values must be multiplied by 10^3 to obtain their true value. Let T be the set of times reported by UKP5 or EDUK2, then the meaning of the columns **avg**, **sd**, **min** and **max**, is, respectively, the arithmetic mean of T, the standard deviation of T, the minimal value of T and the maximal value of T. The time unit of the table values is seconds.

Instance desc.			UKP5				PYAsUKP			
400 inst. per line			Subset-sum. Random c between $[5 \times 10^6 ; 10^7]$							
	n	w_{min}	**avg**	**sd**	**min**	**max**	**avg**	**sd**	**min**	**max**
See section 3.2			0.08	0.20	0.01	1.42	6.39	55.33	0.00	726.34
20 inst. per line			Strong correlation. Random c between $[20\overline{n} ; 100\overline{n}]$							
α	**n**	w_{min}	**avg**	**sd**	**min**	**max**	**avg**	**sd**	**min**	**max**
5	5	10	0.05	0.00	0.05	0.05	2.46	2.81	0.00	6.13
		15	0.07	0.00	0.07	0.09	5.84	2.43	0.00	8.82
		50	0.20	0.06	0.08	0.24	18.35	12.64	0.00	50.58
5	10	10	0.11	0.01	0.10	0.14	0.00	0.00	0.00	0.01
		50	0.49	0.03	0.47	0.60	41.97	33.97	0.00	93.18
		110	1.07	0.02	1.05	1.13	147.60	114.39	0.00	342.86
-5	5	10	0.06	0.00	0.06	0.07	5.98	4.02	0.00	11.99
		15	0.09	0.00	0.08	0.10	10.37	6.73	0.00	21.00
		50	0.21	0.05	0.09	0.24	39.31	30.16	0.00	89.44
-5	10	10	0.19	0.01	0.17	0.21	13.13	12.61	0.00	33.00
		50	0.54	0.02	0.52	0.59	82.97	71.22	0.00	206.74
		110	1.08	0.02	1.07	1.13	261.61	246.21	0.00	721.89
200 inst. per line			Postponed periodicity. Random c between $[w_{max} ; 2 \times 10^6]$							
	n	w_{min}	**avg**	**sd**	**min**	**max**	**avg**	**sd**	**min**	**max**
	20	20	1.42	0.31	0.55	2.77	17.00	17.05	0.01	63.96
	50	20	10.20	1.28	7.91	14.98	208.61	210.72	0.03	828.89
	20	50	1.59	0.32	0.96	2.99	27.68	22.79	0.02	100.96
	50	50	6.86	1.23	4.46	11.78	233.58	187.91	2.65	682.95
500 inst. per line			No collective dominance. Random c between $[w_{max} ; 1000\overline{n}]$							
	n	w_{min}	**avg**	**sd**	**min**	**max**	**avg**	**sd**	**min**	**max**
	5	n	0.05	0.01	0.03	0.10	0.78	0.59	0.00	2.66
	10	n	0.49	0.15	0.21	1.10	3.38	2.80	0.00	12.31
	20	n	0.99	0.19	0.63	2.02	13.08	12.80	0.01	62.12
	50	n	4.69	1.22	3.51	13.18	119.18	131.22	0.04	667.42
qtd inst. per line			SAW. Random c between $[w_{max} ; 10\overline{n}]$							
qtd	**n**	w_{min}	**avg**	**sd**	**min**	**max**	**avg**	**sd**	**min**	**max**
200	10	10	0.11	0.01	0.10	0.16	1.88	1.24	0.01	4.73
500	50	5	0.74	0.08	0.66	1.98	4.79	4.22	0.02	17.78
200	50	10	1.01	0.03	0.97	1.27	10.44	9.02	0.03	38.69
200	100	10	14.13	2.96	9.95	21.94	60.58	54.08	0.05	192.04

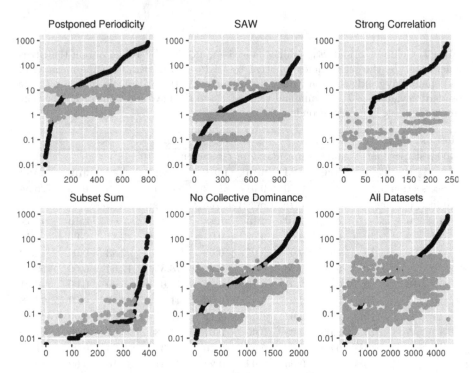

Fig. 1. The times used by UKP5 and PYAsUKP for each instance of each class. The black dots represent PYAsUKP times. The gray dots represent UKP5 times. The y axis is the time used to solve an UKP instance, in seconds. The x axis is the instance index when the instances are sorted by the time PYAsUKP took to solve it. Note that the y axis is in logarithmic scale.

none of its weaknesses, we found anomalous that this typical B&B behavior was present in PYAsUKP. We executed PYAsUKP with the -*nobb* flag, that disables the use of B&B. The PYAsUKP with disabled B&B had a performance worse than the one with B&B. For the presented classes, the ratios $\frac{no\text{-}B\&B\ avg\ time}{B\&B\ avg\ time}$ by instance class are: Subset-sum: 5.70; Strong correlation: 2.47; Postponed periodicity: 2.61; No collective dominance: 4.58; SAW: 4.07. For almost every individual instance no-B&B was worse than B&B (and when no-B&B was better this was by a small relative difference). Based on this evidence, we conclude that the PYAsUKP implementation of the EDUK2 DP-phase is responsible for the larger maximal PYAsUKP times (the time seems exponential but it is instead pseudo-polynomial with a big constant).

Looking back at the first statement of this section, we can now conclude that for instances that are hard for B&B, UKP5 clearly outperforms PYAsUKP DP by a big constant factor. Even considering the instances that PYAsUKP solves almost instantly (because of B&B), UKP5 is about 47 times faster than PYAsUKP, in average. If we ignored the advantage given by B&B (giving UKP5 a B&B phase, or removing the one used on EDUK2) this gap would be even greater.

We also compared our results with CPLEX. In [12] the authors presented results for CPLEX version 10.5, and showed that EDUK2 outperformed CPLEX. However, CPLEX efficiency has grown a lot in the last versions. Due to this, we run CPLEX 12.5. For the instances tested, UKP5 outperformed CPLEX 12.5 considerably. For the presented classes, the ratios $\frac{CPLEX\ avg\ time}{UKP5\ avg\ time}$ by instance class are: Subset-sum: 258.11; Strong correlation: 64.14; Postponed periodicity: 12.18; No collective dominance: 16.23; SAW: 120.14. Moreover, we set a time limit of 1,000 s and a memory limit of 2 GB for CPLEX, while every UKP5 and PYAsUKP run finished before these limits. The ratios above were computed considering 1,000 s for the instances that reached the time limit. However, from 4540 instances, in 402 runs the CPLEX reached the time limit. In 8 instances CPLEX reached the memory limit. We did not compare UKP5 with MTU2 since PYAsUKP already outperformed it, as shown in [12]. However, in a future work we intend to reimplement MTU2 to allow the comparison on the hard instances where it presented overflow problems.

The average UKP5 implementation memory consumption was greater than the PYAsUKP memory consumption. For each instance class, the UKP5-to-PYAsUKP memory consumption ratio was: Subset-sum: 10.09; Strong correlation: 2.84; Postponed periodicity: 1.62; No collective dominance: 12.41; SAW: 1.31. However, note that the UKP5 memory consumption worst case is $n + 2 \times c$ (pseudo-polynomial on n and c). The UKP5 consumed at most $\approx$1.6GB solving an instance.

4 Conclusion and Final Remarks

In this work we present UKP5, a new algorithm to solve the Unbounded Knapsack Problem based on dynamic programming. UKP5 outperformed PYAsUKP, the only known implementation of EDUK2, the state-of-the-art algorithm for solving the problem. When computing the speedups calculated as the ratio of times between the two algorithms, UKP5 is two orders of magnitude faster on average, considering the 4540 tested instances.

The core idea of UKP5 is to apply five improvements over a previously proposed dynamic programming algorithm. An analysis on the individual performance impact caused by each one of the five UKP5 improvements (see Sect. 2) will be presented in an extended version of this paper. Future works on the UKP5 should consider the following unanswered questions: PYAsUKP shows that the addition of a B&B phase before the DP can give good results, how could we apply the same idea to UKP5 and how would be the results? How is the performance of UKP5 applied in real-world instances generated by the column generation iterations for BPP and CSP?

Acknowledgments. We are very thankful to Vincent Poirriez for providing us the codes of a stable version of PYAsUKP, and answering our questions about the paper [12]. We are thankful to the CNPq (Conselho Nacional de Desenvolvimento Científico e Tecnológico) for the financial support.

References

1. Andonov, R., Poirriez, V., Rajopadhye, S.: Unbounded knapsack problem: dynamic programming revisited. Eur. J. Oper. Res. **123**(2), 394–407 (2000)
2. Belov, G., Scheithauer, G.: A branch-and-cut-and-price algorithm for one-dimensional stock cutting and two-dimensional two-stage cutting. Eur. J. Oper. Res. **171**(1), 85–106 (2006)
3. Delorme, M., Iori, M., Martello, S.: Bin packing and cutting stock problems: mathematical models and exact algorithms. In: Decision Models for Smarter Cities (2014)
4. Garfinkel, R.S., Nemhauser, G.L.: Integer Programming, vol. 4. Wiley, New York (1972)
5. Gilmore, P.C., Gomory, R.E.: A linear programming approach to the cutting-stock problem. Oper. Res. **9**(6), 849–859 (1961)
6. Gilmore, P.C., Gomory, R.E.: A linear programming approach to the cutting stock problem-Part II. Oper. Res. **11**(6), 863–888 (1963)
7. Hu, T.C.: Integer programming and network flows. Technical report, DTIC Document (1969)
8. Huang, P.H., Tang, K.: A constructive periodicity bound for the unbounded knapsack problem. Oper. Res. Lett. **40**(5), 329–331 (2012)
9. Iida, H.: Two topics in dominance relations for the unbounded knapsack problem. Open Appl. Math. J. **2**(1), 16–19 (2008)
10. Martello, S., Toth, P.: An exact algorithm for large unbounded knapsack problems. Oper. Res. Lett. **9**(1), 15–20 (1990)
11. Pisinger, D.: Algorithms for knapsack problems (1995)
12. Poirriez, V., Yanev, N., Andonov, R.: A hybrid algorithm for the unbounded knapsack problem. Discrete Optim. **6**(1), 110–124 (2009)

Lempel-Ziv Decoding in External Memory

Djamal Belazzougui[1], Juha Kärkkäinen[2(✉)], Dominik Kempa[2],
and Simon J. Puglisi[2]

[1] CERIST, Algiers, Algeria
dbelazzougui@cerist.dz
[2] Department of Computer Science, Helsinki Institute for Information
Technology HIIT, University of Helsinki, Helsinki, Finland
{juha.karkkainen,dominik.kempa,simon.puglisi}@cs.helsinki.fi

Abstract. Simple and fast decoding is one of the main advantages of
LZ77-type text encoding used in many popular file compressors such as
gzip and *7zip*. With the recent introduction of external memory algorithms for Lempel–Ziv factorization there is a need for external memory
LZ77 decoding but the standard algorithm makes random accesses to the
text and cannot be trivially modified for external memory computation.
We describe the first external memory algorithms for LZ77 decoding,
prove that their I/O complexity is optimal, and demonstrate that they
are very fast in practice, only about three times slower than in-memory
decoding (when reading input and writing output is included in the time).

1 Introduction

The Lempel–Ziv (LZ) factorization [18] is a partitioning of a text string into a
minimal number of phrases consisting of substrings with an earlier occurrence in
the string and of single characters. In LZ77 encoding [20] the repeated phrases are
replaced by a pointer to an earlier occurrence (called the source of the phrase).
It is a fundamental tool for data compression [6,7,15,17] and today it lies at the
heart of popular file compressors (e.g. *gzip* and *7zip*), and information retrieval
systems (see, e.g., [6,10]). Recently the factorization has become the basis for
several compressed full-text self-indexes [5,8,9,16]. Outside of compression, LZ
factorization is a widely used algorithmic tool in string processing: the factorization lays bare the repetitive structure of a string, and this can be used to design
efficient algorithms [2,12–14].

One of the main advantages of LZ77 encoding as a compression technique
is a fast and simple decoding: simply replace each pointer to a source by a
copy of the source. However, this requires a random access to the earlier part of
the text. Thus the recent introduction of external memory algorithms for LZ77
factorization [11] raises the question: Is fast LZ77 decoding possible when the text
length exceeds the RAM size? In this paper we answer the question positively
by describing the first external memory algorithms for LZ77 decoding.

This research is partially supported by Academy of Finland through grant 258308
and grant 250345 (CoECGR).

© Springer International Publishing Switzerland 2016
A.V. Goldberg and A.S. Kulikov (Eds.): SEA 2016, LNCS 9685, pp. 63–74, 2016.
DOI: 10.1007/978-3-319-38851-9_5

In LZ77 compression, the need for external memory algorithms can be avoided by using an encoding window of limited size. However, a longer encoding window can improve the compression ratio [6]. Even with a limited window size, decompression on a machine with a small RAM may require an external memory algorithm if the compression was done on a machine with a large RAM. Furthermore, in applications such as text indexing and string processing limiting the window size is not allowed. While most of these applications do not require decoding, a fast decoding algorithm is still useful for checking the correctness of the factorization.

Our Contribution. We show that in the standard external memory model of computation [19] the I/O complexity of decoding an LZ77-like encoding of a string of length n over an alphabet of size σ is $\Theta\left(\frac{n}{B\log_\sigma n}\log_{M/B}\frac{n}{B\log_\sigma n}\right)$, where M is the RAM size and B is the disk block size in units of $\Theta(\log n)$ bits. The lower bound is shown by a reduction from permuting and the upper bound by describing two algorithms with this I/O complexity.

The first algorithm uses the powerful tools of external memory sorting and priority queues while the second one relies on plain disk I/O only. Both algorithms are relatively simple and easy to implement. Our implementation uses the STXXL library [4] for sorting and priority queues.

Our experiments show that both algorithms scale well for large data but the second algorithm is much faster in all cases. This shows that, while external memory sorting and priority queues are extremely useful tools, they do have a significant overhead when their full power is not needed. The faster algorithm (using a very modest amount of RAM) is only 3–4 times slower than an in-memory algorithm that has enough RAM to perform the decoding in RAM (but has to read the input from disk and write the output to disk).

Our algorithms do not need a huge amount of disk space in addition to the input (factorization) and output (text), but we also describe and implement a version, which can reduce the additional disk space to less than 3 % of total disk space usage essentially with no effect on runtime.

2   Basic Definitions

Strings. Throughout we consider a string $X = X[1..n] = X[1]X[2]\ldots X[n]$ of $|X| = n$ symbols drawn from the alphabet $[0..\sigma-1]$ for $\sigma = n^{\mathcal{O}(1)}$. For $1 \le i \le j \le n$ we write $X[i..j]$ to denote the *substring* $X[i]X[i+1]\ldots X[j]$ of X. By $X[i..j)$ we denote $X[i..j-1]$.

LZ77. The *longest previous factor* (LPF) at position i in string X is a pair $\mathsf{LPF}[i] = (p_i, \ell_i)$ such that $p_i < i$, $X[p_i..p_i+\ell_i) = X[i..i+\ell_i)$, and ℓ_i is maximized. In other words, $X[i..i+\ell_i)$ is the longest prefix of $X[i..n]$ which also occurs at some position $p_i < i$ in X. There may be more than one potential value of p_i, and we do not care which one is used.

The LZ77 factorization (or LZ77 parsing) of a string X is a greedy, left-to-right parsing of X into longest previous factors. More precisely, if the jth LZ factor (or *phrase*) in the parsing is to start at position i, then $\mathsf{LZ}[j] = \mathsf{LPF}[i] = (p_i, \ell_i)$ (to represent the jth phrase), and then the $(j+1)$th phrase starts at position $i + \ell_i$. The exception is the case $\ell_i = 0$, which happens iff $\mathsf{X}[i]$ is the leftmost occurrence of a symbol in X. In this case $\mathsf{LZ}[j] = (\mathsf{X}[i], 0)$ (to represent $\mathsf{X}[i..i]$) and the next phrase starts at position $i + 1$. This is called a *literal phrase* and the other phrases are called *repeat phrases*. For a repeat phrases, the substring $\mathsf{X}[p_i..p_i + \ell_i)$ is called the *source* of the phrase $\mathsf{X}[i..i + \ell_i)$. We denote the number of phrases in the LZ77 parsing of X by z.

LZ77-type Factorization. There are many variations of LZ77 parsing. For example, the original LZ77 encoding [20] had only one type of phrase, a (potentially empty) repeat phrase always followed by a literal character. Many compressors use parsing strategies that differ from the greedy strategy described above to optimize compression ratio after entropy compression or to speed up compression or decompression. The algorithms described in this paper can be easily adapted for most of them. For purposes of presentation and analysis we make two assumptions about the parsing:

- All phrases are either literal or repeat phrases as described above.
- The total number of repeat phrases, denoted by z_{rep}, is $\mathcal{O}(n/\log_\sigma n)$.

We call this an *LZ77-type factorization*. The second assumption holds for the greedy factorization [18] and can always be achieved by replacing too short repeat phrases with literal phrases. We also assume that the repeat phrases are encoded using $\mathcal{O}(\log n)$ bits and the literal phrases using $\mathcal{O}(\log \sigma)$ bits. Then the size of the whole encoding is never more than $\mathcal{O}(n \log \sigma)$ bits.

3   On I/O Complexity of LZ Decoding

Given an LZ77-type factorization of a string encoded as described above, the task of LZ77 decoding is to recover the original string. In this section, we obtain a lower bound on the I/O complexity of LZ decoding by a reduction from permuting.

We do the analysis using the standard external memory model [19] with RAM size M and disk block size B, both measured in units of $\Theta(\log n)$ bits. We are primarily interested in the I/O complexity, i.e., the number of disk blocks moved between RAM and disk.

Given a sequence $\bar{x} = x_1, x_2, \ldots, x_n$ of n objects of size $\Theta(\log n)$ bits each and a permutation $\pi[1..n]$ of $[1..n]$, the task of permuting is to obtain the permuted sequence $\bar{y} = y_1, y_2, \ldots, y_n = x_{\pi[1]}, x_{\pi[2]}, \ldots, x_{\pi[n]}$. Under the mild assumption that $B \log(M/B) = \Omega(\log(n/B))$, the I/O complexity of permuting is $\Theta\left(\frac{n}{B} \log_{M/B} \frac{n}{B}\right)$, the same as the I/O complexity of sorting [1].

We show now that permuting can be reduced to LZ decoding. Let X be the string obtained from the sequence $\bar{x}$ by encoding each x_i as a string of length

$h = \Theta(\log_\sigma n)$ over the alphabet $[0..\sigma)$. Let Y be the string obtained in the same way from the sequence $\bar{y}$. Form an LZ77-type factorization of XY by encoding the first half using literal phrases and the second half using repeat phrases so that the substring representing y_i is encoded by the phrase $(h\pi[i] + 1 - h, h)$. This LZ factorization is easy to construct in $\mathcal{O}(n/B)$ I/Os given $\bar{x}$ and π. By decoding the factorization we obtain XY and thus $\bar{y}$.

Theorem 1. *The I/O complexity of decoding an LZ77-type factorization of a string of length n over an alphabet of size σ is*

$$\Omega\left(\frac{n}{B\log_\sigma n}\log_{M/B}\frac{n}{B\log_\sigma n}\right).$$

Proof. The result follows by the above reduction from permuting a sequence of $\Theta(n/\log_\sigma n)$ objects.   □

For comparison, the worst case I/O complexity of naive LZ decoding is $\mathcal{O}(n/\log_\sigma n)$.

4   LZ Decoding Using EM Sorting and Priority Queue

Our first algorithm for LZ decoding relies on the powerful tools of external memory sorting and external memory priority queues.

We divide the string X into $\lceil n/b \rceil$ segments of size exactly b (except the last segment can be smaller). The segments must be small enough to fit in RAM and big enough to fill at least one disk block. If a phrase or its source overlaps a segment boundary, the phrase is split so that all phrases and their sources are completely inside one segment. The number of phrases increases by at most $\mathcal{O}(z_{\text{rep}} + n/b)$ because of the splitting.

After splitting, the phrases are divided into three sequences. The sequence R_{far} contains repeat phrases with the source more than b positions before the phrase (called far repeat phrases) and the sequence R_{near} the other repeat phrases (called near repeat phrases). The sequence L contains all the literal phrases. The repeat phrases are represented by triples (p, q, ℓ), where p is the starting position of the source, q is the starting position of the phrase and ℓ is the length. The literal phrases are represented by pairs (q, c), where q is the phrase position and c is the character. The sequence R_{far} of far repeat phrases is sorted by the source position. The other two sequences are not sorted, i.e., they remain ordered by the phrase position.

During the computation, we maintain an external memory priority queue Q that stores already recovered far repeat phrases. Each such phrase is represented by a triple (q, ℓ, s), where q and ℓ are as above and s is the phrase as a literal string. The triples are extracted from the queue in the ascending order of q. The maximum length of phrases stored in the queue is bounded by a parameter $\ell_{\max}$. Longer phrases are split into multiple phrases before inserting them into the queue.

The string X is recovered one segment at a time in left-to-right order and each segment is recovered one phrase at a time in left-to-right order. A segment recovery is done in a (RAM) array $Y[0..b)$ of size b. At any moment in time, for some $i \in [0..b]$, $Y[0..i)$ contains the already recovered prefix of the current segment and $Y[i..b)$ contains the last $b - i$ characters of the preceding segment. The next phrase starting at $Y[i]$ is recovered in one of three ways depending on its type:

- A literal phrase is obtained as the next phrase in the sequence L.
- A near repeat phrase is obtained as the next phrase in the sequence R_{near}. The source of the phrase either starts in $Y[0..i)$ or is contained in $Y[i..b)$, and is easily recovered in both cases.
- A far repeat phrase is obtained from the priority queue with the full literal representation.

Once a segment has been fully recovered, we read all the phrases in the sequence R_{far} having the source within the current segment. Since R_{far} is ordered by the source position, this involves a single sequential scan of R_{far} over the whole algorithm. Each such phrase is inserted into the priority queue Q with its literal representation (splitting the phrase into multiple phrases if necessary).

Theorem 2. *A string of length n over an alphabet of size σ can be recovered from its LZ77 factorization in $\mathcal{O}\left(\frac{n}{B \log_\sigma n} \log_{M/B} \frac{n}{B \log_\sigma n}\right)$ I/Os.*

Proof. We set $\ell_{\max} = \Theta(\log_\sigma n)$ and $b = \Theta(B \log_\sigma n)$. Then the objects stored in the priority queue need $\mathcal{O}(\log n + \ell_{\max} \log \sigma) = \mathcal{O}(\log n)$ bits each and the total number of repeat phrases after all splitting is $\mathcal{O}(z_{rep} + n/\log_\sigma n) = \mathcal{O}(n/\log_\sigma n)$. Thus sorting the phrases needs $\mathcal{O}\left(\frac{n}{B \log_\sigma n} \log_{M/B} \frac{n}{B \log_\sigma n}\right)$ I/Os. This is also the I/O complexity of all the external memory priority queue operations [3]. All other processing is sequential and needs $\mathcal{O}\left(\frac{n}{B \log_\sigma n}\right)$ I/Os. □

We have implemented the algorithm using the STXXL library [4] for external memory sorting and priority queues.

5   LZ Decoding Without Sorting or Priority Queue

The practical performance of the algorithm in the previous section is often bounded by in-memory computation rather than I/O, at least on a machine with relatively fast disks. In this section, we describe an algorithm that reduces computation based on the observation that we do not really need the full power of external memory sorting and priority queues.

To get rid of sorting, we replace the sorted sequence R_{far} with $\lceil n/b \rceil$ unsorted sequences $R_1, R_2, \ldots$, where R_i contains all phrases with the source in the ith segment. In other words, sorting R_{far} is replaced with distributing the phrases into $R_1, R_2, \ldots$. If n/b is less than M/B, the distribution can be done in one pass, since we only need one RAM buffer of size B for each segment. Otherwise, we group M/B consecutive segments into a supersegment, distribute the phrases first into supersegments, and then into segments by scanning the supersegment sequences. If necessary, further layers can be added to the segment hierarchy. This operation generates the same amount of I/O as sorting but requires less computation because the segment sequences do not need to be sorted.

In the same way, the priority queue is replaced with $\lceil n/b \rceil$ simple queues. The queue Q_i contains a triple (q, ℓ, s) for each far repeat phrase whose phrase position is within the ith segment. The order of the phrases in the queue is arbitrary. Instead of inserting a recovered far repeat phrase into the priority queue Q it is appended into the appropriate queue Q_i. This requires a RAM buffer of size B for each queue but as above a multi-round distribution can be used if the number of segments is too large. This approach might not reduce the I/O compared to the use of a priority queue but it does reduce computation. Moreover, the simple queue allows the strings s to be of variable sizes and of unlimited length; thus there is no need to split the phrases except at segment boundaries.

Since the queues Q_i are not ordered by the phrase position, we can no more recover a segment in a strict left-to-right order, which requires a modification of the segment recovery procedure. The sequence R_{near} of near repeat phrases is divided into two: R_{prev} contains the phrases with the source in the preceding segment and R_{same} the ones with the source in the same segment.

As before, the recovery of a segment X_j starts with the previous segment in the array $Y[0..b)$ and consists of the following steps:

1. Recover the phrases in R_{prev} (that are in this segment). Note that each source is in the part of the previous segment that is still untouched.
2. Recover the literal phrases by reading them from L.
3. Recover the far repeat phrases by reading them from Q_j (with the full literal representation).
4. Recover the phrases in R_{same}. Note that each source is in the part of the current segment that has been fully recovered.

After the recovery of the segment, we read all the phrases in R_j and insert them into the queues Q_k with their full literal representations.

We want to minimize the number of segments. Thus we choose the segment size to occupy at least half of the available RAM and more if the RAM buffers for the queues Q_k do not require all of the other half. It is easy to see that this algorithm does not generate asymptotically more I/Os than the algorithm of the previous section. Thus the I/O complexity is $\mathcal{O}\left(\frac{n}{B \log_\sigma n} \log_{M/B} \frac{n}{B \log_\sigma n}\right)$. We have implemented the algorithm using standard file I/O (without the help of STXXL).

6   Reducing Disk Space Usage

The algorithm described in the previous section can adapt to a small RAM by using short segments, and if necessary, multiple rounds of distribution. However, reducing the segment size does not affect the disk space usage and the algorithm will fail if it does not have enough disk space to store all the external memory data. In this section, we describe how the disk space usage can be reduced.

The idea is to divide the LZ factorization into parts and to process one part at a time recovering the corresponding part of the text. The first part is processed with the algorithm of the previous section as if it was the full string. To process the later parts, a slightly modified algorithm is needed because, although all the phrases are in the current part, the sources can be in the earlier parts. Thus we will have the R_j queues for all the segments in the current and earlier parts but the Q_j queues only for the current part. The algorithm processes first all segments in the previous parts performing the following steps for each segment X_j:

– Read X_j from disk to RAM.
– Read R_j and for each phrase in R_j create the triple (q, ℓ, s) and write it to the appropriate queue Q_k.

Then the segments of the current part are processed as described in the previous section.

For each part, the algorithm reads all segments in the preceding parts. The number of additional I/Os needed for this is $\mathcal{O}(np/(B \log_\sigma n))$, where p is the number of parts. In other respects, the performance of the algorithm remains essentially the same.

We have implemented this partwise processing algorithm using greedy on-line partitioning. That is, we make each part as large as possible so that the peak disk usage does not exceed a given disk space budget. An estimated peak disk usage is maintained while reading the input. The implementation needs at least enough disk space to store the input (the factorization) and the output (the recovered string) but the disk space needed in addition to that can usually be reduced to a small fraction of the total with just a few parts.

7   Experimental Results

Setup. We performed experiments on a machine equipped with two six-core 1.9 GHz Intel Xeon E5-2420 CPUs with 15 MiB L3 cache and 120 GiB of DDR3 RAM. The machine had 7.2 TiB of disk space striped with RAID0 across four identical local disks achieving a (combined) transfer rate of about 480 MiB/s. The STXXL block size as well as the size of buffers in the algorithm based on plain disk I/O was set to 1 MiB.

The OS was Linux (Ubuntu 12.04, 64bit) running kernel 3.13.0. All programs were compiled using g++ version 4.7.3 with -O3 -DNDEBUG options. The machine had no other significant CPU tasks running and only a single thread of execution

Table 1. Statistics of data used in the experiments. All files are of size 256 GiB. The value of n/z (the average length of a phrase in the LZ77 factorization) is included as a measure of repetitiveness.

Name	σ	n/z
hg.reads	6	52.81
wiki	213	84.26
kernel	229	7767.05
random255	255	4.10

was used for computation. All reported runtimes are wallclock (real) times. In the experiments with a limited amount of RAM, the machine was rebooted with a kernel boot flag so that the unused RAM is unavailable even for the OS.

Datasets. For the experiments we used the following files varying in the number of repetitions and alphabet size (see Table 1 for some statistics):

- hg.reads: a collection of DNA reads (short fragments produced by a sequencing machine) from 40 human genomes[1] filtered from symbols other than $\{A, C, G, T, N\}$ and newline;
- wiki: a concatenation of three different English Wikipedia dumps[2] in XML format dated: 2014-07-07, 2014-12-08, and 2015-07-02;
- kernel: a concatenation of ∼16.8 million source files from 510 versions of Linux kernel[3];
- random255: a randomly generated sequence of bytes.

Experiments. In the first experiment we compare the implementation of the new LZ77 decoding algorithm not using external-memory sorting or priority queue to a straightforward internal-memory LZ77 decoding algorithm that scans the input parsing from disk and decodes the text from left to right. All copying of text from sources to phrases happens in RAM.

We use the latter algorithm as a baseline since it represents a realistic upper bound on the speed of LZ77 decoding. It needs enough RAM to accommodate the output text as a whole, and thus we were only able to process prefixes of test files up to size of about 120 GiB. In the runtime we include the time it takes to read the parsing from disk (we stream the parsing using a small buffer) and to write the output text to disk. The new algorithm, being fully external-memory algorithm, can handle full test instances. The RAM usage of the new algorithm was limited to 3.5 GiB.

The results are presented in Fig. 1. In nearly all cases the new algorithm is about three times slower than the baseline. This is due to the fact that in

---

[1] http://www.1000genomes.org/.
[2] http://dumps.wikimedia.org/.
[3] http://www.kernel.org/.

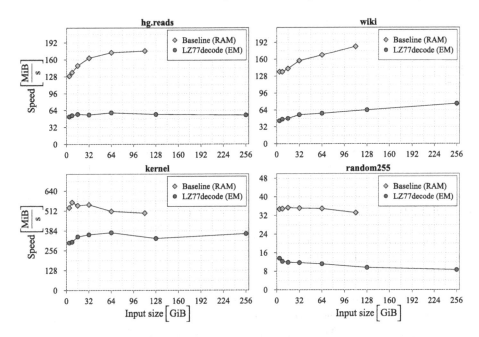

Fig. 1. Comparison of the new external memory LZ77 decoding algorithm based on plain disk I/O ("LZ77decode") with the purely in-RAM decoding algorithm ("Baseline"). The latter represents an upper bound on the speed of LZ77 decoding. The unit of decoding speed is MiB of output text decoded per second.

the external memory algorithm each text symbol in a far repeat phrase is read or written to disk three times: first, when written to a queue Q_j as a part of a recovered phrase, second, when read from Q_j, and third, when we write the decoded text to disk. In comparison, the baseline algorithm transfers each text symbol between RAM and disk once: when the decoded text is written to disk. Similarly, while the baseline algorithm usually needs one cache miss to copy the phrase from the source, the external memory algorithm performs about three cache misses per phrase: when adding the source of a phrase to R_j, when adding a literal representation of a phrase into Q_j, and when copying the symbols from Q_j into their correct position in the text. The exception of the above behavior is the highly repetitive kernel testfile that contains many near repeat phrases, which are processed as efficiently as phrases in the RAM decoding algorithm.

In the second experiment we compare our two algorithms described in Sects. 4 and 5 to each other. For the algorithm based on priority queue we set $\ell_{max} = 16$. The segment size in both algorithms was set to at least half of the available RAM (and even more if it did not lead to multiple rounds of EM sorting/distribution), except in the algorithm based on sorting we also need to allocate some RAM for the internal operations of STXXL priority queue. In all instances we allocate 1 GiB for the priority queue (we did not observe a notable effect on performance from using more space).

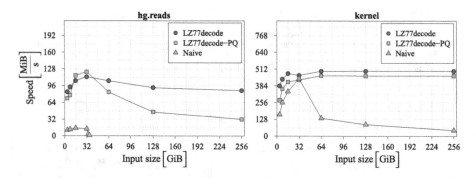

Fig. 2. Comparison of the new external memory LZ77 decoding algorithm based on plain disk I/O ("LZ77decode") to the algorithm implemented using external memory sorting and priority queue ("LZ77decode-PQ"). The comparison also includes the algorithm implementing naive approach to LZ77 decoding in external memory. The speed is given in MiB of output text decoded per second.

In the comparison we also include a naive external-memory decoding algorithm that works essentially the same as baseline RAM algorithm except we do not require that RAM is big enough to hold the text. Whenever the algorithm requests a symbol outside a window, that symbol is accessed from disk. We do not explicitly maintain a window of recently decoded text in RAM, and instead achieve a very similar effect by letting the operating system cache the recently accessed disk pages. To better visualize the differences in performance, all algorithms were allowed to use 32 GiB of RAM.

The results are given in Fig. 2. For highly repetitive input (kernel) there is little difference between the new algorithms, as they both copy nearly all symbols from the window of recently decoded text. The naive algorithm performs much worse, but still finishes in reasonable time due to large average length of phrases (see Table 1).

On the non-repetitive data (hg.reads), the algorithm using external-memory sorting and priority queue clearly gets slower than the algorithm using plain disk I/O as the size of input grows. The difference in constant factors is nearly three for the largest test instance. The naive algorithm maintains acceptable speed only up to a point where the decoded text is larger than available RAM. At this point random accesses to disk dramatically slow down the algorithm.

Note also that the speed of our algorithm in Fig. 2 is significantly higher than in Fig. 1. This is because the larger RAM (32 GiB vs. 3.5 GiB) allows larger segments, and larger segments mean that more of the repeat phrases are near repeat phrases which are faster to process than far repeat phrases.

In the third experiment we explore the effect of the technique described in Sect. 6 aiming at reducing the peak disk space usage of the new algorithm. We executed the algorithm on 32 GiB prefixes of two testfiles using 3.5 GiB of RAM and with varying disk space budgets. As shown in Fig. 3, this technique allows reducing the peak disk space usage to very little over what is necessary to store

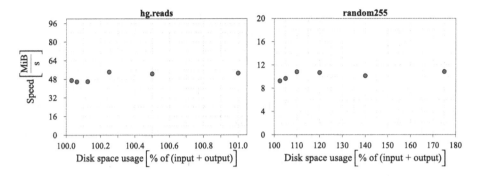

Fig. 3. The effect of disk space budget (see Sect. 6) on the speed of the new external-memory LZ77 decoding algorithm using plain disk I/O. Both testfiles were limited to 32 GiB prefixes and the algorithm was allowed to use 3.5 GiB of RAM. The rightmost data-point on each of the graphs represents a disk space budget sufficient to perform the decoding in one part.

the input parsing and output text and does not have a significant effect on the runtime of the algorithm, even on the incompressible random data.

8   Concluding Remarks

We have described the first algorithms for external memory LZ77 decoding. Our experimental results show that LZ77 decoding is fast in external memory setting too. The state-of-the-art external memory LZ factorization algorithms are more than a magnitude slower than our fastest decoding algorithm, see [11].

References

1. Aggarwal, A., Vitter, J.S.: The input/output complexity of sorting and related problems. Commun. ACM **31**(9), 1116–1127 (1988). doi:10.1145/48529.48535
2. Badkobeh, G., Crochemore, M., Toopsuwan, C.: Computing the maximal-exponent repeats of an overlap-free string in linear time. In: Calderón-Benavides, L., González-Caro, C., Chávez, E., Ziviani, N. (eds.) SPIRE 2012. LNCS, vol. 7608, pp. 61–72. Springer, Heidelberg (2012). doi:10.1007/978-3-642-34109-0_8
3. Brodal, G.S., Katajainen, J.: Worst-case efficient external-memory priority queues. In: Arnborg, S. (ed.) SWAT 1998. LNCS, vol. 1432, pp. 107–118. Springer, Heidelberg (1998). doi:10.1007/BFb0054359
4. Dementiev, R., Kettner, L., Sanders, P.: STXXL: standard template library for XXL data sets. Softw. Pract. Exper. **38**(6), 589–637 (2008). doi:10.1002/spe.844
5. Ferrada, H., Gagie, T., Hirvola, T., Puglisi, S.J.: Hybrid indexes for repetitive datasets. Phil. Trans. R. Soc. A **372** (2014). doi:10.1098/rsta.2013.0137
6. Ferragina, P., Manzini, G.: On compressing the textual web. In: Proceedings of 3rd International Conference on Web Search and Web Data Mining (WSDM), pp. 391–400. ACM (2010). doi:10.1145/1718487.1718536

7. Gagie, T., Gawrychowski, P., Kärkkäinen, J., Nekrich, Y., Puglisi, S.J.: A faster grammar-based self-index. In: Dediu, A.-H., Martín-Vide, C. (eds.) LATA 2012. LNCS, vol. 7183, pp. 240–251. Springer, Heidelberg (2012). doi:10.1007/978-3-642-13089-2_23

8. Gagie, T., Gawrychowski, P., Puglisi, S.J.: Faster approximate pattern matching in compressed repetitive texts. In: Asano, T., Nakano, S., Okamoto, Y., Watanabe, O. (eds.) ISAAC 2011. LNCS, vol. 7074, pp. 653–662. Springer, Heidelberg (2011). doi:10.1007/978-3-642-25591-5_67

9. Gagie, T., Gawrychowski, P., Kärkkäinen, J., Nekrich, Y., Puglisi, S.J.: A faster grammar-based self-index. In: Dediu, A.-H., Martín-Vide, C. (eds.) LATA 2012. LNCS, vol. 7183, pp. 240–251. Springer, Heidelberg (2012). doi:10.1007/978-3-642-28332-1_21

10. Hoobin, C., Puglisi, S.J., Zobel, J.: Relative Lempel-Ziv factorization for efficient storage and retrieval of web collections. Proc. VLDB **5**(3), 265–273 (2011)

11. Kärkkäinen, J., Kempa, D., Puglisi, S.J.: Lempel-Ziv parsing in external memory. In: Proceedings of 2014 Data Compression Conference (DCC), pp. 153–162. IEEE (2014). doi:10.1109/DCC.2014.78

12. Kolpakov, R., Bana, G., Kucherov, G.: MREPS: efficient and flexible detection of tandem repeats in DNA. Nucleic Acids Res. **31**(13), 3672–3678 (2003). doi:10.1093/nar/gkg617

13. Kolpakov, R., Kucherov, G.: Finding maximal repetitions in a word in linear time. In: Proceedings of 40th Annual Symposium on Foundations of Computer Science (FOCS), pp. 596–604. IEEE Computer Society (1999). doi:10.1109/SFFCS.1999.814634

14. Kolpakov, R., Kucherov, G.: Finding approximate repetitions under haamming distance. Theor. Comput. Sci. **303**(1), 135–156 (2003). doi:10.1016/S0304-3975(02)00448-6

15. Kreft, S., Navarro, G.: LZ77-like compression with fast random access. In: Proceedings of 2010 Data Compression Conference (DCC), pp. 239–248 (2010). doi:10.1109/DCC.2010.29

16. Kreft, S., Navarro, G.: Self-indexing based on LZ77. In: Giancarlo, R., Manzini, G. (eds.) CPM 2011. LNCS, vol. 6661, pp. 41–54. Springer, Heidelberg (2011). doi:10.1007/978-3-642-21458-5_6

17. Kuruppu, S., Puglisi, S.J., Zobel, J.: Relative Lempel-Ziv compression of genomes for large-scale storage and retrieval. In: Chavez, E., Lonardi, S. (eds.) SPIRE 2010. LNCS, vol. 6393, pp. 201–206. Springer, Heidelberg (2010). doi:10.1007/978-3-642-16321-0_20

18. Lempel, A., Ziv, J.: On the complexity of finite sequences. IEEE Trans. Inf. Theor. **22**(1), 75–81 (1976). doi:10.1109/TIT.1976.1055501

19. Vitter, J.S.: Algorithms and data structures for external memory. Found. Trends Theoret. Comput. Sci. **2**(4), 305–474 (2006). doi:10.1561/0400000014

20. Ziv, J., Lempel, A.: A universal algorithm for sequential data compression. IEEE Trans. Inf. Theor. **23**(3), 337–343 (1977). doi:10.1109/TIT.1977.1055714

A Practical Method for the Minimum Genus of a Graph: Models and Experiments

Stephan Beyer[1(✉)], Markus Chimani[1], Ivo Hedtke[1(✉)], and Michal Kotrbčík[2]

[1] Institute of Computer Science, University of Osnabrück, Osnabrück, Germany
{stephan.beyer,markus.chimani,ivo.hedtke}@uni-osnabrueck.de
[2] Department of Mathematics and Computer Science,
University of Southern Denmark, Odense, Denmark
kotrbcik@imada.sdu.dk

Abstract. We consider the problem of the minimum genus of a graph, a fundamental measure of non-planarity. We propose the first formulations of this problem as an integer linear program (ILP) and as a satisfiability problem (SAT). These allow us to develop the first working implementations of general algorithms for the problem, other than exhaustive search. We investigate several different ways to speed-up and strengthen the formulations; our experimental evaluation shows that our approach performs well on small to medium-sized graphs with small genus, and compares favorably to other approaches.

1 Introduction

We are concerned with the minimum genus problem, i.e., finding the smallest g such that a given graph $G = (V, E)$ has an embedding in the orientable surface of genus g. As one of the most important measures of non-planarity, the minimum genus of a graph is of significant interest in computer science and mathematics. However, the problem is notoriously difficult from the theoretical, practical, and also structural perspective. Indeed, its complexity was listed as one of the 12 most important open problems in the first edition of Garey and Johnson's book [22]; Thomassen established its NP-completeness in general [36] and for cubic graphs [37]. While the existence of an $\mathcal{O}(1)$-approximation can currently not be ruled out, there was no general positive result beyond a trivial $\mathcal{O}(|V|/g)$-approximation until a recent breakthrough by Chekuri and Sidiropoulos [9]. For graphs with bounded degree, they provide an algorithm that either correctly decides that the genus of the graph G is greater than g, or embeds G in a surface of genus at most $g^{\mathcal{O}(1)} \cdot (\log |V|)^{\mathcal{O}(1)}$. Very recently, Kawarabayashi and Sidiropoulos [27] showed that the bounded degree assumption can be omitted for the related problem of *Euler genus* by providing a $\mathcal{O}(g^{256}(\log |V|)^{189})$-approximation; however, this does not yield an approximation for orientable genus.

M. Chimani—Supported by the German Research Foundation (DFG) project CH 897/2-1.

A.V. Goldberg and A.S. Kulikov (Eds.): SEA 2016, LNCS 9685, pp. 75–88, 2016.
DOI: 10.1007/978-3-319-38851-9_6

Minimum genus is a useful parameter in algorithm design, since, similarly to the planar case, we can take advantage of the topological structure and design faster algorithms for graphs of bounded genus. However, these algorithms typically assume that the input graph is actually embedded in some surface, as for instance in [7,19]. Therefore, without a practical algorithm providing an embedding in a low-genus surface, these algorithms cannot be effectively implemented.

In the mathematical community, the genus of specific graph families is of interest ever since Ringel's celebrated determination of the genus of complete graphs [34]. Such research often combines numerous different approaches, including computer-aided methods, see, e.g., [12,28]. However, in practice it often turns out that even determining the genus of a single relatively small graph can be rather difficult as in [5,12,28,29,31]. One of the reasons is the large problem space—an r-regular graph with n vertices can have $[(r-1)!]^n$ embeddings. It is known that complete graphs have exponentially many embeddings of minimum genus; however, the known constructions are nearly symmetric and the problem becomes much more difficult when the minimum genus does not equal the trivial bound from Euler's formula, see, e.g., [28] for more details. While it is conjectured that the genus distribution of a graph—the number of its embeddings into each orientable surface—is unimodal, very little is known about the structure of the problem space both in theory and practice.

From a slightly different perspective, it has been known for a long time that deciding embeddability in a *fixed* surface is polynomial both for the toroidal [20] and the general case [17,21]. In fact, the minimum genus is fixed-parameter tractable as a result of the Robertson-Seymour theorem, since for every surface there are only finitely many forbidden graph minors, and testing for a fixed minor needs only polynomial time. While there is a direct linear-time algorithm deciding embeddability in a fixed surface [26,30], taking any of these algorithms to practice is very challenging for several reasons. First, the naïve approach of explicitly testing for each forbidden minor is not viable, since the list of forbidden minors is known only for the plane and the projective plane, and the number of minors grows rapidly: for the torus there are already more than 16 000 forbidden minors [8]. Second, Myrvold and Kocay [33] reviewed existing algorithms to evaluate their suitability for implementation in order to compute the complete list of forbidden toroidal minors. Unfortunately, they report that [20] contains a "fatal flaw", which also appears in the algorithm in [21], and that the algorithm in [17] is also "incorrect". Myrvold and Kocay conclude that "There appears to be no way to fix these problems without creating algorithms which take exponential time" [33]. Finally, Mohar's algorithm [30], even in the simpler toroidal case [25], is very difficult to implement correctly (see the discussion in [33]). Consequently, there is currently no correct implementation of any algorithm for the general case of the problem beyond exhaustive search.

It is thus desirable to have an effective and correct implementation of a practical algorithm for the minimum genus. Rather surprisingly, to the best of our knowledge, the approach to obtain practical algorithms via ILP (integer linear program) and SAT (satisfiability) solvers has never been attempted for the minimum genus so far.

Our contribution. We provide the first ILP and SAT formulations for the minimum genus problem, and discuss several different variants both under theoretical and practical considerations. Based thereon, we develop the first implementations of nontrivial general algorithms for the problem. We evaluate these implementations on benchmark instances widely used in the study of non-planarity measures for real-world graphs. In conjunction with suitable algorithmic support via preprocessing and efficient planarity tests, we are for the first time able to tackle general medium-sized, sparse real-world instances with small genus in practice. We also compare our implementations to existing approaches, namely exhaustive search and a tailored algebraic approach for special cases.

2 Minimum Genus ILP and SAT Formulations

Our terminology is standard and consistent with [32]. We consider finite undirected graphs and assume w.l.o.g. that all graphs are simple, connected, and have minimum degree 3. For each nonnegative integer g, there is, up to homeomorphism, a unique *orientable surface of genus g* and this surface is homeomorphic to a sphere with g added handles. An *embedding* of a graph G in a surface S is a representation of G in S without edge crossings; the minimum genus $\gamma(G)$ of a graph G is the minimum genus of an orientable surface into which G has an embedding. When considering embeddings it is often useful to specify the orientation in which we traverse an edge. Therefore, we may speak of two *arcs* (aka. directed edges, halfedges) that correspond to each edge. For a given graph $G = (V, E)$, let $A = \{uv, vu \mid \{u, v\} \in E\}$ denote the arc set arising from E by replacing each undirected edge by its two possible corresponding directed arcs.

A *rotation* at a vertex v is a cyclic order (counter-clockwise) of the neighbors of v. A *rotation system* of a graph G is a set of rotations, one for each vertex of G. Up to mirror images of the surfaces, there is a 1-to-1 correspondence between rotation systems of G and (cellular) embeddings of G into orientable surfaces (see [23, Theorem 3.2.3] and [18,24]). Given a rotation system of G, the corresponding embedding is obtained by *face tracing*: starting with an unused arc uv, move along it from u to v and continue with the arc vw, where w is the vertex after u at the rotation at v. This process stops by computing a face of the embedding when it re-encounters its initial arc. Repeatedly tracing faces eventually finds all faces of the embedding.

Euler's formula asserts that each (cellular) embedding of G in an orientable surface satisfies $|V| - |E| + f = 2 - 2g$, where f is the number of the faces of the embedding, and g is the genus of the underlying surface. It follows that (i) determining the genus of the underlying surface for a given rotation system is essentially equivalent to calculating the number of faces; and (ii) finding the genus of a graph corresponds to maximizing the number of faces over all rotation systems of the graph. See [32] for more details.

In this section, we describe how to reformulate the minimum genus problem as an integer linear program (ILP) or a related problem of Boolean satisfiability (SAT). Generally, such modeling approaches are known for several planarity concepts and non-planarity measures (e.g., crossing number, graph skewness, upward planarity) and often attain surprisingly strong results. However, for the minimum genus problem it is at first rather unclear how to capture the topological nature of the question in simple variables. To the best of our knowledge, there are no known formulations for this problem up to now.

We first describe the basic concepts of both formulations, and later consider possible ways to improve them. For convenience, we write $[k] := \mathbb{Z}_k$; addition and subtraction are considered modulo k.

2.1   ILP Formulation

Our formulation is based on finding an embedding with the largest number of faces. Therefore, it statically simulates the face tracing algorithm. Let $\bar{f}$ be an upper bound on the attainable number of faces; see Sect. 3 on how to obtain a simple linear bound. For each $i \in [\bar{f}]$, we have a binary variable x_i that is 1 iff the i-th face exists and a binary variable c_a^i, for each $a \in A$, that is 1 iff arc a is traversed by the i-th face. For each vertex $v \in V$ and neighbors $u, w \in N(v), u \neq w$, the binary variable $p_{u,w}^v$ is 1 iff w is the successor of u in the rotation at v. The ILP formulation then is:

$$\max \quad \sum_{i=1}^{\bar{f}} x_i \tag{1a}$$

$$\text{s.t} \quad x_i \leq \frac{1}{3} \sum_{a \in A} c_a^i \qquad \forall i \in [\bar{f}] \tag{1b}$$

$$\sum_{i=1}^{\bar{f}} c_a^i = 1 \qquad \forall a \in A \tag{1c}$$

$$\sum_{a \in \delta^-(v)} c_a^i = \sum_{a \in \delta^+(v)} c_a^i \qquad \forall i \in [\bar{f}], v \in V \tag{1d}$$

$$c_{vw}^i \geq c_{uv}^i + p_{u,w}^v - 1 \qquad \forall i \in [\bar{f}], v \in V, u \neq w \in N(v) \tag{1e}$$

$$c_{uv}^i \geq c_{vw}^i + p_{u,w}^v - 1 \qquad \forall i \in [\bar{f}], v \in V, u \neq w \in N(v) \tag{1f}$$

$$\sum_{w \in N(v), u \neq w} p_{u,w}^v = 1 \qquad \forall v \in V, u \in N(v) \tag{1g}$$

$$\sum_{u \in N(v), w \neq u} p_{u,w}^v = 1 \qquad \forall v \in V, w \in N(v) \tag{1h}$$

$$\sum_{u \in U} \sum_{w \in N(v) \setminus U} p_{u,w}^v \geq 1 \quad \forall v \in V, \emptyset \neq U \subsetneq N(v) \tag{1i}$$

$$x_i \in \{0, 1\} \qquad \forall i \in [\bar{f}] \tag{1j}$$

$$c_a^i \in \{0, 1\} \qquad \forall i \in [\bar{f}], a \in A \tag{1k}$$

$$p_{u,w}^v \in \{0, 1\} \qquad \forall v \in V, u \neq w \in N(v). \tag{1l}$$

Constraints (1b) ensure that if a face exists, it traverses *at least* three arcs[1]; inversely, each arc is traversed by exactly one face due to (1c). Equalities (1d) guarantee that at every vertex of a face i, the number of i-traversed incoming

---

[1] For a simple graph, the minimum genus embedding contains no face of length 1 or 2. On the other hand, we cannot be more specific than the lower bound of 3.

and outgoing arcs is identical. Inequalities (1e) and (1f) ensure that arcs uv and vw are both in the same face if w is the successor of u in the rotation at v. Constraints (1g) and (1h) ensure that p^v represents a permutation of the vertices in $N(v)$; (1i) ensures that p^v consists of a single cycle. Observe that maximizing (1a) guarantees that each face index corresponds to at most one facial walk.

2.2 SAT Formulation

To solve the above ILP, we will need to consider its linear relaxation (where the binary variables are replaced by variables in the interval $[0,1]$). It is easy to see that fractional values for the p^v matrices lead to very weak dual bounds. Therefore, we also consider SAT formulations. While general SAT solvers cannot take advantage of algebraically obtained (lower) bounds, state-of-the-art SAT solvers are highly tuned to quickly search a vast solution space by sophisticated branching, backtracking, and learning strategies. This can give them an upper hand over ILP approaches, in particular when the ILP's relaxation is weak.

In contrast to the ILP, a SAT problem has no objective function and simply asks for *some* satisfying variable assignment. In our case, we construct a SAT instance to answer the question whether the given graph allows an embedding with *at least* f faces. To solve the optimization problem, we iterate the process for increasing values of f until reaching unsatisfiability. We use the same notation as before, and construct the SAT formulation around the very same ideas. Each binary variable is now a Boolean variable instead. While a SAT is typically given in conjunctive normal form (CNF), we present it here as a conjunction of separate Boolean formulae (*rules*) for better readability. Their transformation into equisatisfiable CNFs is trivial. The SAT formulation is:

$$\neg(c_a^i \wedge c_a^j) \qquad \forall a \in A, i \neq j \in [f] \tag{2a}$$

$$\bigvee_{a \in A} c_a^i \qquad \forall i \in [f] \tag{2b}$$

$$p_{u,w}^v \rightarrow (c_{uv}^i \leftrightarrow c_{vw}^i) \qquad \forall v \in V, u \neq w \in N(v), i \in [f] \tag{2c}$$

$$\bigvee_{u \in N(v), u \neq w} p_{u,w}^v \qquad \forall v \in V, w \in N(v) \tag{2d}$$

$$\neg(p_{u,w}^v \wedge p_{u',w}^v) \qquad \forall v \in V, w \in N(v), u \neq u' \in N(v)\backslash\{w\} \tag{2e}$$

$$\bigvee_{w \in N(v), w \neq u} p_{u,w}^v \qquad \forall v \in V, u \in N(v) \tag{2f}$$

$$\neg(p_{u,w}^v \wedge p_{u,w'}^v) \qquad \forall v \in V, u \in N(v), w \neq w' \in N(v)\backslash\{u\} \tag{2g}$$

$$\bigvee_{u \in U, w \in N(v)\backslash U} p_{u,w}^v \qquad \forall v \in V, \emptyset \neq U \subsetneq N(v) \tag{2h}$$

Rules (2a) and (2b) enforce that each arc is traversed by exactly one face, cf. (1c). Rule (2c) ensures that the successor is in the same face, cf. (1e)–(1f). Rules (2d)–(2h) guarantee that p^v variables form rotations at v, cf. (1g)–(1i).

2.3   Improvements

There are several potential opportunities to improve upon the above formulations. In pilot studies we investigated their practical ramifications.

Symmetries (ILP). It seems worthwhile to add symmetry-breaking constraints $x_i \geq x_{i+1}$ or even $\sum_{a \in A} c_a^i \geq \sum_{a \in A} c_a^{i+1}$ for all $i \in [f-1]$ to the ILP. Surprisingly, this does not improve the overall running time (and the latter is even worse by orders of magnitude), and we refrain from using these constraints in the following.

Vertices of degree 3 (ILP&SAT). Let $V_3 := \{v \in V \mid \deg(v) = 3\}$. Consider a degree-3 vertex $v \in V_3$ with neighbors u_0, u_1, u_2. The only two possible rotations at v are $u_0 u_1 u_2$ and $u_2 u_1 u_0$. Hence, we can use a single binary/Boolean variable p^v whose assignment represents this choice.

In the ILP, we remove all $p_{u,w}^v$ variables for $v \in V_3$ and replace (1e)–(1i) by

$$c_{vu_{k+1}}^i \geq c_{u_k v}^i + p^v - 1 \quad \forall i \in [f], v \in V_3, k \in [3] \tag{3a}$$

$$c_{u_k v}^i \geq c_{vu_{k+1}}^i + p^v - 1 \quad \forall i \in [f], v \in V_3, k \in [3] \tag{3b}$$

$$c_{vu_k}^i \geq c_{u_{k+1}v}^i - p^v \quad \forall i \in [f], v \in V_3, k \in [3] \tag{3c}$$

$$c_{u_{k+1}v}^i \geq c_{vu_k}^i - p^v \quad \forall i \in [f], v \in V_3, k \in [3], \tag{3d}$$

where u_0, u_1, u_2 denote the arbitrarily but statically ordered neighbors of $v \in V_3$.

In the SAT formulation, we analogously replace (2c) by

$$p^v \to (c_{u_k v}^i \leftrightarrow c_{vu_{k+1}}^i) \quad \forall v \in V_3, k \in [3], i \in [f] \tag{4a}$$

$$\neg p^v \to (c_{u_{k+1}v}^i \leftrightarrow c_{vu_k}^i) \quad \forall v \in V_3, k \in [3], i \in [f]. \tag{4b}$$

As expected, this is faster by orders of magnitude for certain families of graphs, especially for instances with many degree-3 vertices. On the real world *Rome* benchmark set (see Sect. 4), the performance improves by about 10 % for both the ILP and the SAT formulations, compared to their respective formulations with $p_{u,w}^v$ variables.

This idea can be generalized for vertices v of arbitrary degree $d \geq 4$. There are $\varrho := (d-1)!$ different rotations. Instead of using $\mathcal{O}(d^2)$ many variables $p_{u,w}^v$, we introduce $\lceil \log_2 \varrho \rceil$ binary variables and representing the index of the rotation as a binary number. Since this process is coupled with a substantial trade-off of more complicated and weaker constraints, we refrain from using it for $d \geq 4$.

Binary face representations (SAT). Let $i \in [f]$ be a face index, and $\mathbb{B}(i)$ the vector of its binary representation, i.e., $i = \sum_{j=0}^{\ell} 2^j \cdot \mathbb{B}(i)_j$, where $\ell = \lfloor \log_2 f \rfloor$. We define new Boolean variables b_a^j that are true iff arc a is contained in a face i with $\mathbb{B}(i)_j = 1$. In logic formulae, value $\mathbb{B}(i)_j = 1$ is mapped to *true*, 0 to *false*.

By changing the following clauses of the SAT formulation above, we construct a new formulation that asks for a solution with at least f faces, because we do not forbid the usage of binary representations outside of $[f]$.

$$\bigvee_{a \in A} \bigwedge_{j \in [\ell]} (b_a^j \leftrightarrow \mathbb{B}(i)_j) \qquad \forall i \in [f] \qquad (2b')$$

$$p_{u,w}^v \rightarrow (b_{uv}^j \leftrightarrow b_{vw}^j) \qquad \forall v \in V \backslash V_3, u \neq w \in N(v), j \in [\ell] \qquad (2c')$$

$$p^v \rightarrow (b_{u_k v}^j \leftrightarrow b_{vu_{k+1}}^j) \qquad \forall v \in V_3, k \in [3], j \in [\ell] \qquad (4a')$$

$$\neg p^v \rightarrow (b_{u_{k+1}v}^j \leftrightarrow b_{vu_k}^j) \qquad \forall v \in V_3, k \in [3], j \in [\ell] \qquad (4b')$$

This variant achieves a more than 100-fold speedup.

2.4 Exponential vs. Polynomial Size Formulations

Observe that the number of inequalities (1i), or rules (2h) respectively, is exponential in the degree of each vertex v. Therefore, we investigate ways to obtain a polynomial time solution strategy or a polynomially sized formulation.

Efficient Separation. For the ILP we can *separate* violating constraints (also known as row generation) using a well-known separation oracle based on minimum cuts (see, e.g., [13, Sect. 7.4]). While this guarantees that only a polynomial-sized subset of (1i) is used, it is not worthwhile in practice: the separation process requires a comparably large overhead and state-of-the-art ILP solvers offer a lot of speed-up techniques that need to be deactivated to separate constraints on the fly. Overall, this more than doubles the running times compared to a direct inclusion of all (1i), even if we separate only for vertices with large degrees.

Another option is to use different representations for rotation systems. Here we discuss an *ordering* approach and a *betweenness* approach. Both yield polynomial size formulations.

Ordering Reformulation. For the ordering approach we replace the permutation variables with variables that attach vertices to specific positions in the rotation. This is known to be weaker in the realm of ILPs, and we hence concentrate on the SAT formulation. There, we introduce for any $v \in V, u \in N(v)$ a Boolean variable $q_{j,u}^v$ that is true iff u is the j-th vertex in the rotation at v. We do not use the p variables any longer, replace the old permutation rules (2d)–(2h) with rules to ensure that each q^v is a bijective mapping, and change (2c) to $\bigvee_{j \in [\deg(v)]} (q_{j,u}^v \wedge q_{j+1,w}^v) \rightarrow (c_{uv}^i \leftrightarrow c_{vw}^i)$ for all $v \in V$, $u \neq w \in N(v)$, $i \in [f]$. However, the SAT running times thereby increase 50–100-fold.

Betweenness Reformulation. For the betweenness approach we add the variables $r_{x,y,z}^v$ for each triple $x, y, z \in N(v)$. By $r_{x,y,z}^v = 1$ (true, respectively) we denote that y is (somewhere) between x and z in the rotation at v. Here we only describe the usage of the r variables in the SAT formulation. The usage in the ILP is analogous. First of all, the cyclicity of a rotation implies the symmetries $r_{x,y,z}^v \equiv r_{y,z,x}^v \equiv r_{z,x,y}^v \equiv \neg r_{x,z,y}^v \equiv \neg r_{z,y,x}^v \equiv \neg r_{y,x,z}^v$ for all $\{x,y,z\} \subseteq N(v)$. Instead of ensuring that each p^v represents a permutation, we connect the p variables to the new r variables via $p_{u,w}^v \leftrightarrow \bigwedge_{y \in N(v) \setminus \{u,w\}} r_{u,w,y}^v$. The rules to model the betweenness conditions for the neighborhood of a given vertex v are simply $r_{u,w,x}^v \wedge r_{u,x,y}^v \rightarrow r_{u,w,y}^v \wedge r_{w,x,y}^v$ for all $\{u,w,x,y\} \subseteq N(v)$. However, the SAT running times thereby increase 20–50-fold.

Overall, we conclude that the exponential dependencies of the original formulations are not so much of an issue in practice after all, and the overhead and weaknesses of polynomial strategies typically seem not worthwhile. However, if one considers problems with many very high degree vertices where the exponential dependency becomes an issue, the above approaches can be merged very naturally, leading to an overall polynomial model: Let τ be some fixed constant threshold value (to be decided upon experimentally). For vertices v of degree at most τ, we use the original formulation requiring an exponential (in constant τ) number of constraints over p^v. Vertices of degree above τ are handled via the betweenness reformulation.

3   A Minimum Genus Computation Framework

Before deploying any of our approaches on a given graph, we consider several preprocessing steps. Since the genus is additive over biconnected components [1,2], we decompose the input graph G accordingly. We can test $\gamma = 0$ by simply running a linear time planarity test, in our case [4]. Next, we observe that the genus problem is susceptible to *non-planar core reduction* [10]: A *maximal planar 2-component* is defined as a maximal subgraph $S \subset G$ that (i) has only two vertices x, y in common with the rest of the graph, and (ii) $S + (x,y)$ is planar. The (in our case unweighted) non-planar core (*NPC*) of G is obtained (in linear time) by replacing each such maximal planar 2-components by an edge.[2] After these steps we are in general left with a set of simple biconnected (preprocessed) graphs with minimum degree at least 3, for each of which we want to compute the genus.

By Euler's formula, we only have to calculate SAT instances with $f \equiv |E| - |V| \bmod 2$. For increasing number of faces we compute the satisfiability until we get the first unsatisfiable instance. Such an iteration is clearly not necessary

---

[2] In [10], the validity of such a preprocessing is shown for several non-planarity measures, namely crossing number, skewness, coarseness, and thickness. Let H be the NPC of G. We can trivially observe that (A) $\gamma(G) \leq \gamma(H)$, and (B) $\gamma(G) \geq \gamma(H)$. A: Given an optimal solution for H, we can embed each S onto the surface in place of its replacement edge, without any crossings. B: Each replaced component S contains a path connecting its poles that is drawn crossing-free in the optimal embedding of G; we can planarly draw all of S along this path, and then simplify the embedding by replacing this locally drawn S by its replacement edge; this gives a solution for H on the same surface.

in the ILP approach, where our objective function explicitly maximizes f and we only require an upper bound of $\bar{f} = \min\{\lfloor 2|E|/3 \rfloor, |E| - |V|\}$,[3] adjusted for parity.

Table 1. Characteristics of instances and resulting formulations. The graphs from the *Rome* (left table) and *North* (right table) benchmark sets are grouped by their number of vertices in the given ranges. For each group, we give the averages for the following values: number of vertices and percentage of degree-3 vertices in the NPC, upper bound $\bar{f}$ on the number of faces, number of variables and constraints in the ILP formulation.

range	avg. for computation on NPC					range	avg. for computation on NPC				
$|V|$	$|V|$	$\%|V_3|$	$\bar{f}$	#vars	#cons	$|V|$	$|V|$	$\%|V_3|$	$\bar{f}$	#vars	#cons
10–40	12.8	64.2	10.0	616.1	3399.5	10–40	12.6	38.3	17.4	2200.0	102295.9
41–60	18.5	60.3	15.3	1310.7	7639.9	41–60	24.6	40.3	29.9	4916.7	197577.3
61–80	26.8	59.4	22.5	2624.4	15735.1	61–80	32.1	43.5	35.5	7741.7	249864.6
81–100	36.4	58.5	30.9	4718.4	28778.3	81–100	24.3	40.6	34.7	7146.7	632634.6

4 Experimental Evaluation

Our C++ code is compiled with GCC 4.9.2, and runs on a single core of an AMD Opteron 6386 SE with DDR3 Memory @ 1600 MHz under Debian 8.0. We use the ILP solver CPLEX 12.6.1, the SAT solver lingeling (improved version for SMT Competition 2015 by Armin Biere)[4], and the Open Graph Drawing Framework (www.ogdf.net, GPL), and apply a 72 GB memory limit.

Real world graphs. We consider the established *Rome* [16] and *North* [15] benchmark sets of graphs collected from real-world applications. They are commonly used in the evaluation of algorithms in graph drawing and non-planarity measures. We use the ILP and SAT approaches to compute the genera of all 8249 (423) non-planar Rome (North) graphs. Each approach is run with a 30 min time limit for each graph to compute its genus; we omit 10 (*North*) instances that failed due to the memory limitation. Characteristics about the data sets and the resulting formulations can be found in Table 1.

Figure 1(a) shows the success rate (computations finished within the time limit) for the Rome graphs, depending on the number of vertices of the input graph. Both the SAT and ILP approach exhibit comparable numbers, but nearly always, the success rate of the SAT approach is as good or better than the ILP's. However, the differences are statistically not significant. Instances with up to 40 vertices can be solved with a high success rate; our approach degrades heavily

---

[3] First term: each edge lies on at most two faces, each face has size at least 3; second term: Euler's formula with genus at least 1.

[4] The previous version was the winner of the *Sequential Appl. SAT+UNSAT Track* of the SAT competition 2014 [3]. This improved version is even faster.

for graphs with more than 60–70 vertices. However, it is worth noting that even if the genus is not calculated to provable optimality, we obtain highly nontrivial bounds on the genus of the graphs in question.

In Fig. 1(b) we see that, given *any* fixed time limit below 30 min, the SAT approach solves clearly more instances than the ILP approach. Note that the curve that corresponds to the solved SAT instances flattens out very quickly.

When we compare the success rates to the density of the NPC (see Fig. 1(c)), we see the same characteristics as in Fig. 1(a). Both approaches are able to solve instances with density (i.e., $|E|/|V|$) up to 1.6 with a high success rate but are typically not able to obtain provably optimal values for densities above 1.9.

Finally, we compare the average running time of the instances that are solved by both approaches. Out of the 8249 non-planar Rome graphs we are able to solve 2571 with SAT *and* ILP, and additionally 96 (24) more with the SAT (ILP, respectively). Except for very small graphs, the average running time of the SAT approach is always at least one or two orders of magnitude lower than the average running time of the ILP approach, see Fig. 1(d).

Considering the non-planar North graphs, Fig. 1(e) shows that the success rates of both approaches are again comparable. Again, the differences are statistically not significant. However, ten instances could not be solved due to the high memory consumption caused by the exponential number of constraints (1i) and rules (2h). Since the results for the North graphs are analogous to those for the Rome graphs, we omit discussing them in detail.

Generally, we observe that the SAT approach is particularly fast to show the existence of an embedding, but is relatively slow to prove that there is no embedding with a given number of faces. This is of particular interest for non-planar graphs that allow a genus-1 embedding, since there the SAT is quick to find such a solution and need not prove that a lower surface is infeasible. The SAT's behavior in fact suggests an easy heuristical approach: if solving the SAT instance for f faces needs a disproportionally long running time (compared to the previous iterations for lower face numbers), this typically indicates that it is an unsatisfiable instance and $f - 2$ faces is the optimal value.

Comparison to existing genus computations. An evaluation of exhaustive search algorithms for determining the genus distribution of complete graphs was performed in [35]. Fixing the rotation of the first vertex, it is possible to compute the genus of distribution the complete graph K_7 within 896 h of computation (112 h on 8 parallel threads). While *both* our approaches perform significantly better, there is a notable (and w.r.t. to the above evaluations particularly surprising) difference in their performance: the SAT approach needs one hour to find and prove the optimal genus; solving the ILP takes only 30 s.

A circulant $C_n(S)$ is the Cayley graph of $\mathbb{Z}_n$ with generating set S. Conder and Grande [12] recently characterized all circulants with genus 1 and 2. A crucial part of the characterization is the determination of the genus of several sporadic cases where the lower bounds are more problematic. At the same time, these sporadic cases constitute the main obstacle in both obtaining a simpler proof, as well as extending the results to higher genera. By far the most difficult case

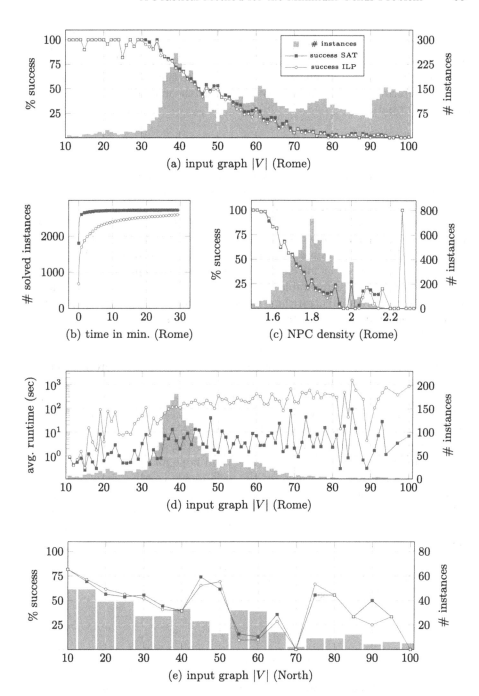

Fig. 1. Rome Graphs: (a) success rate per $|V|$, (b) solved instances per given time, (c) success rate per non-planar core density $|E|/|V|$, (d) average running time per $|V|$ where both approaches were successful. North graphs: (e) success rate per $|V|$.

is proving that the genus of $C_{11}(1, 2, 4)$ is at least 3. The proof takes three pages of theoretical analysis and eventually resorts to a computational verification of three subcases, taking altogether around 85 h using the MAGMA computational algebra system in a nontrivial problem-specific setting. The ILP solver needs 180 h to determine the genus without using any theoretical results or problem-specific information.

5   Conclusion

The minimum genus problem is very difficult from the mathematical, algorithmic, and practical perspective—the problem space is large and seems not to be well-structured, the existing algorithms are error-prone and/or very difficult to implement, and only little progress was made on the (practice-oriented) algorithmic side. In this paper we have presented the first ILP and SAT formulations, together with several variants and alternative reformulations, for the problem, and investigated them in an experimental study. Our approach leads to the first (even easily!) implementable general-purpose minimum genus algorithms. Besides yielding practical algorithms for small to medium-sized graphs and small genus, one of the further advantages of our approach is that the formulations are adaptable and can be modified to tackle other related problems of interest. For example, the existence of polyhedral embeddings [32], or embeddings with given face lengths, say 5 and 6 as in the case of fullerenes (graph-theoretic models of carbon molecules), see [14].

On the negative side, our implementations cannot deal with too large graphs without resorting to extensive computational resources. However, this is not very surprising considering the difficulty of the problem—a fast exact algorithm could be used to solve several long-standing open problems, such as completing the list of forbidden toroidal minors. We also see—and hope for—certain similarities to the progress on exact algorithms for the well-known crossing number problem: while the first published report [6] was only capable of solving Rome graphs with 30–40 vertices, it led to a series of improvements that culminated in the currently strongest variant [11] which is capable to tackle even the largest Rome graphs.

Acknowledgements. We thank Armin Biere for providing the most recent version (as of 2015-06-05) of the lingeling SAT solver.

References

1. Archdeacon, D.: The orientable genus is nonadditive. J. Graph Theor. **10**(3), 385–401 (1986)
2. Battle, J., Harary, F., Kodama, Y., Youngs, J.W.T.: Additivity of the genus of a graph. Bull. Amer. Math. Soc. **68**, 565–568 (1962)
3. Belov, A., Diepold, D., Heule, M.J., Järvisalo, M. (eds.): Proceedings of SAT Competition 2014: Solver and Benchmark Descriptions. No. B-2014-2 in Series of Publications B, Department Of Computer Science, University of Helsinki (2014)

4. Boyer, J.M., Myrvold, W.J.: On the cutting edge: simplified $O(n)$ planarity by edge addition. J. Graph Algorithms Appl. **8**(2), 241–273 (2004)
5. Brin, M.G., Squier, C.C.: On the genus of $Z_3 \times Z_3 \times Z_3$. Eur. J. Comb. **9**(5), 431–443 (1988)
6. Buchheim, C., Ebner, D., Jünger, M., Klau, G.W., Mutzel, P., Weiskircher, R.: Exact crossing minimization. In: Healy, P., Nikolov, N.S. (eds.) GD 2005. LNCS, vol. 3843, pp. 37–48. Springer, Heidelberg (2006)
7. Cabello, S., Chambers, E.W., Erickson, J.: Multiple-source shortest paths in embedded graphs. SIAM J. Comput. **42**(4), 1542–1571 (2013)
8. Chambers, J.: Hunting for torus obstructions. M.Sc. thesis, University of Victoria (2002)
9. Chekuri, C., Sidiropoulos, A.: Approximation algorithms for euler genus and related problems. In: Proceedings of FOCS 2013, pp. 167–176 (2013)
10. Chimani, M., Gutwenger, C.: Non-planar core reduction of graphs. Disc. Math. **309**(7), 1838–1855 (2009)
11. Chimani, M., Mutzel, P., Bomze, I.: A new approach to exact crossing minimization. In: Halperin, D., Mehlhorn, K. (eds.) ESA 2008. LNCS, vol. 5193, pp. 284–296. Springer, Heidelberg (2008)
12. Conder, M., Grande, R.: On embeddings of circulant graphs. Electron. J. Comb. **22**(2), P2.28 (2015)
13. Cook, W.J., Cunningham, W.H., Pulleyblank, W.R., Schrijver, A.: Combinatorial Optimization. Wiley-Interscience Series in Discrete Mathematics and Optimization. Wiley, New York (1998)
14. Deza, M., Fowler, P.W., Rassat, A., Rogers, K.M.: Fullerenes as tilings of surfaces. J. Chem. Inf. Comput. Sci. **40**(3), 550–558 (2000)
15. Di Battista, G., Garg, A., Liotta, G., Parise, A., Tamassia, R., Tassinari, E., Vargiu, F., Vismara, L.: Drawing directed acyclic graphs: an experimental study. Int. J. Comput. Geom. Appl. **10**(6), 623–648 (2000)
16. Di Battista, G., Garg, A., Liotta, G., Tamassia, R., Tassinari, E., Vargiu, F.: An experimental comparison of four graph drawing algorithms. Comput. Geom. **7**(5–6), 303–325 (1997)
17. Djidjev, H., Reif, J.: An efficient algorithm for the genus problem with explicit construction of forbidden subgraphs. In: Proceedings of STOC 1991, pp. 337–347. ACM (1991)
18. Edmonds, J.: A combinatorial representation for polyhedral surfaces. Not. Amer. Math. Soc. **7**, 646 (1960)
19. Erickson, J., Fox, K., Nayyeri, A.: Global minimum cuts in surface embedded graphs. In: Proceedings of SODA 2012, pp. 1309–1318. SIAM (2012)
20. Filotti, I.S.: An efficient algorithm for determining whether a cubic graph is toroidal. In: Proceedings of STOC 1978, pp. 133–142. ACM (1978)
21. Filotti, I.S., Miller, G.L., Reif, J.: On determining the genus of a graph in $O(V^{O(G)})$ steps. In: Proceedings of STOC 1979, pp. 27–37. ACM (1979)
22. Garey, M.R., Johnson, D.S.: Computers and Intractability. A Guide to the theory of NP-completeness. Bell Telephone Laboratories, New York (1979)
23. Gross, J.L., Tucker, T.W.: Topological Graph Theory. Wiley-Interscience Series in Discrete Mathematics and Optimization. Wiley, New York (1987)
24. Heffter, L.: Ueber das Problem der Nachbargebiete. Math. Ann. **38**, 477–508 (1891)
25. Juvan, M., Marinček, J., Mohar, B.: Embedding graphs in the torus in linear time. In: Balas, E., Clausen, J. (eds.) IPCO 1995. LNCS, vol. 920, pp. 360–363. Springer, Heidelberg (1995)

26. Kawarabayashi, K., Mohar, B., Reed, B.: A simpler linear time algorithm for embedding graphs into an arbitrary surface and the genus of graphs of bounded tree-width. In: Proceedings of FOCS 2008, pp. 771–780 (2008)
27. Kawarabayashi, K., Sidiropoulos, A.: Beyond the euler characteristic: approximating the genus of general graphs. In: Proceedings of STOC 2015. ACM (2015)
28. Kotrbčík, M., Pisanski, T.: Genus of cartesian product of triangles. Electron. J. Comb. **22**(4), P4.2 (2015)
29. Marušič, D., Pisanski, T., Wilson, S.: The genus of the GRAY graph is 7. Eur. J. Comb. **26**(3–4), 377–385 (2005)
30. Mohar, B.: Embedding graphs in an arbitrary surface in linear time. In: Proceedings of STOC 1996, pp. 392–397. ACM (1996)
31. Mohar, B., Pisanski, T., Škoviera, M., White, A.: The cartesian product of 3 triangles can be embedded into a surface of genus 7. Disc. Math. **56**(1), 87–89 (1985)
32. Mohar, B., Thomassen, C.: Graphs on Surfaces. Johns Hopkins Studies in the Mathematical Sciences. Johns Hopkins University Press, Baltimore (2001)
33. Myrvold, W., Kocay, W.: Errors in graph embedding algorithms. J. Comput. Syst. Sci. **77**(2), 430–438 (2011)
34. Ringel, G.: Map Color Theorem. Springer, Heidelberg (1974)
35. Schmidt, P.: Algoritmické vlastnosti vnorení grafov do plôch. B.Sc. thesis, Comenius University (2012). In Slovak
36. Thomassen, C.: The graph genus problem is NP-complete. J. Algorithms **10**, 568–576 (1989)
37. Thomassen, C.: The graph genus problem is NP-complete for cubic graphs. J. Comb. Theor. Ser. B **69**, 52–58 (1997)

Compact Flow Diagrams for State Sequences

Kevin Buchin[1]([✉]), Maike Buchin[2], Joachim Gudmundsson[3], Michael Horton[3], and Stef Sijben[2]

[1] Department of Mathematics and Computer Science,
TU Eindhoven, Eindhoven, The Netherlands
`k.a.buchin@tue.nl`
[2] Department of Mathematics, Ruhr-Universität Bochum, Bochum, Germany
`{Maike.Buchin,Stef.Sijben}@ruhr-uni-bochum.de`
[3] School of Information Technologies, The University of Sydney, Sydney, Australia
`{joachim.gudmundsson,michael.horton}@sydney.edu.au`

Abstract. We introduce the concept of compactly representing a large number of state sequences, e.g., sequences of activities, as a flow diagram. We argue that the flow diagram representation gives an intuitive summary that allows the user to detect patterns among large sets of state sequences. Simplified, our aim is to generate a small flow diagram that models the flow of states of all the state sequences given as input. For a small number of state sequences we present efficient algorithms to compute a minimal flow diagram. For a large number of state sequences we show that it is unlikely that efficient algorithms exist. More specifically, the problem is $W[1]$-hard if the number of state sequences is taken as a parameter. We thus introduce several heuristics for this problem. We argue about the usefulness of the flow diagram by applying the algorithms to two problems in sports analysis. We evaluate the performance of our algorithms on a football data set and generated data.

1 Introduction

Sensors are tracking the activity and movement of an increasing number of objects, generating large data sets in many application domains, such as sports analysis, traffic analysis and behavioural ecology. This leads to the question of how large sets of sequences of activities can be represented compactly. We introduce the concept of representing the "flow" of activities in a compact way and argue that this is helpful to detect patterns in large sets of state sequences.

To describe the problem we start by giving a simple example. Consider three objects (people) and their sequences of states, or activities, during a day. The set of state sequences $\mathcal{T} = \{\tau_1, \tau_2, \tau_3\}$ are shown in Fig. 1(a). As input we are also given a set of criteria $\mathcal{C} = \{C_1, \ldots, C_k\}$, as listed in Fig. 1(b). Each criterion is a Boolean function on a single subsequence of states, or a set of subsequences of states. For example, in the given example the criterion C_1 = "eating" is true for Person 1 at time intervals 7–8 am and 7–9 pm, but false for all other time intervals. Thus, a criterion partitions a sequence of states into subsequences, called *segments*. In each segment the criterion is either true or false. A *segmentation* of

© Springer International Publishing Switzerland 2016
A.V. Goldberg and A.S. Kulikov (Eds.): SEA 2016, LNCS 9685, pp. 89–104, 2016.
DOI: 10.1007/978-3-319-38851-9_7

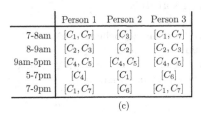

	Person 1	Person 2	Person 3
7-8am	breakfast	gym	breakfast
8-9am	cycle to work	drive to work	cycle to work
9am-5pm	work	work	work
5-7pm	study	dinner	shop
7-9pm	dinner	shop	dinner

(a)

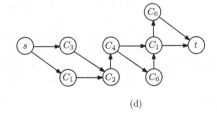

C_1: Eating {breakfast,dinner}
C_2: Commuting {cycle/drive to work}
C_3: Exercising {gym,cycle to work}
C_4: Working or studying
C_5: Working for at least 4 hours
C_6: Shopping
C_7: At least 2 people eating simultaneously

(b)

	Person 1	Person 2	Person 3
7-8am	$[C_1, C_7]$	$[C_3]$	$[C_1, C_7]$
8-9am	$[C_2, C_3]$	$[C_2]$	$[C_2, C_3]$
9am-5pm	$[C_4, C_5]$	$[C_4, C_5]$	$[C_4, C_5]$
5-7pm	$[C_4]$	$[C_1]$	$[C_6]$
7-9pm	$[C_1, C_7]$	$[C_6]$	$[C_1, C_7]$

(c)

(d)

Fig. 1. The input is (a) a set $T = \{\tau_1, \ldots, \tau_m\}$ of sequences of states and (b) a set of criteria $C = \{C_1, \ldots, C_k\}$. (c) The criteria partition the states into a segmentation. (d) A valid flow diagram for T according to C.

T is a partition of each sequence in T into true segments, which is represented by the corresponding sequence of criteria. If a criterion C is true for a set of subsequences, we say they *fulfil* C. Possible segments of T according to the set C are shown in Fig. 1(c). The aim is to summarize segmentations of all sequences efficiently; that is, build a flow diagram $\mathcal{F}$, starting at a start state s and ending at an end state t, with a small number of nodes such that for each sequence of states τ_i, $1 \le i \le m$, there exists a segmentation according to C which appears as an s–t path in $\mathcal{F}$. A possible flow diagram is shown in Fig. 1(d). This flow diagram for T according to C can be validated by going through a segmentation of each object while following a path in $\mathcal{F}$ from s to t. For example, for Person 1 the s–t path $s \to C_1 \to C_2 \to C_4 \to C_1 \to t$ is a valid segmentation.

Now we give a formal description of the problem. A *flow diagram* is a node-labelled DAG containing a source node s and sink node t, and where all other nodes are labelled with a criterion. Given a set T of sequences of states and a set of criteria C, the goal is to construct a flow diagram with a minimum number of nodes, such that a segmentation of each sequence of states in T is represented, that is, included as an s–t path, in the flow diagram. Furthermore (when criteria depend on multiple state sequences, e.g. C_7 in Fig. 1) we require that the segmentations represented in the flow diagram are consistent, i.e. can be jointly realized. The *Flow Diagram problem* thus requires the segmentations of each sequence of states and the minimal flow diagram of the segmentations to be computed. It can be stated as:

Problem 1. Flow Diagram (FD)

Instance: A set of sequences of states $T = \{\tau_1, \ldots, \tau_m\}$, each of length at most n, a set of criteria $C = \{C_1, \ldots, C_k\}$ and an integer $\lambda > 2$.

Question: Is there a flow diagram $\mathcal{F}$ with $\leq \lambda$ nodes, such that for each $\tau_i \in \mathcal{T}$, there exists a segmentation according to $\mathcal{C}$ which appears as an s–t path in $\mathcal{F}$?

Even the small example above shows that there can be considerable space savings by representing a set of state sequences as a flow diagram. This is not a lossless representation and comes at a cost. The flow diagram represents the sequence of flow between states, however, the information about an individual sequence of states is lost. As we will argue in Sect. 3, paths representing many segments in the obtained flow diagrams show interesting patterns. We will give two examples. First we consider segmenting the morphology of formations of a defensive line of football players during a match (Fig. 4). The obtained flow diagram provides an intuitive summary of these formations. The second example models attacking possessions as state sequences. The summary given by the flow diagram gives intuitive information about differences in attacking tactics.

Properties of Criteria. The efficiency of the algorithms will depend on properties of the criteria on which the segmentations are based. Here we consider four cases: (i) general criteria without restrictions; (ii) monotone decreasing and independent criteria; (iii) monotone decreasing and dependent criteria; and (iv) fixed criteria. To illustrate the properties we will again use the example in Fig. 1.

A criterion C is *monotone decreasing* [8] for a given sequence of states τ that fulfils C, if all subsequences of τ also fulfil C. For example, if C_4 is fulfilled by a sequence τ then any subsequence τ' of τ will also fulfil C_4. This is in contrast to criterion C_5 which is not monotone decreasing.

A criterion C is *independent* if checking whether a subsequence τ' of a sequence $\tau_i \in \mathcal{T}$ fulfils C can be achieved without reference to any other sequences $\tau_j \in \mathcal{T}, i \neq j$. Conversely, C is *dependent* if checking that a subsequence τ' of τ_i requires reference to other state sequences in $\mathcal{T}$. In the above example C_4 is an example of an independent criterion while C_7 is a dependent criterion since it requires that at least two objects fulfil the criterion at the same time.

Related Work. To the best of our knowledge compactly representing sequences of states as flow diagrams has not been considered before. The only related work we are aware of comes from the area of trajectory analysis. Spatial trajectories are a special case of state sequences. A spatial trajectory describes the movement of an object through space over time, where the states are location points, which may also include additional information such as heading, speed, and temperature. For a single trajectory a common way to obtain a compact representation is *simplification* [10]. Trajectory simplification asks to determine a subset of the data that represents the trajectory well in terms of the location over time. If the focus is on characteristics other than the location, then *segmentation* [1,2,8] is used to partition a trajectory into a small number of subtrajectories, where each subtrajectory is homogeneous with respect to some characteristic. This allows a trajectory to be compactly represented as a sequence of characteristics.

For multiple trajectories other techniques apply. A large set of trajectories might contain very unrelated trajectories, hence *clustering* may be used. Clustering on complete trajectories will not represent information about interesting parts of trajectories; for this clustering on subtrajectories is needed [6,12]. A set of trajectories that forms different groups over time may be captured by a *grouping structure* [7]. These approaches also focus on location over time.

For the special case of spatial trajectories, a flow diagram can be illustrated by a simple example: trajectories of migrating geese, see [9]. The individual trajectories can be segmented into phases of activities such as directed flight, foraging and stop overs. This results in a flow diagram containing a path for the segmentation of each trajectory. More complex criteria can be imagined that depend on a group of geese, or frequent visits to the same area, resulting in complex state sequences that are hard to analyze without computational tools.

Results, Organization and Hardness. In Sect. 2 we present algorithms for the Flow Diagram problem using criteria with the properties described above. These algorithms only run in polynomial time if the number of state sequences m is constant. Below we observe that this is essentially the best we can hope for by showing that the problem is $W[1]$-hard.

Theorem 2. *The FD problem is NP-hard. This even holds when only two criteria are used or when the length of every state sequence is 2. Furthermore, for any $0 < c < 1/4$, the FD problem cannot be approximated within factor of $c \log m$ in polynomial time unless $NP \subset DTIME(m^{\text{polylog } m})$.*

Also for bounded m the running times of our algorithms is rather high. Again, we can show that there are good reasons for this.

Theorem 3. *The FD problem parameterized in the number of state sequences is $W[1]$-hard even when the number of criteria is constant.*

Both theorems are proved in the longer version of this paper [5]. Unless $W[1] = FPT$, this rules out the existence of algorithms with time complexity of $O(f(m) \cdot (nk)^c)$ for some constant c and any computable function $f(m)$, where m, n and k are the number of state sequences, the length of the state sequences and the number of criteria, respectively. To obtain flow diagrams for larger groups of state sequences we propose two heuristics for the problem in Sect. 2. We experimentally evaluate the algorithms and heuristics in Sect. 3.

2   Algorithms

In this section, we present algorithms that compute a smallest flow diagram representing a set of m state sequences of length n for a set of k criteria. First, we present an algorithm for the general case, followed by a more efficient algorithm for the case of monotone increasing and independent criteria, and then two heuristic algorithms. The algorithm for monotone increasing and dependent criteria, and the proofs omitted in this section are in the extended version of this paper [5].

2.1   General Criteria

Next, we present a dynamic programming algorithm for finding a smallest flow diagram. Recall that a node v in the flow diagram represents a criterion C_j that is fulfilled by a contiguous segment in some of the state sequences. Let $\tau[i,j]$, $i \leq j$, denote the subsequence of τ starting at the ith state of τ and ending at the jth state, where $\tau[i,i]$ is the empty sequence. Construct an $(n+1)^m$ grid of vertices, where a vertex with coordinates $(x_1, \ldots, x_m)$, $0 \leq x_1, \ldots, x_m \leq n$, represents $(\tau_1[0, x_1], \ldots, \tau_m[0, x_m])$. Construct a *prefix graph* G as follows:

There is an edge between two vertices $v = (x_1, \ldots, x_m)$ and $v' = (x'_1, \ldots, x'_m)$, labeled by some criterion C_j, if and only if, for every i, $1 \leq i \leq m$, one of the following two conditions is fulfilled: (1) $x_i = x'_i$, or (2) all remaining $\tau_i[x_i + 1, x'_i]$ jointly fulfil C_j. Consider the edge between $(x_1, x_2) = (1, 0)$ and $(x'_1, x'_2) = (1, 1)$ in Fig. 2(b). Here $x_1 = x'_1$ and $\tau_2[x_2 + 1, x'_2]$ fulfils C_2.

Finally, define v_s to be the vertex in G with coordinates $(0, \ldots, 0)$ and add an additional vertex v_t outside the grid, which has an incoming edge from $(n, \ldots, n)$. This completes the construction of the prefix graph G.

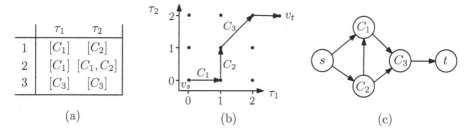

(a)   (b)   (c)

Fig. 2. (a) A segmentation of $\mathcal{T} = \{\tau_1, \tau_2\}$ according to $\mathcal{C} = \{C_1, C_2, C_3\}$. (b) The prefix graph G of the segmentation, omitting all but four of the edges. (c) The resulting flow diagram generated from the highlighted path in the prefix graph.

Now, a path in G from v_s to a vertex v represents a valid segmentation of some prefix of each state sequence, and defines a flow diagram that describes these segmentations in the following way: the empty path represents the flow diagram consisting only of the start node s. Every edge of the path adds one new node to the flow diagram, labeled by the criterion that the segments fulfil. Additionally, for each node the flow diagram contains an edge from every node representing a previous segment, or from s if the node is the first in a segmentation. For a path leading from v_s to v_t, the target node t is added to the flow diagram, together with its incoming edges. This ensures that the flow diagram represents valid segmentations and that each node represents at least one segment. An example of this construction is shown in Fig. 2.

Hence the length of a path (where length is the number of edges on the path) equals the number of nodes of the corresponding flow diagram, excluding s and t. Thus, we find an optimal flow diagram by finding a shortest v_s–v_t path in G.

Lemma 4. *A smallest flow diagram for a given set of state sequences is represented by a shortest v_s–v_t path in G.*

Recall that G has $(n+1)^m$ vertices. Each vertex has $O(k(n+1)^m)$ outgoing edges, thus, G has $O(k(n+1)^{2m})$ edges in total. To decide if an edge is present in G, check if the nonempty segments the edge represents fulfil the criterion. Thus, we need to perform $O(k(n+1)^{2m})$ of these checks. There are m segments of length at most n, and we assume the cost for checking this is $T(m,n)$. Thus, the cost of constructing G is $O(k(n+1)^{2m} \cdot T(m,n))$, and finding the shortest path requires $O(k(n+1)^{2m})$ time.

Theorem 5. *The algorithm described above computes a smallest flow diagram for a set of m state sequences, each of length at most n, and k criteria in $O((n+1)^{2m}k \cdot T(m,n))$ time, where $T(m,n)$ is the time required to check if a set of m subsequences of length at most n fulfils a criterion.*

2.2 Monotone Decreasing and Independent Criteria

If all criteria are decreasing monotone and independent, we can use ideas similar to those presented in [8] to avoid constructing the full graph. From a given vertex with coordinates $(x_1, \ldots, x_m)$, we can greedily move as far as possible along the sequences, since the monotonicity guarantees that this never leads to a solution that is worse than one that represents shorter segments. For a given criterion C_j, we can compute for each τ_i independently the maximum x'_i such that $\tau_i[x_i+1, x'_i]$ fulfils C_j. This produces coordinates $(x'_1, \ldots, x'_m)$ for a new vertex, which is the optimal next vertex using C_j. By considering all criteria we obtain k new vertices. However, unlike the case with a single state sequence, there is not necessarily one vertex that is better than all others (i.e. largest ending position), since there is no total order on the vertices. Instead, we consider all vertices that are not dominated by another vertex, where a vertex p dominates a vertex p' if each coordinate of p is at least as large as the corresponding coordinate of p', and at least one of p's coordinates is larger.

Let V_i be the set of vertices of G that are reachable from v_s in exactly i steps, and define $M(V) := \{v \in V \mid \text{no vertex } u \in V \text{ dominates } v\}$ to be the set of *maximal vertices* of a vertex set V. Then a shortest v_s–v_t path through G can be computed by iteratively computing $M(V_i)$ for increasing i, until a value of i is found for which $v_t \in M(V_i)$. Observe that $|M(V)| = O((n+1)^{m-1})$ for any set V of vertices in the graph. Also note that $V_0 = M(V_0) = v_s$.

Lemma 6. *For each $i \in \{1, \ldots, \ell - 1\}$, every vertex in $M(V_i)$ is reachable in one step from a vertex in $M(V_{i-1})$. Here, ℓ is the distance from v_s to v_t.*

$M(V_i)$ is computed by computing the farthest reachable vertex for each $v \in M(V_{i-1})$ and criterion, thus yielding a set D_i of $O((n+1)^{m-1}k)$ vertices. This set contains $M(V_i)$ by Lemma 6, so we now need to remove all vertices that are dominated by some other vertex in the set to obtain $M(V_i)$.

We find $M(V_i)$ using a copy of G. Each vertex may be marked as being in D_i or dominated by a vertex in D_i. We process the vertices of D_i in arbitrary order. For a vertex v, if it is not yet marked, we mark it as being in D_i. When a vertex is newly marked, we mark its $\leq m$ immediate neighbours dominated by it as being dominated. After processing all vertices, the grid is scanned for the vertices still marked as being in D_i. These vertices are exactly $M(V_i)$.

When computing $M(V_i)$, $O((n+1)^{m-1}k)$ vertices need to be considered, and the maximum distance from v_s to v_t is $m(n+1)$, so the algorithm considers $O(mk(n+1)^m)$ vertices. We improve this bound by a factor m using the following:

Lemma 7. *The total size of all D_i, for $0 \leq i \leq \ell - 1$, is $O(k(n+1)^m)$.*

Using this result, we compute all $M(V_i)$ in $O((k+m)(n+1)^m)$ time, since $O(k(n+1)^m)$ vertices are marked directly, and each of the $(n+1)^m$ vertices is checked at most m times when a direct successor is marked. One copy of the grid can be reused for each $M(V_i)$, since each vertex of D_{i+1} dominates at least one vertex of $M(V_i)$ and is thus not yet marked while processing D_j for any $j \leq i$.

Since the criteria are independent, the farthest reachable point for a given starting point and criterion can be precomputed for each state sequence separately. Using the monotonicity we can traverse each state sequence once per criterion and thus need to test only $O(nmk)$ times whether a subsequence fulfils a criterion.

Theorem 8. *The algorithm described above computes a smallest flow diagram for m state sequences of length n with k independent and monotone decreasing criteria in $O(mnk \cdot T(1,n) + (k+m)(n+1)^m)$ time, where $T(1,n)$ is the time required to check if a subsequence of length at most n fulfils a criterion.*

2.3 Heuristics

The hardness results presented in the introduction indicate that it is unlikely that the performance of the algorithms will be acceptable in practical situations, except for very small inputs. As such, we investigated heuristics that may produce usable results that can be computed in reasonable time.

We consider heuristics for monotone decreasing and independent criteria. These are based on the observation that by limiting V_i, the vertices that are reachable from v_s in i steps, to a fixed size, the complexity of the algorithm can be controlled. Given that every path in a prefix graph represents a valid flow diagram, any path chosen in the prefix graph will be valid, though not necessarily optimal. In the worst case, a vertex that advances along a single state sequence a single time-step (i.e. advancing only one state) will be selected, and for each vertex, all k criteria must be evaluated, so $O(kmn)$ vertices may be processed by the algorithm. We consider two strategies for selecting the vertices in V_i to retain:

(1) For each vertex in V_i, determine the number of state sequences that are advanced in step i and retain the top q vertices [*sequence heuristic*].

(2) For each vertex in V_i, determine the number of time-steps that are advanced in all state sequences in step i and retain the top q vertices [*time-step heuristic*].

In our experiments we use $q = 1$ since any larger value would immediately give an exponential worst-case running time.

3   Experiments

The objectives of the experiments were twofold: to determine whether compact and useful flow diagrams could be produced in real application scenarios; and to empirically investigate the performance of the algorithms on inputs of varying sizes. We implemented the algorithms described in Sect. 2 using the Python programming language. For the first objective, we considered the application of flow diagrams to practical problems in football analysis in order to evaluate their usefulness. For the second objective, the algorithms were run on generated datasets of varying sizes to investigate the impact of different parameterisations on the computation time required to produce the flow diagram and the complexity of the flow diagram produced.

3.1   Tactical Analysis in Football

Sports teams will apply tactics to improve their performance, and computational methods to detect, analyse and represent tactics have been the subject of several recent research efforts [4,11,14,16–18]. Two manifestations of team tactics are in the persistent and repeated occurrence of spatial formations of players, and in *plays* — a coordinated sequence of actions by players. We posited that flow diagrams would be a useful tool for compactly representing both these manifestations, and we describe the approaches used in this section.

The input for the experiments is a database containing player trajectory and match event data from four home matches of the Arsenal Football Club from the 2007/08 season, provided by Prozone Sports Limited [15]. For each player and match, there is a trajectory comprising a sequence of timestamped location points in the plane, sampled at 10 Hz and accurate to 10 cm. The origin of the coordinate system coincides with the centre point of the football pitch and the longer side of the pitch is parallel to the x-axis — i.e. the pitch is oriented so the goals are to the left and right. In addition, for each match, there is a log of all the match events, comprising the type, time-stamp and location of each event.

Defensive Formations. The spatial formations of players in football matches are known to characterize a team's tactics [3], and a compact representation of how formations change over time would be a useful tool for analysis. We investigated whether a flow diagram could provide such a compact representation of the defensive formation of a team, specifically to show how the formation evolves during a phase of play. In our match database, all the teams use a formation of four defensive players who orient themselves in line across the pitch. Broadly speaking, the ideal is for the formation to be "flat", i.e. the players are positioned in a line parallel to the y-axis. However the defenders will react to changes circumstances, for example in response to opposition attacks, possibly causing the formation to deform. We constructed the following flow diagram to analyse the defensive formations used in the football matches in our database.

For each match in the database, the trajectories of the four defensive players were re-sampled at one-second intervals to extract the point-locations of the four defenders. The samples were partitioned into sequences $\mathcal{T} = \{\tau_1, \ldots, \tau_m\}$ corresponding to phases such that a single team was in possession of the ball, and where the phase began with a goal kick event, or the goalkeeper kicks or throws the ball from hand. Let $\tau_i[j]$ be the j-th state in the i-th state sequence. Each $\tau_i[j] = (p_1, p_2, p_3, p_4)$, where p_i is the location of a player in the plane, such that the locations are ordered by their y-coordinate: $y(p_i) \leq y(p_{i+1}) : i \in \{1, 2, 3\}$.

The criteria used to summarise the formations were derived from those presented by Kim et al. [13]. The angles between pairs of adjacent players (along the defensive line) were used to compute the formation criteria, see Fig. 3. The scheme in Kim et al. was extended to allow multiple criteria to be applied where the angle between pairs of players is close to $10°$. The reason for this was to facilitate compact results by allowing for smoothing of small variations in contiguous time-steps.

The criteria C applied to each state is a triple (x_1, x_2, x_3), computed as follows. Given two player positions p and q as points in the plane such that $y(p) \leq y(q)$, let p' be an arbitrary point on the interior of the half-line from p in the direction of the positive y-axis, and let $\angle p'pq$ be the angle induced by these points, and thus denotes the angle between the two player's positions relative to the goal-line. Let $R(-1) = [-90°, -5°)$, $R(0) = (-15°, +15°)$, and $R(1) = (+5°, +90°]$ be three angular ranges. Thus, $C = \{(x_1, x_2, x_3) : x_1, x_2, x_3 \in \{-1, 0, 1\}\}$ is the set of available criteria.

Each state sequence $\tau_i \in \mathcal{T}$ is segmented according to the criteria set C. A given state $\tau_i[j] = (p_1, p_2, p_3, p_4)$ may satisfy the criteria (and thus have the formation) (x_1, x_2, x_3) if $\angle p'_i p_i p_{i+1} \in R(x_i)$ for all $i \in \{1, 2, 3\}$.

The criteria are monotone decreasing and independent, and we ran the corresponding algorithm using randomly selected sets of the state sequences as input. The size m of the input was increased until the running time exceeded a threshold of $6\,\mathrm{h}$. The algorithm successfully processed up to $m = 12$ state sequences, having a total of 112 assigned segments. The resulting flow diagram, Fig. 4, has a total complexity of 12 nodes and 27 edges.

We believe that the flow diagram provides an intuitive summary of the defensive formation, and several observations are apparent. There appears to be a preference amongst the teams for the right-back to position himself in advance of the right centre-half (i.e. the third component of the triple is $+1$). Furthermore, the $(0, 0, 0)$ triple, corresponding to a "flat back four" is not present in the diagram. This is typically considered the ideal formation for teams that utilise the offside trap, and thus may suggest that the defences here are not employing this tactic. These observations were apparent to the authors as laymen, and we would expect that a domain expert would be able to extract further useful insights from the flow diagrams.

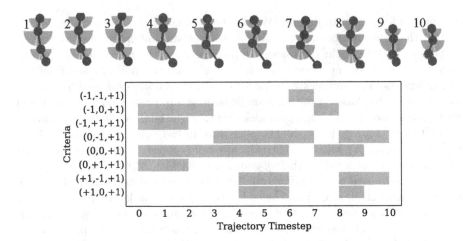

Fig. 3. Segmentation of a single state sequence τ_i. The formation state sequence is used to compute the segmentation representation, where segments corresponding to criteria span the state sequence *(bottom)*. The representation of this state sequence in the movement flow diagram is shaded in Fig. 4.

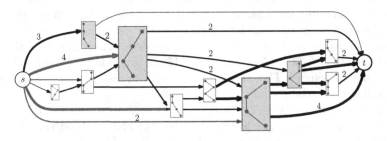

Fig. 4. Flow diagram for formation morphologies of twelve defensive possessions. The shaded nodes are the segmentation of the state sequence in Fig. 3.

Attacking Plays. In this second experiment, we used a different formulation to produce flow diagrams to summarise phases of attack. During a match, the team in possession of the ball regularly attempts to reach a position where they can take a shot at goal. Teams will typically use a variety of tactics to achieve such a position, e.g. teams can vary the intensity of an attack by pushing forward, moving laterally, making long passes, or retreating and regrouping. We modelled attacking possessions as state sequences, segmented according to criteria representing the attacking intensity and tactics employed, and computed flow diagrams for the possessions. In particular, we were interested in determining whether differences in tactics employed by teams when playing at home or away [4] are apparent in the flow diagrams.

We focus on *ball events*, where a player touches the ball, e.g. passes, touches, dribbles, headers, and shots at goal. The event sequence for each match was partitioned into sequences $\mathcal{T} = \{\tau_1, \ldots, \tau_m\}$ such that each τ_i is an event sequence where a single team was in possession, and $\mathcal{T}$ includes only the sequences that end with a shot at goal. Let $\tau_i[j]$ be a tuple (p, t, e) where p is the location in the plane where an event of type $e \in \{touch, pass, dribble, header, shot, clearance\}$ occurred at time t. We are interested in the movement of the ball between an event state $\tau_i[j]$ and the next event state $\tau_i[j+1]$, in particular, let $d_x(\tau_i[j])$ (resp. $d_y(\tau_i[j])$) be the distance in the x-direction (resp. y-direction) between state $\tau_i[j]$ and the next state. Similarly, let $v_x(\tau_i[j])$ (resp. $v_y(\tau_i[j])$) be the velocity of the ball in the x-direction (resp. y-direction) between $\tau_i[j]$ and its successor state. Let $\angle \tau_i[j]$ be the angle defined by the location of $\tau_i[j]$, $\tau_i[j + 1]$ and a point on the interior of the half-line from the location of $\tau_i[j]$ in the positive y-direction.

Criteria were defined to characterise the movement of the ball — relative to the goal the team is attacking — between event states in the possession sequence. The criteria $\mathcal{C} = \{C_1, \ldots, C_8\}$ were defined as follows.

C_1: *Backward movement (BM):* $v_x(\tau_i[j]) < 1$ — a sub-sequence of passes or touches that move in a defensive direction.

C_2: *Lateral movement (LM):* $-5 < v_x(\tau_i[j]) < 5$ — passes or touches that move in a lateral direction.

C_3: *Forward movement (FM):* $-1 < v_x(\tau_i[j]) < 12$ — passes or touches that move in an attacking direction, at a velocity in the range achievable by a player sprinting, i.e. approximately $12\,\mathrm{m/s}$.

C_4: *Fast forward movement (FFM):* $8 < v_x(\tau_i[j])$ — passes or touches moving in an attacking direction at a velocity generally in excess of maximum player velocity.

C_5: *Long ball (LB):* $30 < d_x(\tau_i[j])$ — a single pass travelling 30 m in the attacking direction.

C_6: *Cross-field bal (CFB):* $20 < d_y(\tau_i[j]) \wedge \angle\tau_i[j] \in [-10, 10] \cup [170, 190]$ — a single pass travelling 20 m in the cross-field direction with an angle within $10°$ of the y-axis.

C_7: *Shot resulting in goal (SG):* a successful shot resulting in a goal.

C_8: *Shot not resulting in goal (SNG):* a shot that does not produce a goal.

For a football analyst, the first four criteria are simple movements, and are not particularly interesting. The last four events are significant: the long ball and cross-field ball change the locus of attack; and the shot criteria represent the objective of an attack.

The possession state sequences for the home and visiting teams were segmented according to the criteria and the time-step heuristic algorithm was used to compute the flow diagrams. The home-team input consisted of 66 sequences covered by a total of 866 segments, and resulted in a flow diagram with 25 nodes and 65 edges, see Fig. 5. Similarly, the visiting-team input consisted of 39 state sequences covered by 358 segments and the output flow diagram complexity was 22 nodes and 47 edges, as shown in Fig. 6.

At first glance, the differences between these flow diagrams may be difficult to appreciate, however closer inspection reveals several interesting observations. The s–t paths in the home-team flow diagram tend to be longer than those in the visiting team's, suggesting that the home team tends to retain possession of

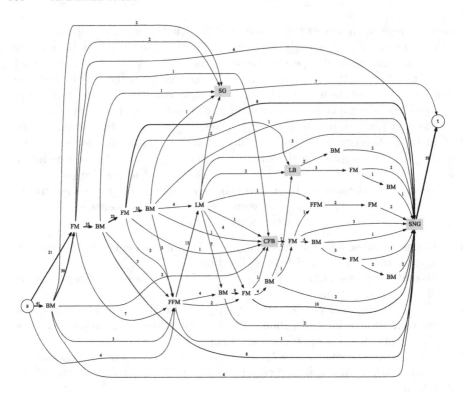

Fig. 5. Flow diagrams produced for home team. The edge weights are the number of possessions that span the edge, and the nodes with grey background are event types that are significant.

the ball for longer, and varies the intensity of attack more often. Moreover, the nodes for cross-field passes and long-ball passes tend to occur earlier in the s–t paths in the visiting team's flow diagram. These are both useful tactics as they alter the locus of attack, however they also carry a higher risk. This suggests that the home team is more confident in its ability to maintain possession for long attack possessions, and will only resort to such risky tactics later in a possession. Furthermore, the tactics used by the team in possession are also impacted by the defensive tactics. As Bialkowski et al. [4] found, visiting teams tend to set their defence deeper, i.e. closer to the goal they are defending. When the visiting team is in possession, there is thus likely to be more space behind the home team's defensive line, and the long ball may appear to be a more appealing tactic. The observations made from these are consistent with our basic understanding of football tactics, and suggest that the flow diagrams are interpretable in this application domain.

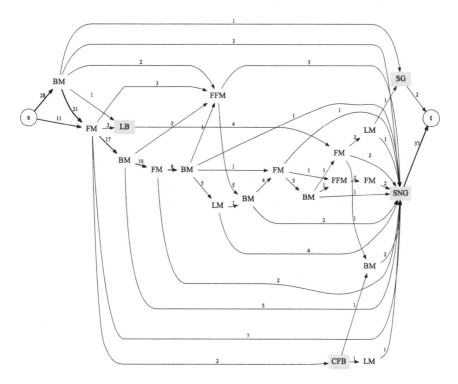

Fig. 6. Flow diagrams produced for visiting team. The edge weights are the number of possessions that span the edge, and the nodes with grey background are event types that are significant.

3.2   Performance Testing

In the second experiment, we used a generator that outputs synthetic state sequences and segmentations, and tested the performance of the algorithms on inputs of varying sizes.

The segmentations were generated using Markov-Chain Monte-Carlo sampling. Nodes representing the criteria set of size k were arranged in a ring and a Markov chain constructed, such that each node had a transition probability of 0.7 to remain at the node, 0.1 to move to the adjacent node, and 0.05 to move to the node two places away. Segmentations were computed by sampling the Markov chain starting at a random node. Thus, simulated datasets of arbitrary size m, state sequence length n, criteria set size k were generated.

We performed two tests on the generated segmentations. In the first, experiments were run on the four algorithms described in Sect. 2 with varying configurations of m, n and k to investigate the impact of input size on the algorithm's performance. The evaluation metric used was the CPU time required to generate the flow diagram for the input. In the second test, we compared the total complexity of the output flow diagram produced by the two heuristic algorithms with

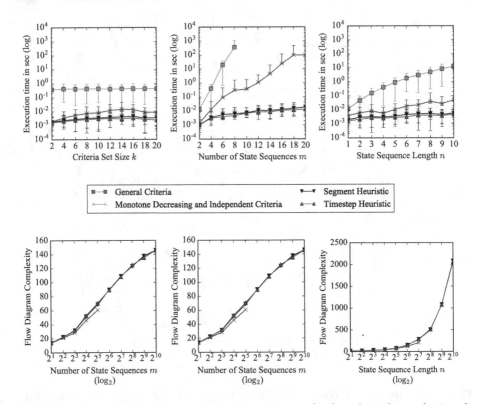

Fig. 7. Runtime statistics for generating flow diagram *(top)*, and total complexity of flow diagrams produced *(bottom)*. Default values of $m = 4$, $n = 4$ and $k = 10$ were used. The data points are the mean value and the error bars delimit the range of values over the five trials run for each input size.

the baseline complexity of the flow diagram produced by the exact algorithm for monotone increasing and independent criteria.

We repeated each experiment five times with different input sequences for each trial, and the results presented are the mean values of the metrics over the trials. Limits were set such that the process was terminated if the CPU time exceeded 1 h, or the memory required exceeded 8 GB.

The results of the first test showed empirically that the exact algorithms have time and storage complexity consistent with the theoretical worst-case bounds, Fig. 7 *(top)*. The heuristic algorithms were subsequently run against larger test data sets to examine the practical limits of the input sizes, and were able to process larger input — for example, an input of $k = 128$, $m = 32$ and $n = 1024$ was tractable — the trade-off is that the resulting flow diagrams were suboptimal, though correct, in terms of their total complexity.

For the second test, we investigated the complexity of the flow diagram induced by inputs of varying parameterisations when using the heuristic algorithms. The objective was to examine how close the complexity was to the optimal complexity produced using an exact algorithm. The inputs exhibited monotone decreasing and independent criteria, and thus the corresponding algorithm was used to produce the baseline. Figure 7 *(bottom)* summarises the results for varying input parameterisations. The complexity of the flow diagrams produced by the two heuristic algorithms are broadly similar, and increase at worst linearly as the input size increases. Moreover, while the complexity is not optimal it appears to remain within a constant factor of the optimal, suggesting that the heuristic algorithms could produce usable flow diagrams for inputs where the exact algorithms are not tractable.

4   Concluding Remarks

We introduced flow diagrams as a compact representation of a large number of state sequences. We argued that this representation gives an intuitive summary allowing the user to detect patterns among large sets of state sequences, and gave several algorithms depending on the properties of the segmentation criteria. These algorithms only run in polynomial time if the number of state sequences m is constant, which is the best we can hope for given the problem is $W[1]$-hard. As a result we considered two heuristics capable of processing large data sets in reasonable time, however we were unable to give an approximation bound. We tested the algorithms experimentally to assess the utility of the flow diagram representation in a sports analysis context, and also analysed the performance of the algorithms of inputs of varying parameterisations.

References

1. Alewijnse, S.P.A., Buchin, K., Buchin, M., Kölzsch, A., Kruckenberg, H., Westenberg, M.: A framework for trajectory segmentation by stable criteria. In: Proceedings of 22nd ACM SIGSPATIAL/GIS, pp. 351–360. ACM (2014)
2. Aronov, B., Driemel, A., van Kreveld, M.J., Löffler, M., Staals, F.: Segmentation of trajectories for non-monotone criteria. In: Proceedings of 24th ACM-SIAM SODA, pp. 1897–1911 (2013)
3. Bialkowski, A., Lucey, P., Carr, G.P.K., Yue, Y., Sridharan, S., Matthews, I.: Identifying team style in soccer using formations learned from spatiotemporal tracking data. In: ICDM Workshops, pp. 9–14. IEEE (2014)
4. Bialkowski, A., Lucey, P., Carr, P., Yue, Y., Matthews, I.: Win at home and draw away: automatic formation analysis highlighting the differences in home and away team behaviors. In: Proceedings of 8th Annual MIT Sloan Sports Analytics Conference (2014)
5. Buchin, K., Buchin, M., Gudmundsson, J., Horton, M., Sijben, S.: Compact flow diagrams for state sequences. CoRR, abs/1602.05622 (2016)
6. Buchin, K., Buchin, M., Gudmundsson, J., Löffler, M., Luo, J.: Detecting commuting patterns by clustering subtrajectories. Int. J. Comput. Geom. Appl. **21**(3), 253–282 (2011)
7. Buchin, K., Buchin, M., van Kreveld, M., Speckmann, B., Staals, F.: Trajectory grouping structure. In: Dehne, F., Solis-Oba, R., Sack, J.-R. (eds.) WADS 2013. LNCS, vol. 8037, pp. 219–230. Springer, Heidelberg (2013)

8. Buchin, M., Driemel, A., van Kreveld, M., Sacristan, V.: Segmenting trajectories: a framework and algorithms using spatiotemporal criteria. J. spat. inf. sci. **3**, 33–63 (2011)

9. Buchin, M., Kruckenberg, H., Kölzsch, A.: Segmenting trajectories based on movement states. In: Proceedings of 15th SDH, pp. 15–25. Springer (2012)

10. Cao, H., Wolfson, O., Trajcevski, G.: Spatio-temporal data reduction with deterministic error bounds. VLDB J. **15**(3), 211–228 (2006)

11. Gudmundsson, J., Wolle, T.: Football analysis using spatio-temporal tools. Comput. Environ. Urban Syst. **47**, 16–27 (2014)

12. Han, C.-S., Jia, S.-X., Zhang, L., Shu, C.-C.: Sub-trajectory clustering algorithm based on speed restriction. Comput. Eng. **37**(7), 219–221 (2011)

13. Kim, H.-C., Kwon, O., Li, K.-J.: Spatial and spatiotemporal analysis of soccer. In: Proceedings of 19th ACM SIGSPATIAL/GIS, pp. 385–388. ACM (2011)

14. Lucey, P., Bialkowski, A., Carr, G.P.K., Morgan, S., Matthews, I., Sheikh, Y.: Representing and discovering adversarial team behaviors using player roles. In: Proceedings of IEEE Conference on Computer Vision and Pattern Recognition (CVPR 2013), Portland, pp. 2706–2713. IEEE, June 2013

15. Prozone Sports Ltd: Prozone Sports - Our technology (2015). http://prozonesports.stats.com/about/technology/

16. Van Haaren, J., Dzyuba, V., Hannosset, S., Davis, J.: Automatically discovering offensive patterns in soccer match data. In: Fromont, E., De Bie, T., van Leeuwen, M. (eds.) IDA 2015. LNCS, vol. 9385, pp. 286–297. Springer, Heidelberg (2015). doi:10.1007/978-3-319-24465-5_25

17. Wang, Q., Zhu, H., Hu, W., Shen, Z., Yao, Y.: Discerning tactical patterns for professional soccer teams. In: Proceedings of the 21th ACM SIGKDD International Conference on Knowledge Discovery and Data Mining - KDD 2015, Sydney, pp. 2197–2206. ACM Press, August 2015

18. Wei, X., Sha, L., Lucey, P., Morgan, S., Sridharan, S.: Large-scale analysis of formations in soccer. In: 2013 International Conference on Digital Image Computing: Techniques and Applications (DICTA), Hobart, pp. 1–8. IEEE, November 2013

Practical Dynamic Entropy-Compressed Bitvectors with Applications

Joshimar Cordova$^{(\boxtimes)}$ and Gonzalo Navarro

Department of Computer Science, CeBiB — Center of Biotechnology
and Bioengineering, University of Chile, Santiago, Chile
{jcordova,gnavarro}@dcc.uchile.cl

Abstract. Compressed data structures provide the same functionality as their classical counterparts while using entropy-bounded space. While they have succeeded in a wide range of static structures, which do not undergo updates, they are less mature in the dynamic case, where the theory-versus-practice gap is wider. We implement compressed dynamic bitvectors B using $|B|H_0(B)+o(|B|)$ or $|B|H_0(B)(1+o(1))$ bits of space, where H_0 is the zero-order empirical entropy, and supporting queries and updates in $\mathcal{O}(w)$ time on a w-bit word machine. This is the first implementation that provably achieves compressed space and is also practical, operating within microseconds. Bitvectors are the basis of most compressed data structures; we explore applications to sequences and graphs.

1 Introduction

Compact data structures have emerged as an attractive solution to reduce the significant memory footprint of classical data structures, which becomes a more relevant problem as the amount of available data grows. Such structures aim at representing the data within almost its entropy space while supporting a rich set of operations on it. Since their beginnings [12], several compact structures have been proposed to address a wide spectrum of applications, with important success stories like ordinal trees with full navigation in less than 2.5 bits [1], range minimum queries in 2.1 bits per element [7], and full-text indexes using almost the space of the compressed text [15], among others. Most of the major practical solutions are implemented in the *Succinct Data Structures Library* [10], which offers solid C++ implementations and extensive test datasets.

Most of these implemented structures, however, are static, that is, they do not support updates to the data once they are built. While dynamic variants exist for many compact data structures, they are mostly theoretical and their practicality is yet to be established.

At the core of many compact structures lay simple bitvectors supporting two important queries: counting the number of bits b up to a given position (*rank*) and finding the position of the i-th occurrence of bit b (*select*). Such bitvectors enable well-known compact structures like sequences, two-dimensional

---

Funded by Basal Funds FB0001 and with Fondecyt Grant 1-140796, Conicyt, Chile.

A.V. Goldberg and A.S. Kulikov (Eds.): SEA 2016, LNCS 9685, pp. 105–117, 2016.
DOI: 10.1007/978-3-319-38851-9_8

grids, graphs, trees, etc. Supporting insertion and deletion of bits in the bitvectors translates into supporting insertion and deletions of symbols, points, edges, and nodes, in those structures. Very recent work [16] shows that dynamic bitvectors are practical and that compression can be achieved for skewed frequencies of 0 s and 1 s, provided that the underlying dynamic memory allocation is handled carefully. Furthermore, the authors implement the *compressed RAM* [13] and show that it is practical by storing in it a directed graph.

In this paper we build on a theoretical proposal [17] to present the first practical dynamic bitvector representations whose size is provably entropy-bounded. A first variant represents $B[1,n]$ in $nH_0(B) + o(n)$ bits, where H_0 denotes the zero-order empirical entropy. For bitvectors with few 1 s, a second variant that uses $nH_0(B)(1+o(1))$ bits is preferable. Both representations carry out updates and *rank/select* queries in time $\mathcal{O}(w)$ on a w-bit machine. In practice, the times are just a few microseconds and the compression obtained is considerable. Instead of using our structure to implement a compressed RAM, we use our bitvectors to implement (a) a practical dynamic wavelet matrix [5] to handle sequences of symbols and two-dimensional grids, and (b) a compact dynamic graph that achieves considerable space savings with competitive edge-insertion times.

Along the way we also describe how we handle the dynamic memory allocation with the aim of reducing RAM fragmentation, and unveil a few related practical results that had not been mentioned in the literature.

2   Basic Concepts

Given a sequence $S[1,n]$ over the alphabet $[1,\sigma]$, $access(S,i)$ returns the character $S[i]$, $rank_c(S,i)$ returns the number of occurrences of character c in $S[1,i]$ and $select_c(S,j)$ returns the position of the j-th occurrence of c. The (empirical) zero-order entropy of S is defined as $H_0(S) = \sum_{1 \leq c \leq \sigma} \frac{n_c}{n} \lg \frac{n}{n_c}$, where c occurs n_c times in S, and is a lower bound on the average code length for any compressor that assigns fixed (variable-length) codes to symbols. When $\sigma = 2$ we refer to the sequence as a bitvector $B[1,n]$ and the entropy becomes $H_0(B) = \frac{m}{n} \lg \frac{n}{m} + \frac{n-m}{n} \lg \frac{n}{n-m}$, where $m = n_1$. The entropy decreases when m is closer to 0 or n. In the first case, another useful formula is $H_0(B) = \frac{m}{n}(\lg \frac{n}{m} + \mathcal{O}(1))$.

Dynamism is supported by the operations $insert(S,i,c)$, which inserts the character c before position i in S and moves characters $S[i,n]$ one position to the right; $delete(S,i)$, which removes character $S[i]$ and moves the characters $S[i+1,n]$ one position to the left; and $modify(S,i,c)$, which sets $S[i] = c$.

Uncompressed (or plain) bitvector representations use $n + o(n)$ bits, and can answer queries in $\mathcal{O}(1)$ time [3]. Compressed representations reduce the space to $nH_0(B) + o(n)$ bits while retaining the constant query times [24]. Dynamic bitvectors cannot be that fast, however: queries require $\Omega(\lg n / \lg \lg n)$ time if the updates are to be handled in $\mathcal{O}(\text{polylog}\, n)$ time [8]. Dynamic plain bitvectors with optimal times $\mathcal{O}(\lg n / \lg \lg n)$ for all the operations exist [23]. Mäkinen and Navarro [17] presented the first dynamic bitvectors using compressed space,

$nH_0(B) + o(n)$ bits, and $\mathcal{O}(\lg n)$ times. It is possible to improve the times to the optimal $\mathcal{O}(\lg n/\lg \lg n)$ within compressed space [21], but the solutions are complicated and unlikely to be practical.

A crucial aspect of the dynamic bitvectors is memory management. When insertions/deletions occur in the bit sequence, the underlying memory area needs to grow/shrink appropriately. The classical solution, used in most of the theoretical results, is the allocator presented by Munro [18]. Extensive experiments [16] showed that this allocator can have a drastic impact on the actual memory footprint of the structure: the standard allocator provided by the operating system may waste up to 25 % of the memory due to fragmentation.

The first implementation of compact dynamic structures we know of is that of Gerlang [9]. He presents dynamic bitvectors and wavelet trees [11], and uses them to build a compact dynamic full-text index. However, memory management is not considered and bitvectors $B[1, n]$ use $\mathcal{O}(n)$ bits of space, $3.5n$–$14n$ in practice. A more recent implementation [25] has the same problems and thus is equally unattractive. Brisaboa et al. [2] also explore plain dynamic bitvectors; they use a B-tree-like structure where leaves store blocks of bits. While their query/update times are competitive, the space reported should be read carefully as they do not consider memory fragmentation. In the context of compact dynamic ordinal trees, Joannou and Raman [14] present a practical study of dynamic Range Min-Max trees [21]. Although the space usage is not clear, the times are competitive and almost as fast as the static implementations [1].

There also exist open-source libraries providing compact dynamic structures. The *ds-vector* library [22] provides dynamic bitvectors and wavelet trees, but their space overhead is large and their wavelet tree is tailored to byte sequences; memory fragmentation is again disregarded. The compressed data structures framework *Memoria* [26] offers dynamic compact bitvectors and ordinal trees, among other structures. A custom memory allocator is provided to reduce fragmentation, but unfortunately the library is not in a stable state yet (as confirmed by the author of the library).

Klitzke and Nicholson [16] revisit dynamic bitvectors. They present the first practical implementation of the memory allocation strategy of Munro [18] tailored to using compact data structures, and show that it considerably reduces memory fragmentation without incurring in performance penalties. They present plain dynamic bitvectors $B[1, n]$ using only $1.03n$ bits. For bitvectors with $m \ll n$ 1 s, they build on general-purpose compressors lz4 and lz4hc to reduce the space up to $0.06n$. However, they lack theoretical guarantees on the compression achieved. While their work is the best practical result in the literature, the code and further technical details are unfortunately unavailable due to legal issues (as confirmed by the first author).

3 Dynamic Entropy-Compressed Bitvectors

In this section we present engineered dynamic bitvectors that achieve zero-order entropy compression. These are based on the ideas of Mäkinen and Navarro [17], but are modified to be more practical. The following general scheme underlies almost all practical results to date and is used in this work as well. The bitvector $B[1, n]$ is partitioned into *chunks* of contiguous bits and a balanced search tree (we use AVLs) is built where the leaves store these chunks. The actual partition strategy and storage used in the leaves vary depending on the desired compression. Each internal node v of the balanced tree stores two fields: v.ones (v.length) is the number of 1 s (total number of bits) present in the left subtree of v. The field v.length is used to find a target position i in B: if $i \leq$ v.length we descend to the left child, otherwise we descend to the right child and i becomes $i -$ v.length. This is used to answer *access/rank* queries and also to find the target leaf where an update will take place (for *rank* we add up the v.ones field whenever we go right). The field v.ones is used to answer $select_1(B, j)$ queries: if $j \leq$ v.ones the answer is in the left subtree; otherwise we move to the right child, add v.length to the answer, and j becomes $j -$ v.ones. For $select_0(B, j)$ we proceed analogously, replacing v.ones by v.length $-$ v.ones. The leaves are sequentially scanned, taking advantage of locality. Section 3.2 assumes the tree is traversed according to these rules.

3.1   Memory Management

Although Klitzke and Nicholson [16] present and study a practical implementation of Munro's allocator [18], the technical details are briefly mentioned and the implementation is not available. We then provide an alternative implementation with its details. In Sect. 5, both implementations are shown to be comparable.

Munro's allocator is tailored to handle small *blocks* of bits, in particular blocks whose size lies in the range $[L, 2L]$ for some $L = $ polylog n. It keeps $L + 1$ linked lists, one for each possible size, with $L + 1$ pointers to the heads of the lists. Each list l_i consists of fixed-length *cells* of $2L$ bits where the blocks of i bits are stored contiguously. In order to allocate a block of i bits we check if there is enough space in the head cell of l_i, otherwise a new cell of $2L$ bits is allocated and becomes the head cell. To deallocate a block we fill its space with the last block stored in the head cell of list l_i; if the head cell no longer stores any block it is deallocated and returned to the OS. Finally, given that we move blocks to fill the gaps left by deallocation, *back pointers* need to be stored from each block to the external structure that *points* to the block, to update the pointers appropriately. Note that in the original proposal a block may span up to two cells and a cell may contain pieces of up to three different blocks.

Implementation. Blocks are fully stored in a single cell to improve locality. As in the previous work [16], we only allocate blocks of *bytes*: L is chosen as a multiple of 8 and we only handle blocks of size $L, L+8, L+16, \ldots, 2L$, rounding the requested sizes to the next multiple of 8. The cells occupy $T = 2L/8$ bytes and are allocated using the default allocator provided by the system. Doing increments of 8 bits has two benefits: the total number of allocations is reduced and the memory pointers returned by our allocator are byte-aligned. The head pointers and lists l_i are implemented verbatim. The *back pointers* are implemented using a folklore idea: when allocating a block of l bytes we instead allocate $l + w/8$ bytes and store in the first w bits the address of the *pointer* to the block, so that when moving blocks to fill gaps the pointer can be modified. This creates a strong binding between the external structure and the block, which can be pointed only from one place. This restriction can be alleviated by storing the pointer in our structure, in an immutable memory area, and let the external structures point to the pointer. This requires that the external structures know that the handle they have for the block is not a pointer to the data but a pointer to the pointer. In this sense, the memory allocator is not completely transparent.

As a further optimization, given that our dynamic bitvectors are based on search trees, we will be constantly (de)allocating very small structures representing the nodes of the trees (eg. 4 words for a AVL node). We use another folklore strategy for these structures: given that modern operating systems usually provide 8 MB of *stack* memory for a running process, we implement an allocator on top of that memory, avoiding the use of the heap area for these tiny structures; (de)allocation simply moves the end of the stack.

3.2   Entropy-Based Compression

Our first variant builds on the compression format of Raman et al. [17,24], modified to be practical. We partition the bitvector B into chunks of $\Theta(w^2)$ bits and these become the leaves of an AVL tree. We store the chunks using the *(class, offset)* encoding (c, o) [24]: a chunk is further partitioned into blocks of $b = w/2$ bits; the class of a block is the number of 1 s it contains and its offset is the index of the block among all possible blocks of the same class when sorted lexicographically. A class component requires $\lg w$ bits, while the offset of a block of class k requires $\lg \binom{b}{k}$ bits. All class/offset components are concatenated in arrays C/O, which are stored using our custom memory allocator. The overall space of this encoding is $nH_0(B) + o(n)$ bits [24]. The space overhead of the AVL tree is $\mathcal{O}(n/w)$ bits, since there are $\mathcal{O}(n/w^2)$ nodes, each requiring $\Theta(w)$ bits. Since $w = \Omega(\lg n)$, this overhead is $o(n)$. It is important to notice that while leaves represent $\Theta(w^2)$ *logical* bits, the *actual* space used by the (c, o) encoding may be considerably smaller. In practice we choose a parameter L', and all leaves will store a number of physical bytes in the range $[L', 2L']$.

To answer $access(B, i)/select(B, j)$ queries we navigate, using the AVL tree, to the leaf containing the target position and then decode the blocks sequentially until the desired position is found. A block is decoded in constant time using a lookup table that, given a pair (c, o), returns the original b bits of the block. This table has $2^{w/2}$ entries, which is small and can be regarded as program size, since it does not depend on the data. Note that we only need to decode the last block; for the previous ones the class component is sufficient to determine how much to advance in array O. For $rank_1(B, i)$ we also need to add up the class components (i.e., number of 1 s) up to the desired block. Again, this only requires accessing array C, while O is only read to decode the last block. We spend $\mathcal{O}(\lg n)$ time to navigate the tree, $\mathcal{O}(w)$ time to traverse the blocks in the target leaf, and $\mathcal{O}(w)$ time to process the last block bitwise. Thus queries take $\mathcal{O}(w)$ time. In practice we set $b = 15$, hence the class components require 4 bits (and can be read by pairs from each single byte of C), the (uncompressed) blocks are 16-bit integers, and the decoding table overhead (which is shared by all the bitvectors) is only 64 KB.

To handle updates we navigate towards the target leaf and proceed to decompress, update, and recompress all the blocks to the right of the update position. If the number of physical bytes stored in a leaf grows beyond $2L$ we split it in two leaves and add a new internal node to be tree; if it shrinks beyond L we move a single bit from the left or right sibling leaf to the current leaf. If this is not possible (because both siblings store L physical bytes) we merge the current leaf with one of its siblings; in either case we perform rotations on the internal nodes of the tree appropriately to restore the AVL invariant.

Recompressing a block is done using an encoding lookup table that, given a block of b bits, returns the associated (c, o) encoding. This adds other 64 KB of memory. To avoid overwriting memory when the physical leaf size grows, recompression is done by reading the leaf data and writing the updated version in a separate memory area, which is later copied back to the leaf.

3.3   Compression of Very Sparse Bitvectors

When the number m of 1 s in B is very low, the $o(n)$ term may be significant compared to $nH_0(B)$. In this case we seek a structure whose space depends mainly on m. We present our second variant (also based on Mäkinen and Navarro [17]) that requires only $m \lg \frac{n}{m} + \mathcal{O}(m \lg \lg \frac{n}{m})$ bits, while maintaining the $\mathcal{O}(w)$-time complexities. This space is $nH_0(B)(1 + o(1))$ bits if $m = o(n)$.

The main building blocks is Elias δ-codes [6]. Given a positive integer x, let $|x|$ denote the length of its binary representation (eg. $|7| = 3$). The δ-code for x is obtained by writing $||x|| - 1$ zeros followed by the binary representation of $|x|$ and followed by the binary representation of x without the leading 1 bit. For example $\delta(7) = 01111$ and $\delta(14) = 00100110$. It follows easily that the length of the code $\delta(x)$ is $|\delta(x)| = \lg x + 2 \lg \lg x + \mathcal{O}(1)$ bits.

We partition B into chunks containing $\Theta(w)$ 1 s. We build an AVL tree where leaves store the chunks. A chunk is stored using δ-codes for the distance between pairs of consecutive 1 s. This time the overhead of the AVL tree is $\mathcal{O}(m)$ bits. By using the Jensen inequality on the lengths of the δ-codes it can be shown [17] that the overall space of the leaves is $m \lg \frac{n}{m} + \mathcal{O}(m \lg \lg \frac{n}{m})$ bits and the redundancy of the AVL tree is absorbed in the second term. In practice we choose a constant M and leaves store a number of 1 s in the range $[M, 2M]$. Within this space we now show how to answer queries and handle updates in $\mathcal{O}(w)$ time.

To answer $access(i)$ we descend to the target leaf and start decoding the δ-codes sequentially until the desired position is found. Note that each δ-code represents a run of 0 s terminated with a 1, so as soon as the current run contains the position i we return the answer. To answer $rank(i)$ we increase the answer by 1 per δ-code we traverse. Finally, to answer $select_1(j)$, when we reach the target leaf looking for the j-th local 1-bit we decode the first j codes and add their sum (since they represent the lengths of the runs). Instead, $select_0(j)$ is very similar to the *access* query.

To handle the insertion of a 0 at position i in a leaf we sequentially search for the δ-code that contains position i. Let this code be $\delta(x)$; we then replace it by $\delta(x+1)$. To insert a 1, let $i' \leq x+1$ be the local offset inside the run $0^{x-1}1$ (represented by the code $\delta(x)$) where the insertion will take place. We then replace $\delta(x)$ by $\delta(i')\delta(x-i'+1)$ if $i' \leq x$ and by $\delta(x)\delta(1)$ otherwise. In either case (inserting a 1 or a 0) we copy the remaining δ-codes to the right of the insertion point. Deletions are handled analogously; we omit the description. If, after an update, the number of 1 s of a leaf lies outside the interval $[M, 2M]$ we move a run from a neighbor leaf or perform a split/merge just as in the previous solution and then perform tree rotations to restore the AVL invariant.

The times for the queries and updates are $\mathcal{O}(w)$ provided that δ-codes are encoded/decoded in constant time. To decode a δ-code we need to find the highest 1 in a word (as this will give us the necessary information to decode the rest). Encoding a number x requires efficiently computing $|x|$ (the length of its binary representation), which is also the same problem. Modern CPUs provide special support for this operation; otherwise we can use small precomputed tables. The rest of the encoding/decoding process is done with appropriate bitwise operations. Furthermore, the local encoding/decoding is done on sequential memory areas, which is cache-friendly.

4   Applications

4.1   Dynamic Sequences

The wavelet matrix [5] is a compact structure for sequences $S[1, n]$ over a fixed alphabet $[1, \sigma]$, providing support for $access(i), rank_c(i)$ and $select_c(i)$ queries. The main idea is to store $\lg \sigma$ bitvectors B_i defined as follows: let $S_1 = S$ and $B_1[j] = 1$ iff the most significant bit of $S_1[j]$ is set. Then S_2 is obtained by moving to the front all characters $S_1[j]$ with $B_1[j] = 0$ and moving the rest to the back (the internal order of front and back symbols is retained). Then $B_2[j] = 1$ iff the second most significant bit of $S_2[j]$ is set, we create S_3 by shuffling S_2 according to B_2, and so on. This process is repeated $\lg \sigma$ times. We also store $\lg \sigma$ numbers $z_j = rank_0(B_j, n)$. The *access/rank/select* queries on this structure reduce to $\mathcal{O}(\lg \sigma)$ analogous queries on the bitvectors B_j, thus the times are $\mathcal{O}(\lg \sigma)$ and the final space is $n \lg \sigma + o(n \lg \sigma)$ (see the article [5] for more details).

Our results in Sect. 3 enable a dynamic implementation of wavelet matrices with little effort. The insertion/deletion of a character at position i is implemented by the insertion/deletion of a single bit in each of the bitvectors B_j. For insertion of c, we insert the highest bit of c in $B_1[i]$. If the bit is a 0, we increase z_1 by one and change i to $rank_0(B_1, i)$; otherwise we change i to $z_1 + rank_1(B_1, i)$. Then we continue with B_2, and so on. Deletion is analogous. Hence all query and update operations require $\lg \sigma \; \mathcal{O}(w)$-time operations on our dynamic bitvectors. By using our uncompressed dynamic bitvectors, we maintain a dynamic string $S[1, n]$ over a (fixed) alphabet $[1, \sigma]$ in $n \lg \sigma + o(n \lg \sigma)$ bits, handling queries and updates in $\mathcal{O}(w \lg \sigma)$ time. An important result [11] states that if the bitvectors B_j are compressed to their zero-order entropy $nH_0(B_j)$, then the overall space is $nH_0(S)$. Hence, by switching to our compressed dynamic bitvectors (in particular, our first variant) we immediately achieve $nH_0(S) + o(n \lg \sigma)$ bits and the query/update times remain $\mathcal{O}(w \lg \sigma)$.

4.2   Dynamic Graphs and Grids

The wavelet matrix has a wide range of applications [19]. One is directed graphs. Let us provide dynamism to the compact structure of Claude and Navarro [4]. Given a directed graph $G(V, E)$ with $n = |V|$ vertices and $e = |E|$ edges, consider the adjacency list $G[v]$ of each node v. We concatenate all the adjacency lists in a single sequence $S[1, e]$ over the alphabet $[1, n]$ and build the dynamic wavelet matrix on S. Each outdegree d_v of vertex v is written as 10^{d_v} and appended to a bitvector $B[1, n + e]$. The final space is $e \lg n(1 + o(1)) + \mathcal{O}(n)$ bits.

This representation allows navigating the graph. The outdegree of vertex v is computed as $select_1(B, v + 1) - select_1(B, v) - 1$. The j-th neighbor of vertex v is $access(S, select_1(B, v) - v + j)$. The edge (v, u) exists iff $rank_u(S, select_1(B, v + 1) - v - 1) - rank_u(S, select_1(B, v) - v) = 1$. The main advantage of this representation is that it also enables *backwards* navigation of the graph without doubling the space: the indegree of vertex v is $rank_v(S, e)$ and the j-th *reverse* neighbor of v is $select_0(B, select_v(S, j)) - select_v(S, j)$.

To insert an edge (u, v) we insert a 0 at position $select_1(B, u) + 1$ to increment the indegree of u, and then insert in S the character v at position $select_1(B, u) - u + 1$. Edge deletion is handled in a similar way. We thus obtain $\mathcal{O}(w \lg n)$ time to update the edges. Unfortunately, the wavelet matrix does not allow changing the alphabet size. Despite this, providing edge dynamism is sufficient in several applications where an upper bound on the number of vertices is known.

This same structure is useful to represent two-dimensional $n \times n$ grids with e points, where we can insert and delete points. It is actually easy to generalize the grid size to any $c \times r$. Then the space is $n \lg r(1 + o(1)) + \mathcal{O}(n + c)$ bits. The static wavelet matrix [5] can count the number of points in a rectangular area in time $\mathcal{O}(\lg r)$, and report each such point in time $\mathcal{O}(\lg r)$ as well. On our dynamic variant, times become $\mathcal{O}(w \lg r)$, just like the time to insert/delete points.

5   Experimental Results and Discussion

The experiments were run on a server with 4 Intel Xeon cores (each at 2.4 GHz) and 96 GB RAM running Linux version 3.2.0-97. All implementations are in C++.

We first reproduce the memory fragmentation *stress* test [16] using our allocator of Sect. 3.1. The experiment initially creates n chunks holding C bytes. Then it performs C steps. In the i-th step n/i chunks are randomly chosen and their memory area is expanded to $C + i$ bytes. We set $C = 2^{11}$ and use the same settings [16] for our custom allocator: the cell size T is set to 2^{16} and L is set to 2^{11}. Table 1 shows the results. The memory consumption is measured as the

Resident Set Size (RSS),[1] which is the actual amount of physical RAM retained by a running process. `Malloc` represents the default allocator provided by the operating system and `custom` is our implementation. Note that for all the tested values of n our allocator holds less RAM memory, and in particular for $n = 2^{24}$ (i.e., $nC = 32\,\text{GB}$) it saves up to 12 GB. In all cases the CPU times of our allocator are faster than the default `malloc`. This shows that our implementation is competitive with the previous one [16], which reports similar space and time.

Table 1. Memory consumption measured as RSS in GBs and CPU time (seconds) for the RAM fragmentation test.

$\lg n$	malloc RSS	custom RSS	malloc time	custom time
18	0.889	0.768	0.668	0.665
19	1.777	1.478	1.360	1.325
20	3.552	2.893	2.719	2.635
21	7.103	5.727	5.409	5.213
22	14.204	11.392	10.811	10.446
23	28.407	22.725	21.870	21.163
24	56.813	45.388	45.115	43.081

Having established that our allocator enables considerable reductions in RAM fragmentation, we study our practical compressed bitvectors. We generate random bitvectors of size $n = 50 \cdot 2^{23}$ (i.e., 50 MB) and set individual bits to 1 with probability p. We consider skewed frequencies $p = 0.1, 0.01$, and 0.001. Preliminary testing showed that, for our variant of Sect. 3.2, setting the range of physical leaf sizes to $[2^{11}, 2^{12}]$ bytes provided the best results. Table 2 gives our results for the compression achieved and the time for queries and updates (averaging insertions and deletions). We achieve 0.3–0.4 bits of redundancy over the entropy, which is largely explained by the component c of the pairs (c, o): these add $\lg(b + 1)/b = 4/15 = 0.27$ bits of redundancy, whereas the O array adds up to $nH_0(B)$. The rest of the redundancy is due to the AVL tree nodes and the space wasted by our memory allocator. For the very sparse bitvectors ($p = 0.001$), the impact of this fixed redundancy is very high.

Operation times are measured by timing 10^5 operations on random positions of the bitvectors. The queries on our first variant take around $1\,\mu s$ (and even less for *access*), whereas the update operations take 8–$15\,\mu s$. The operations become faster as more compression is achieved.

For the very sparse bitvectors ($p = 0.001$) we also test our variant of Sect. 3.3. Preliminary testing showed that enforcing leaves to handle a number of 1s in the range $[128, 256]$ provided the best results. The last row of Table 2 shows the compression and timing results for this structure. As promised in theory, the

---

[1] Measured with https://github.com/mpetri/mem_monitor.

Table 2. Memory used (measured as RSS, in MB, in bits per bit, and in redundancy over $H_0(B)$) and timing results (in microseconds) for our compressed dynamic bitvectors. The first three rows refer to the variant of Sect. 3.2, and the last to Sect. 3.3.

p	MB	Bits/n	$-H_0(B)$	Updates	Access	Rank	Select
0.1	38.57	0.77	0.30	15.08	0.80	1.10	1.20
0.01	21.27	0.43	0.35	10.77	0.60	0.90	1.10
0.001	19.38	0.39	0.38	8.50	0.70	0.90	1.00
*0.001	1.50	0.03	0.02	5.26	1.38	1.47	1.35

Table 3. Memory usage (MBs) and times (in seconds) for the online construction and a breadth-first traversal of the *DBLP* graph to find its weakly connected components. The data for previous work [16] is a rough approximation.

Structure	RSS	Ratio	Build time	Ratio	BFS time	Ratio
std::vector	22.20	1.00	7.40	1.00	0.06	1.00
Wavelet matrix	4.70	0.21	12.42	1.68	9.34	155.67
Previous [16]		0.30		9.00		30.00

compression achieved by this representation is remarkable, achieving 0.02 bits of redundancy (its space is much worse on the higher values of p, however). The query times become slightly over $1\,\mu s$ and the update times are around $5\,\mu s$.

Finally, we present a single application for the dynamic wavelet matrix and graphs. We find the weakly connected components of a sample of the *DBLP* social graph stored using the dynamic representation of Sect. 4.2 with plain dynamic bitvectors, that is, they are stored verbatim. We use range $[2^{10}, 2^{11}]$ bytes for the leaf sizes.

The sample dataset consists of 317,080 vertices and 1,049,866 edges taken from https://snap.stanford.edu/data/com-DBLP.html, with edge directions assigned at random. We build the graph by successive insertions of the edges. Table 3 shows the memory consumption, the construction time (i.e., inserting all the edges), and the time to perform a breadth-first search of the graph. Our baseline is a representation of graphs based on adjacency lists implemented using the std::vector class from the STL library in C++, where each directed edge (u, v) is also stored as (v, u) to enable backwards navigation. Considerable space savings are achieved using the dynamic wavelet matrix, 5-fold with respect to the baseline. The edge insertion times are very competitive, only 70 % slower than the baseline. The time to perform a full traversal of the graph, however, is two orders of magnitude slower.

We now briefly make an informal comparison between our results and the best previous work [16] by extrapolating some numbers.[2] For bitvectors with

---

[2] A precise comparison is not possible since their results are not available. We use their plots as a reference.

density $p = 0.1$ our first variant achieves 77 % compression compared to their 85 %. For $p = 0.01$ ours achieves 43 % compared to their 35 %, and for $p = 0.001$ our second variant achieves 3 % compared to their 6 %. In terms of running times our results handle queries in about $1\,\mu s$ and updates in 8–$15\,\mu s$, while their most practical variant, based on lz4, handles queries and updates in around 10–$25\,\mu s$. These results are expected since the encodings we used ((c, o) pairs and δ-codes) are tailored to answer *rank/select* queries without the need of full decompression. Finally, they also implement a compressed dynamic graph (based on compressed RAM and not on compact structures). The rough results (extrapolated from their own comparison against std::vector) are shown in the last line of Table 3: they use 50 % more space and 5 times more construction time than our implementation, but their BFS time is 5 times faster.

6   Conclusions

We have presented the first practical entropy-compressed dynamic bitvectors with good space/time theoretical guarantees. The structures solve queries in around a microsecond and handle updates in 5–$15\,\mu s$. An important advantage compared with previous work [16] is that we do not need to fully decompress the bit chunks to carry out queries, which makes us an order of magnitude faster. Another advantage over previous work is the guaranteed zero-order entropy space, which allows us using bitvectors for representing sequences in zero-order entropy, and full-text indexes in high-order entropy space [15].

Several improvements are possible. For example, we reported times for querying random positions, but many times we access long contiguous areas of a sequence. Those can be handled much faster by remembering the last accessed AVL tree node and block. In the (c, o) encoding, we would access a new byte of C every 30 operations, and decode a new block of O every 15, which would amount to at least an order-of-magnitude improvement in query times. For δ-encoded bitvectors, we would decode a new entry every n/m operations on average.

Another improvement is to allow for faster queries when updates are less frequent, tending to the fast static query times in the limit. We are studying policies to turn an AVL subtree into static when it receives no updates for some time. This would reduce, for example, the performance gap for the BFS traversal in our graph application once it is built, if further updates are infrequent.

Finally, there exist theoretical proposals [20] to represent dynamic sequences that obtain the optimal time $\mathcal{O}(\lg n/\lg\lg n)$ for all the operations. This is much better than the $\mathcal{O}(w\lg\sigma)$ time we obtain with dynamic wavelet matrices. An interesting future work path is to try to turn that solution into a practical implementation. It has the added benefit of allowing us to update the alphabet, unlike wavelet matrices.

Our implementation of dynamic bitvectors and the memory allocator are available at https://github.com/jhcmonroy/dynamic-bitvectors.

References

1. Arroyuelo, D., Cánovas, R., Navarro, G., Sadakane, K.: Succinct trees in practice. In: Proceedings of the 12th ALENEX, pp. 84–97 (2010)
2. Brisaboa, N., de Bernardo, G., Navarro, G.: Compressed dynamic binary relations. In: Proceedings of the 22nd DCC, pp. 52–61 (2012)
3. Clark, D.: Compact PAT Trees. Ph.D. thesis, Univ. Waterloo, Canada (1996)
4. Claude, F., Navarro, G.: Extended compact web graph representations. In: Elomaa, T., Mannila, H., Orponen, P. (eds.) Ukkonen Festschrift 2010. LNCS, vol. 6060, pp. 77–91. Springer, Heidelberg (2010)
5. Claude, F., Navarro, G., Ordóñez, A.: The wavelet matrix: an efficient wavelet tree for large alphabets. Inf. Syst. **47**, 15–32 (2015)
6. Elias, P.: Universal codeword sets and representations of the integers. IEEE Trans. Inf. Theor. **21**(2), 194–203 (1975)
7. Ferrada, H., Navarro, G.: Improved range minimum queries. In: Proceedings of the 26th DCC, pp. 516–525 (2016)
8. Fredman, M., Saks, M.: The cell probe complexity of dynamic data structures. In: Proceedings of the 21st STOC, pp. 345–354 (1989)
9. Gerlang, W.: Dynamic FM-Index for a Collection of Texts with Application to Space-efficient Construction of the Compressed Suffix Array. Master's thesis, Univ. Bielefeld, Germany (2007)
10. Gog, S., Beller, T., Moffat, A., Petri, M.: From theory to practice: plug and play with succinct data structures. In: Gudmundsson, J., Katajainen, J. (eds.) SEA 2014. LNCS, vol. 8504, pp. 326–337. Springer, Heidelberg (2014)
11. Grossi, R., Gupta, A., Vitter, J.S.: High-order entropy-compressed text indexes. In: Proceedings of the 14th SODA, pp. 841–850 (2003)
12. Jacobson, G.: Space-efficient static trees and graphs. In: Proceedings of the 30th FOCS, pp. 549–554 (1989)
13. Jansson, J., Sadakane, K., Sung, W.-K.: CRAM: Compressed Random Access Memory. In: Czumaj, A., Mehlhorn, K., Pitts, A., Wattenhofer, R. (eds.) ICALP 2012, Part I. LNCS, vol. 7391, pp. 510–521. Springer, Heidelberg (2012)
14. Joannou, S., Raman, R.: Dynamizing succinct tree representations. In: Klasing, R. (ed.) SEA 2012. LNCS, vol. 7276, pp. 224–235. Springer, Heidelberg (2012)
15. Kärkkäinen, J., Puglisi, S.J.: Fixed block compression boosting in FM-indexes. In: Grossi, R., Sebastiani, F., Silvestri, F. (eds.) SPIRE 2011. LNCS, vol. 7024, pp. 174–184. Springer, Heidelberg (2011)
16. Klitzke, P., Nicholson, P.K.: A general framework for dynamic succinct and compressed data structures. In: Proceedings of the 18th ALENEX, pp. 160–173 (2016)
17. Mäkinen, V., Navarro, G.: Dynamic entropy-compressed sequences and full-text indexes. ACM Trans. Algorithms **4**(3), 32–38 (2008)
18. Munro, J.I.: An implicit data structure supporting insertion, deletion, and search in o(log2 n) time. J. Comput. Syst. Sci. **33**(1), 66–74 (1986)
19. Navarro, G.: Wavelet trees for all. J. Discrete Algorithms **25**, 2–20 (2014)
20. Navarro, G., Nekrich, Y.: Optimal dynamic sequence representations. SIAM J. Comput. **43**(5), 1781–1806 (2014)
21. Navarro, G., Sadakane, K.: Fully-Functional static and dynamic succinct trees. ACM Trans. Algorithms **10**(3), 16 (2014)
22. Okanohara, D.: Dynamic succinct vector library. https://code.google.com/archive/p/ds-vector/. Accessed 30 Jan 2016

23. Raman, R., Raman, V., Rao, S.S.: Succinct dynamic data structures. In: Dehne, F., Sack, J.-R., Tamassia, R. (eds.) WADS 2001. LNCS, vol. 2125, p. 426. Springer, Heidelberg (2001)
24. Raman, R., Raman, V., Satti, S.R.: Succinct indexable dictionaries with applications to encoding k-ary trees, prefix sums and multisets. ACM Trans. Algorithms **3**(4), 43 (2007)
25. Salson, M.: Dynamic fm-index library. http://dfmi.sourceforge.net/. Accessed 30 Jan 2016
26. Smirnov, V.: Memoria library. https://bitbucket.org/vsmirnov/memoria/. Accessed 30 Jan 2016

Accelerating Local Search for the Maximum Independent Set Problem

Jakob Dahlum[1], Sebastian Lamm[1], Peter Sanders[1], Christian Schulz[1],
Darren Strash[1(✉)], and Renato F. Werneck[2]

[1] Institute of Theoretical Informatics,
Karlsruhe Institute of Technology, Karlsruhe, Germany
{dahlum,lamm}@ira.uka.de,
{sanders,christian.schulz,strash}@kit.edu
[2] San Francisco, USA
rwerneck@acm.org

Abstract. Computing high-quality independent sets quickly is an important problem in combinatorial optimization. Several recent algorithms have shown that kernelization techniques can be used to find exact maximum independent sets in medium-sized sparse graphs, as well as high-quality independent sets in huge sparse graphs that are intractable for exact (exponential-time) algorithms. However, a major drawback of these algorithms is that they require significant preprocessing overhead, and therefore cannot be used to find a high-quality independent set quickly.

In this paper, we show that performing simple kernelization techniques in an online fashion significantly boosts the performance of local search, and is much faster than pre-computing a kernel using advanced techniques. In addition, we show that cutting high-degree vertices can boost local search performance even further, especially on huge (sparse) complex networks. Our experiments show that we can drastically speed up the computation of large independent sets compared to other state-of-the-art algorithms, while also producing results that are very close to the best known solutions.

Keywords: Maximum independent set · Minimum vertex cover · Local search · Kernelization · Reduction

1 Introduction

The maximum independent set problem is a classic NP-hard problem [13] with applications spanning many fields, such as classification theory, information retrieval, computer vision [11], computer graphics [29], map labeling [14] and routing in road networks [20]. Given a graph $G = (V, E)$, our goal is to compute a maximum cardinality set of vertices $\mathcal{I} \subseteq V$ such that no vertices in $\mathcal{I}$ are adjacent to one another. Such a set is called a *maximum independent set* (MIS).

© Springer International Publishing Switzerland 2016
A.V. Goldberg and A.S. Kulikov (Eds.): SEA 2016, LNCS 9685, pp. 118–133, 2016.
DOI: 10.1007/978-3-319-38851-9_9

1.1   Previous Work

Since the MIS problem is NP-hard, all known exact algorithms for these problems take exponential time, making *large* graphs infeasible to solve in practice. Instead, heuristic algorithms such as local search are used to efficiently compute high-quality independent sets. For many practical instances, some local search algorithms even quickly find exact solutions [3,16].

Exact Algorithms. Much research has been devoted to reducing the base of the exponent for exact branch-and-bound algorithms. One main technique is to apply *reductions*, which remove or modify subgraphs that can be solved simply, reducing the graph to a smaller instance. Reductions have consistently been used to reduce the running time of exact MIS algorithms [31], with the current best polynomial-space algorithm having running time $O(1.2114^n)$ [7]. These algorithms apply reductions during recursion, only branching when the graph can no longer be reduced [12]. This resulting graph is called a *kernel*.

Relatively simple reduction techniques are known to be effective at reducing graph size in practice [1,8]. Recently, Akiba and Iwata [2] showed that more advanced reduction rules are also highly effective, finding an exact minimum vertex cover (and by extension, an exact maximum independent set) on a corpus of large social networks with up to 3.2 million vertices in less than a second. However, their algorithm still requires $O(1.2210^n)$ time in the worst case, and its running time has exponential dependence on the kernel size. Since much larger graph instances have consistently large kernels, they remain intractable in practice [24]. Even though small benchmark graphs with up to thousands of vertices have been solved exactly with branch-and-bound algorithms [28,30, 32], many similarly-sized instances remain unsolved [8]. Even a graph on 4,000 vertices was only recently solved exactly, and it required hundreds of machines in a MapReduce cluster [33]. Heuristic algorithms are clearly still needed in practice, even for small graphs.

Heuristic Approaches. There are a wide range of heuristics and local search algorithms for the complementary maximum clique problem [6,15–17,19,27]. These algorithms work by maintaining a single solution and attempt to improve it through node deletions, insertions, swaps, and *plateau* search. Plateau search only accepts moves that do not change the objective function, which is typically achieved through *node swaps*—replacing a node by one of its neighbors. Note that a node swap cannot directly increase the size of the independent set. A very successful approach for the maximum clique problem has been presented by Grosso et al. [16]. In addition to plateau search, it applies various diversification operations and restart rules. The iterated local search algorithm of Andrade et al. [3] is one of the most successful local search algorithms in practice. On small benchmark graphs requiring hours of computation to solve with exact algorithms, their algorithm often finds optimal solutions in milliseconds. However, for huge complex networks such as social networks and Web graphs, it is consistently outperformed by other methods [23,24]. We give further details of this algorithm in Sect. 2.1.

To solve these largest and intractable graphs, Lamm et al. [24] proposed ReduMIS, an algorithm that uses reduction techniques combined with an evolutionary approach. It finds the exact MIS for many of the benchmarks used by Akiba and Iwata [2], and consistently finds larger independent sets than other heuristics. Its major drawback is the significant preprocessing time it takes to apply reductions and initialize its evolutionary algorithm, especially on larger instances. Thus, though ReduMIS finds high-quality independent sets faster than existing methods, it is still slow in practice on huge complex networks. However, for many of the applications mentioned above, a near-optimal independent set is not needed in practice. The main goal then is to quickly compute an independent set of sufficient quality. Hence, to find high-quality independent sets faster, we need a different approach.

1.2 Our Results

We develop an advanced local search algorithm that quickly computes large independent sets by combining iterated local search with reduction rules that reduce the size of the search space without losing solution quality. By running local search on the kernel, we significantly boost its performance, especially on huge sparse networks. In addition to exact kernelization techniques, we also apply inexact reductions that remove high-degree vertices from the graph. In particular, we show that cutting a small percentage of high-degree vertices from the graph minimizes performance bottlenecks of local search, while maintaining high solution quality. Experiments indicate that our algorithm finds large independent sets much faster than existing state-of-the-art algorithms, while still remaining competitive with the best solutions reported in literature.

2 Preliminaries

Let $G = (V = \{0, \ldots, n-1\}, E)$ be an undirected graph with $n = |V|$ nodes and $m = |E|$ edges. The set $N(v) = \{u : \{v, u\} \in E\}$ denotes the open neighborhood of v. We further define the open neighborhood of a set of nodes $U \subseteq V$ to be $N(U) = \cup_{v \in U} N(v) \setminus U$. We similarly define the closed neighborhood as $N[v] = N(v) \cup \{v\}$ and $N[U] = N(U) \cup U$. A graph $H = (V_H, E_H)$ is said to be a *subgraph* of $G = (V, E)$ if $V_H \subseteq V$ and $E_H \subseteq E$. We call H an *induced* subgraph when $E_H = \{\{u, v\} \in E : u, v \in V_H\}$. For a set of nodes $U \subseteq V$, $G[U]$ denotes the subgraph induced by U.

An *independent set* is a set $\mathcal{I} \subseteq V$, such that all nodes in $\mathcal{I}$ are pairwise nonadjacent. An independent set is *maximal* if it is not a subset of any larger independent set. The *maximum independent set problem* is that of finding the maximum cardinality independent set among all possible independent sets. Such a set is called a *maximum independent set* (MIS).

Finally, we note the maximum independent set problem is equivalent to the *maximum clique* and *minimum vertex cover* problems. We see this equivalence as follows: Given a graph $G = (V, E)$ and an independent set $\mathcal{I} \in V$, $V \setminus \mathcal{I}$ is a vertex cover and $\mathcal{I}$ is a clique in the complement graph (the graph containing all edges missing in G). Thus, algorithms for any of these problems can also solve the maximum independent set problem.

2.1   The ARW Algorithm

We now review the local search algorithm by Andrade et al. [3] (ARW) in more detail, since we use this algorithm in our work. For the independent set problem, Andrade et al. [3] extended the notion of swaps to (j, k)-swaps, which remove j nodes from the current solution and insert k nodes. The authors present a fast linear-time implementation that, given a maximal solution, can find a $(1, 2)$-swap or prove that no $(1, 2)$-swap exists. One *iteration* of the ARW algorithm consists of a perturbation and a local search step. The ARW *local search* algorithm uses $(1, 2)$-swaps to gradually improve a single current solution. The simple version of the local search iterates over all nodes of the graph and looks for a $(1, 2)$-swap. By using a data structure that allows insertion and removal operations on nodes in time proportional to their degree, this procedure can find a valid $(1, 2)$-swap in $\mathcal{O}(m)$ time, if it exists.

A *perturbation step*, used for diversification, forces nodes into the solution and removes neighboring nodes as necessary. In most cases a single node is forced into the solution; with a small probability the number of forced nodes f is set to a higher value (f is set to $i + 1$ with probability $1/2^i$). Nodes to be forced into a solution are picked from a set of random candidates, with priority given to those that have been outside the solution for the longest time. An even faster incremental version of the algorithm (which we use here) maintains a list of *candidates*, which are nodes that may be involved in $(1, 2)$-swaps. It ensures a node is not examined twice unless there is some change in its neighborhood. Furthermore, an external memory version of this algorithm by Liu et al. [25] runs on graphs that do not fit into memory on a standard machine. The ARW algorithm is efficient in practice, finding the exact maximum independent sets orders of magnitude faster than exact algorithms on many benchmark graphs.

3   Techniques for Accelerating Local Search

First, we note that while local search techniques such as ARW perform well on huge uniformly sparse mesh-like graphs, they perform poorly on complex networks, which are typically scale-free. We first discuss *why* local search performs poorly on huge complex networks, then introduce the techniques we use to address these shortcomings.

The first performance issue is related to vertex selection for perturbation. Many vertices are *always* in some MIS. These include, for example, vertices with degree one. However, ARW treats such vertices like any other. During a perturbation step, these vertices may be forced out of the current solution, causing extra searching that may not improve the solution.

The second issue is that high-degree vertices may slow ARW down significantly. Most internal operations of ARW (including (1,2)-swaps) require traversing the adjacency lists of multiple vertices, which takes time proportional to their degree. Although high-degree vertices are only scanned if they have at most one solution neighbor (or belong to the solution themselves), this happens often in complex networks.

A third issue is caused by the particular implementation. When performing an (1,2)-swap involving the insertion of a vertex v, the original ARW implementation (as tested by Andrade et al. [3]) picks a pair of neighbors u, w of v at random among all valid ones. Although this technically violates that $O(m)$ worst-case bound (which requires the first such pair to be taken), the effect is minimal on the small-degree networks. On large complex networks, this can become a significant bottleneck.

To deal with the third issue, we simply modified the ARW code to limit the number of valid pairs considered to a small constant (100). Addressing the first two issues requires more involved techniques (*kernelization* and *high-degree vertex cutting*, respectively), as we discuss next.

3.1   Exact Kernelization

First, we investigate kernelization, a technique known to be effective in practice for finding an exact minimum vertex cover (and hence, a maximum independent set) [1,2]. In kernelization, we repeatedly apply reductions to the input graph G until it cannot be reduced further, producing a *kernel* $\mathcal{K}$. Even simple reduction rules can significantly reduce the graph size. Indeed, in some cases $\mathcal{K}$ may be empty—giving an exact solution without requiring any additional steps. We note that this is the case for many of the graphs in the experiments by Akiba and Iwata [2]. Furthermore, any solution of $\mathcal{K}$ can be extended to a solution of the input.

The size of the kernel depends entirely on the structure of the input graph. In many cases, the kernel can be too large, making it intractable to find an exact maximum independent set in practice (see Sect. 4). In this case "too large" can mean a few thousand vertices. However, for many graphs, the kernel is still significantly smaller than the input graph, and even though it is intractable for exact algorithms, local search algorithms such as ARW have been shown to find the exact MIS quickly on small benchmark graphs. It therefore stands to reason that ARW would perform better on a small kernel.

Reductions. We now briefly describe the reduction rules that we consider. Each of these exact reductions allow us to choose vertices that are in some MIS by following simple rules. If an MIS is found on the kernel graph $\mathcal{K}$, then each reduction may be undone, producing an MIS in the original graph.

Reductions of Akiba and Iwata [2]. First, we briefly describe the reductions used by Akiba and Iwata [2]. Akiba and Iwata use a full suite of advanced reduction rules, which they show can efficiently solve the minimum vertex cover problem exactly for a variety of real-world instances. We consider all of their reductions here. Refer to Akiba and Iwata [2] for a thorough discussion, including implementation details.

Pendant vertices: Any vertex v of degree one, called a *pendant*, is in some MIS; therefore, v and its neighbor u can be removed from G.

Vertex folding: For a vertex v with degree two whose neighbors u and w are not adjacent, either v is in some MIS, or both u and w are in some MIS. Therefore, we can contract u, v, and w to a single vertex v' and decide which vertices are in the MIS later.

Linear Programming: A well-known [26] linear programming relaxation for the MIS problem with a half-integral solution (i.e., using only values 0, 1/2, and 1) can be solved using bipartite matching: maximize $\sum_{v \in V} x_v$ such that $\forall (u,v) \in E$, $x_u + x_v \leq 1$ and $\forall v \in V$, $x_v \geq 0$. Vertices with value 1 must be in the MIS and can thus be removed from G along with their neighbors. We use an improved version [18] that computes a solution whose half-integral part is minimal.

Unconfined [34]: Though there are several definitions of *unconfined* vertex in the literature, we use the simple one from Akiba and Iwata [2]. A vertex v is *unconfined* when determined by the following simple algorithm. First, initialize $S = \{v\}$. Then find a $u \in N(S)$ such that $|N(u) \cap S| = 1$ and $|N(u) \setminus N[S]|$ is minimized. If there is no such vertex, then v is confined. If $N(u) \setminus N[S] = \emptyset$, then v is unconfined. If $N(u) \setminus N[S]$ is a single vertex w, then add w to S and repeat the algorithm. Otherwise, v is confined. Unconfined vertices can be removed from the graph, since there always exists an MIS $\mathcal{I}$ that contains no unconfined vertices.

Twin [34]: Let u and v be vertices of degree three with $N(u) = N(v)$. If $G[N(u)]$ has edges, then add u and v to $\mathcal{I}$ and remove u, v, $N(u)$, $N(v)$ from G. Otherwise, some vertices in $N(u)$ may belong to some MIS $\mathcal{I}$. We still remove u, v, $N(u)$ and $N(v)$ from G, and add a new gadget vertex w to G with edges to u's two-neighborhood (vertices at a distance 2 from u). If w is in the computed MIS, then none of u's two-neighbors are $\mathcal{I}$, and therefore $N(u) \subseteq \mathcal{I}$. Otherwise, if w is not in the computed MIS, then some of u's two-neighbors are in $\mathcal{I}$, and therefore u and v are added to $\mathcal{I}$.

Alternative: Two sets of vertices A and B are set to be *alternatives* if $|A| = |B| \geq 1$ and there exists an MIS $\mathcal{I}$ such that $\mathcal{I} \cap (A \cup B)$ is either A or B. Then we remove A and B and $C = N(A) \cap N(B)$ from G and add edges from each $a \in N(A) \setminus C$ to each $b \in N(B) \setminus C$. Then we add either A or B to $\mathcal{I}$, depending on which neighborhood has vertices in $\mathcal{I}$. Two structures are detected as alternatives. First, if $N(v) \setminus \{u\}$ induces a complete graph, then $\{u\}$ and $\{v\}$ are alternatives (a *funnel*). Next, if there is a cordless 4-cycle $a_1 b_1 a_2 b_2$ where each vertex has at least degree three. Then sets $A = \{a_1, a_2\}$ and $B = \{b_1, b_2\}$ are alternatives when $|N(A) \setminus B| \leq 2$, $|N(A) \setminus B| \leq 2$, and $N(A) \cap N(B) = \emptyset$.

Packing [2]: Given a non-empty set of vertices S, we may specify a *packing constraint* $\sum_{v \in S} x_v \leq k$, where x_v is 0 when v is in some MIS $\mathcal{I}$ and

1 otherwise. Whenever a vertex v is excluded from $\mathcal{I}$ (i.e., in the unconfined reduction), we remove x_v from the packing constraint and decrease the upper bound of the constraint by one. Initially, packing constraints are created whenever a vertex v is excluded or included into the MIS. The simplest case for the packing reduction is when k is zero: all vertices must be in $\mathcal{I}$ to satisfy the constraint. Thus, if there is no edge in $G[S]$, S may be added to $\mathcal{I}$, and S and $N(S)$ are removed from G. Other cases are much more complex. Whenever packing reductions are applied, existing packing constraints are updated and new ones are added.

The Reduction of Butenko et al. [8]. We now describe one last reduction that was not included in the exact algorithm by Akiba and Iwata [2], but was shown by Butenko et al. [8] to be highly effective on medium-sized graphs derived from error-correcting codes.

Isolated Vertex Removal: The most relevant reduction for our purposes is the *isolated vertex removal*. If a vertex v forms a single clique C with all its neighbors, then v is called *isolated* (*simplicial* is also used in the literature) and is always contained in some MIS. To see this, at most one vertex from C may is an MIS. Either it is v or, if a neighbor of v is in an MIS, then we select v instead (See Fig. 1).

Fig. 1. An isolated vertex v, in a single clique of five vertices.

When this reduction is applied in practice, vertices with degree three or higher are often excluded—as checking all pairwise adjacencies of v's neighbors can be expensive, especially in sparse representations. Degree zero and pendant vertices can be checked purely by the number of neighbors, and triangles can be detected by storing neighbors in increasing order by vertex number and performing a single binary search to check if v's neighbors are adjacent.

3.2 Inexact Reductions: Cutting High-Degree Vertices

To further boost local search, we investigate removing (cutting) high-degree vertices outright. This is a natural strategy: intuitively, vertices with very high degree are unlikely to be in a large independent set (consider a maximum independent set of graphs with few high-degree vertices, such as a star graph, or scale-free networks). In particular, many reduction rules show that low-degree vertices are in some MIS, and applying them results in a small kernel [24]. Thus, high-degree vertices are left behind. This is especially true for huge complex networks considered here, which generally have few high-degree vertices.

Besides intuition, there is much additional evidence to support this strategy. In particular, the natural greedy algorithm that repeatedly selects low-degree vertices to construct an independent set is typically within 1 %–10 % of the maximum independent set size for sparse graphs [3]. Moreover, several successful algorithms make choices that favor low-degree vertices. ReduMIS [24] forces low-degree vertices into an independent set in a multi-level algorithm, giving high-quality independent sets as a result. Exact branch-and-bound algorithms order vertices so that vertices of high-degree are considered first during search. This reduces the search space size initially, at the cost of finding poor initial independent sets. In particular, optimal and near-optimal independent sets are typically found after high-degree vertices have been evaluated and excluded from search; however, it is then much slower to find the remaining solutions, since only low-degree vertices remain in the search. This slowness can be observed in the experiments of Batsyn et al. [5], where better initial solutions from local search significantly speed up exact search.

We consider two strategies for removing high-degree vertices from the graph. When we cut by *absolute degree*, we remove the vertices with degree higher than a threshold. In *relative degree* cutting, we iteratively remove highest-degree vertices and their incident edges from the graph. This is the mirror image of the greedy algorithm that repeatedly selects smallest-degree vertices in the graph to be in an independent set until the graph is empty. We stop when a fixed fraction of all vertices is removed. This better ensures that clusters of high-degree vertices are removed, leaving high-degree vertices that are isolated from one another, which are more likely to be in large independent sets.

3.3   Putting Things Together

We use reductions and cutting in two ways. First, we explore the standard technique of producing a kernel in advance, and then run ARW on the kernel. Second, we investigate applying reductions online as ARW runs.

Preprocessing. Our first algorithm (KerMIS) uses exact reductions in combination with relative degree cutting. It uses the full set of reductions from Akiba and Iwata [2], as described in Sect. 3. Note that we do not include isolated vertex removal, as it was not included in their reductions. After computing a kernel, we then cut 1 % of the highest-degree vertices using relative degree cutting, breaking ties randomly. We then run ARW on the resulting graph.

Online. Our second approach (OnlineMIS) applies a set of simple reductions on the fly. For this algorithm, we use only the isolated vertex removal reduction (for degrees zero, one, and two), since it does not require the graph to be modified— we can just mark isolated vertices and their neighbors as removed during local search. In more detail, we first perform a quick *single pass* when computing the initial solution for ARW. We force isolated vertices into the initial solution, and mark them and their neighbors as removed. Note that this does not result in a kernel, as this pass may create more isolated vertices. We further mark the top 1 % of high-degree vertices as removed during this pass. As local search continues, whenever we check if a vertex can be inserted into the solution, we check if it is isolated and update the solution and graph similarly to the single pass. Thus, OnlineMIS kernelizes the graph online as local search proceeds.

4   Experimental Evaluation

4.1   Methodology

We implemented our algorithms (OnlineMIS, KerMIS), including the kernelization techniques, using C++ and compiled all code using gcc 4.6.3 with full optimizations turned on (-O3 flag). We further compiled the original implementations of ARW and ReduMIS using the same settings. For ReduMIS, we use the same parameters as Lamm et al. [24] (convergence parameter $\mu = 1,000,000$, reduction parameter $\lambda = 0.1 \cdot |\mathcal{I}|$, and cutting percentage $\eta = 0.1 \cdot |\mathcal{K}|$). For all instances, we perform three independent runs of each algorithm. For small instances, we run each algorithm sequentially with a five-hour wall-clock time limit to compute its best solution. For huge graphs, with tens of millions of vertices and at least one billion edges, we use a time limit of 10 h. Each run was performed on a machine that is equipped with four Octa-Core Intel Xeon E5-4640 processors running at 2.4 GHz. It has 512 GB local memory, 4×20 MB L3-Cache and $4 \times 8 \times 256$ KB L2-Cache.

We consider social networks, autonomous systems graphs, and Web graphs taken from the 10th DIMACS Implementation Challenge [4], and two additional large Web graphs, webbase-2001 [22] and wikilinks [21]. We also include road networks from Andrade et al. [3] and meshes from Sander et al. [29]. The graphs europe and USA-road are large road networks of Europe [9] and the USA [10]. The instances as-Skitter-big, web-Stanford and libimseti are the hardest instances from Akiba and Iwata [2]. We further perform experiments on huge instances with billions of edges taken from the Laboratory of Web Algorithmics [22]: it-2004, sk-2005, and uk-2007.

4.2   Accelerated Solutions

We now illustrate the speed improvement over existing heuristic algorithms. First, we measure the speedup of OnlineMIS over other high-quality heuristic search algorithms. In particular, in Table 1, we report the maximum speedup that OnlineMIS compared with the state-of-the-art competitors. We compute the maximum speedup for an instance as follows. For each solution size i, we compute the speedup $s_{\mathsf{Alg}}^i = t_{\mathsf{Alg}}^i / t_{\mathsf{OnlineMIS}}^i$ of OnlineMIS over algorithm Alg for that solution size. We then report the maximum speedup $s_{\mathsf{Alg}}^{\max} = \max_i s_{\mathsf{Alg}}^i$ for the instance.

As can be seen in Table 1, OnlineMIS always has a maximum speedup greater than 1 over every other algorithm. We first note that OnlineMIS is significantly faster than ReduMIS and KerMIS. In particular, on 14 instances, OnlineMIS achieves a maximum speedup of over 100 over ReduMIS. KerMIS performs only slightly better than ReduMIS in this regard, with OnlineMIS achieving similar speedups on 12 instances. Though, on meshes, KerMIS fairs especially poorly. On these instances, OnlineMIS always finds a better solution than KerMIS (instances marked with an *), and on the bunny and feline instances, OnlineMIS achieves a maximum speedup of over 10,000 against KerMIS. Furthermore, on

Table 1. For each graph instance, we give the number of vertices n and the number of edges m. We further give the maximum speedup for OnlineMIS over other heuristic search algorithms. For each solution size i, we compute the speedup $s^i_{\text{Alg}} = t^i_{\text{Alg}}/t^i_{\text{OnlineMIS}}$ of OnlineMIS over algorithm Alg for that solution size. We then report the maximum speedup $s^{\max}_{\text{Alg}} = \max_i s^i_{\text{Alg}}$ for the instance. When an algorithm never matches the final solution quality of OnlineMIS, we give the highest non-infinite speedup and give an *. A '∞' indicates that all speedups are infinite.

Graph			Maximum Speedup of OnlineMIS		
Name	n	m	$s^{\max}_{\text{ARW}}$	$s^{\max}_{\text{KerMIS}}$	$s^{\max}_{\text{ReduMIS}}$
Huge instances:					
it-2004	41 291 594	1 027 474 947	4.51	221.26	266.30
sk-2005	50 636 154	1 810 063 330	356.87*	201.68	302.64
uk-2007	105 896 555	1 154 392 916	11.63*	108.13	122.50
Social networks and Web graphs:					
amazon-2008	735 323	3 523 472	43.39*	13.75	50.75
as-Skitter-big	1 696 415	11 095 298	355.06*	2.68	7.62
dewiki-2013	1 532 354	33 093 029	36.22*	632.94	1 726.28
enwiki-2013	4 206 785	91 939 728	51.01*	146.58	244.64
eu-2005	862 664	22 217 686	5.52	62.37	217.39
hollywood-2011	2 180 759	114 492 816	4.35	5.51	11.24
libimseti	220 970	17 233 144	15.16*	218.30	1 118.65
ljournal-2008	5 363 260	49 514 271	2.51	3.00	5.33
orkut	3 072 441	117 185 082	1.82*	478.94*	8 751.62*
web-Stanford	281 903	1 992 636	50.70*	29.53	59.31
webbase-2001	118 142 155	854 809 761	3.48	33.54	36.18
wikilinks	25 890 800	543 159 884	3.88	11.54	11.89
youtube	1 134 890	543 159 884	6.83	1.83	7.29
Road networks:					
europe	18 029 721	22 217 686	5.57	12.79	14.20
USA-road	23 947 347	28 854 312	7.17	24.41	27.84
Meshes:					
buddha	1 087 716	1 631 574	1.16	154.04*	976.10*
bunny	68 790	103 017	3.26	16 616.83*	526.14
dragon	150 000	225 000	2.22*	567.39*	692.60*
feline	41 262	61 893	2.00*	13 377.42*	315.48
gameguy	42 623	63 850	3.23	98.82*	102.03
venus	5 672	8 508	1.17	∞	157.78*

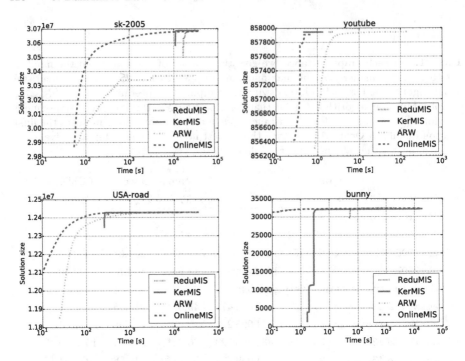

Fig. 2. Convergence plots for sk-2005 (top left), youtube (top right), USA-road (bottom left), and bunny (bottom right).

the venus mesh graph, KerMIS never matches the quality of a single solution from OnlineMIS, giving infinite speedup. ARW is the closest competitor, where OnlineMIS only has 2 maximum speedups greater than 100. However, on a further 6 instances, OnlineMIS achieves a maximum speedup over 10, and on 11 instances ARW fails to match the final solution quality of OnlineMIS, giving an effective infinite maximum speedup.

We now give several representative *convergence plots* in Fig. 2, which illustrate the early solution quality of OnlineMIS compared to ARW, the closest competitor. We construct these plots as follows. Whenever an algorithm finds a new large independent set I at time t, it reports a tuple $(t, |I|)$; the convergence plots show average values over all three runs. In the non-mesh instances, OnlineMIS takes a early lead over ARW, though solution quality converges over time. Lastly, we give the convergence plot for the bunny mesh graph. Reductions and high-degree cutting aren't effective on meshes, thus ARW and OnlineMIS have similar initial solution sizes.

4.3 Time to High-Quality Solutions

We now look at the time it takes an algorithm to find a high-quality solution. We first determine the largest independent set found by any of the four algorithms, which represent the best-known solutions [24], and compute how long it takes

Table 2. For each algorithm, we give the average time t_{avg} to reach 99.5 % of the best solution found by any algorithm. The fastest such time for each instance is marked in bold. We also give the size of the largest solution found by any algorithm and list the algorithms (abbreviated by first letter) that found this largest solution in the time limit. A '-' indicates that the algorithm did not find a solution of sufficient quality.

Graph Name	OnlineMIS t_{avg}	ARW t_{avg}	KerMIS t_{avg}	ReduMIS t_{avg}	Best IS Size	Best IS Algorithms
Huge instances:						
it-2004	**86.01**	327.35	7 892.04	9 448.18	25 620 285	R
sk-2005	**152.12**	-	10 854.46	16 316.59	30 686 766	K
uk-2007	**403.36**	3 789.74	23 022.26	26 081.36	67 282 659	K
Social networks and Web graphs:						
amazon-2008	**0.76**	1.26	5.81	15.23	309 794	K, R
as-Skitter-big	**1.26**	2.70	2.82	8.00	1 170 580	K, R
dewiki-2013	**4.10**	7.88	898.77	2 589.32	697 923	K
enwiki-2013	**10.49**	19.26	856.01	1 428.71	2 178 457	K
eu-2005	**1.32**	3.11	29.01	95.65	452 353	R
hollywood-2011	**1.28**	1.46	7.06	14.38	523 402	O, A, K, R
libimseti	**0.44**	0.45	50.21	257.29	127 293	R
ljournal-2008	**3.79**	8.30	10.20	18.14	2 970 937	K, R
orkut	**42.19**	49.18	2 024.36	-	839 086	K
web-Stanford	**1.58**	8.19	3.57	7.12	163 390	R
webbase-2001	**144.51**	343.86	2 920.14	3 150.05	80 009 826	R
wikilinks	**34.40**	85.54	348.63	358.98	19 418 724	R
youtube	**0.26**	0.81	0.48	1.90	857 945	A, K, R
Road networks:						
europe	**28.22**	75.67	91.21	101.21	9 267 811	R
USA-road	**44.21**	112.67	259.33	295.70	12 428 105	R
Meshes:						
buddha	**26.23**	26.72	119.05	1 699.19	480 853	A
bunny	**3.21**	9.22	-	70.40	32 349	R
dragon	**3.32**	4.90	5.18	97.88	66 502	A
feline	**1.24**	1.27	-	39.18	18 853	R
gameguy	15.13	**10.60**	60.77	12.22	20 727	R
venus	**0.32**	0.36	-	6.52	2 684	O, A, R

Table 3. For each algorithm, we include average solution size and average time t_{avg} to reach it within a time limit (5 hours for normal graphs, 10 hours for huge graphs). Solutions in italics indicate the larger solution between ARW and OnlineMIS local search, bold marks the largest overall solution. A '-' in our indicates that the algorithm did not find a solution in the time limit.

Graph Name	OnlineMIS Avg.	t_{avg}	ARW Avg.	t_{avg}	KerMIS Avg.	t_{avg}	ReduMIS Avg.	t_{avg}
Huge instances:								
it-2004	25 610 697	35 324	25 612 993	33 407	25 619 988	35 751	**25 620 246**	35 645
sk-2005	*30 680 869*	34 480	30 373 880	11 387	**30 686 684**	34 923	30 684 867	35 837
uk-2007	*67 265 560*	35 982	67 101 065	8 702	**67 282 347**	35 663	67 278 359	35 782
Social networks and Web graphs:								
amazon-2008	*309 792*	6 154	309 791	12 195	309 793	818	**309 794**	153
as-Skitter-big	*1 170 560*	7 163	1 170 548	14 017	**1 170 580**	4	**1 170 580**	9
dewiki-2013	*697 789*	17 481	697 669	16 030	**697 921**	14 070	697 798	17 283
enwiki-2013	*2 178 255*	13 612	2 177 965	17 336	**2 178 436**	17 408	2 178 327	17 697
eu-2005	452 296	11 995	452 311	22 968	452 342	5 512	**452 353**	2 332
hollywood-2011	**523 402**	33	**523 402**	101	**523 402**	9	**523 402**	17
libimseti	*127 288*	8 250	127 284	9 308	**127 292**	102	**127 292**	16 747
ljournal-2008	2 970 236	428	2 970 887	16 571	**2 970 937**	36	**2 970 937**	41
orkut	*839 073*	17 764	839 001	17 933	839 004	19 765	806 244	34 197
web-Stanford	*163 384*	5 938	163 382	10 924	163 388	35	**163 390**	12
webbase-2001	79 998 332	35 240	80 002 845	35 922	80 009 041	30 960	**80 009 820**	31 954
wikilinks	19 404 530	21 069	19 416 213	34 085	19 418 693	23 133	**19 418 724**	854
youtube	857 914	< 1	**857 945**	93	**857 945**	< 1	**857 945**	2
Road networks:								
Europe	9 267 573	15 622	9 267 587	28 450	9 267 804	27 039	**9 267 809**	115
USA-road	12 426 557	10 490	12 426 582	31 583	12 427 819	32 490	**12 428 099**	4 799
Meshes:								
buddha	480 795	17 895	**480 808**	17 906	480 592	16 695	479 905	17 782
bunny	32 283	13 258	32 287	13 486	32 110	14 185	**32 344**	1 309
dragon	*66 501*	15 203	66 496	14 775	66 386	16 577	66 447	3 456
feline	*18 846*	15 193	18 844	10 547	18 732	15 055	**18 851**	706
gameguy	20 662	6 868	20 674	12 119	20 655	7 467	**20 727**	191
venus	**2 684**	507	**2 684**	528	2 664	9	2 683	74

each algorithm to find an independent set within 99.5% of this size. The results are shown in Table 2. With a single exception, OnlineMIS is the fastest algorithm to be within 99.5% of the target solution. In fact, OnlineMIS finds such a solution at least twice as fast as ARW in 14 instances, and it is almost 10 times faster on the largest instance, uk-2007. Further, OnlineMIS is orders of magnitude faster than ReduMIS (by a factor of at least 100 in seven cases). We also see that KerMIS is faster than ReduMIS in 19 cases, but much slower than OnlineMIS for all instances. It does eventually find the largest independent set (among all algorithms) for 10 instances. This shows that the full set of reductions is not always necessary, especially when the goal is to get a high-quality solution quickly. It also justifies our choice of cutting: the solution quality of KerMIS rivals (and sometimes even improves) that of ReduMIS.

4.4   Overall Solution Quality

Next, we show that OnlineMIS has high solution quality when given a time limit for searching (5 hours for normal graphs, 10 hours for huge graphs). Although long-run quality is not the goal of the OnlineMIS algorithm, in 11 instances OnlineMIS finds a larger independent set than ARW, and in four instances OnlineMIS finds the largest solution in the time limit. As seen in Table 3, OnlineMIS also finds a solution within 0.1% of the best solution found by any algorithm for all graphs. However, in general OnlineMIS finds lower-quality solutions than ReduMIS, which we believe is from high-degree cutting removing vertices in large independent sets. Nonetheless, as this shows, even when cutting out 1% of the vertices, the solution quality remains high.

On eight instances, KerMIS finds a better solution than ReduMIS. However, kernelization and cutting take a long time (over three hours for sk-2005, 10 h for uk-2007), and therefore KerMIS is much slower to get to a high-quality solution than OnlineMIS. Thus, our experiments show that the full set of reductions is not always necessary, especially when the goal is to get a high-quality solution quickly. This also further justifies our choice of cutting, as the solution quality of KerMIS remains high. On the other hand, instances as-Skitter-big, ljournal-2008, and youtube are solved quickly with advanced reduction rules.

5   Conclusion and Future Work

We have shown that applying reductions on the fly during local search leads to high-quality independent sets quickly. Furthermore, cutting few high-degree vertices has little effect on the quality of independent sets found during local search. Lastly, by kernelizing with advanced reduction rules, we can further speed up local search for high-quality independent sets in the long-run—rivaling the current best heuristic algorithms for complex networks. Determining which reductions give a desirable balance between high-quality results and speed is an interesting topic for future research. While we believe that OnlineMIS gives a nice balance, it is possible that further reductions may achieve higher-quality results even faster.

References

1. Faisal Abu-Khzam, N., Michael Fellows, R., Michael Langston, A., Suters, H.W.: Crown structures for vertex cover kernelization. Theor. Comput. Syst. **41**(3), 411–430 (2007)
2. Akiba, T., Iwata, Y.: Branch-and-reduce exponential/FPT algorithms in practice: A case study of vertex cover. Theor. Comput. Sci. **609**, 211–225 (2016). Part 1
3. Andrade, D.V., Resende, M.G.C., Werneck, R.F.: Fast local search for the maximum independent set problem. J. Heuristics **18**(4), 525–547 (2012)
4. Bader, D.A., Meyerhenke, H., Sanders, P., Schulz, C., Kappes, A., Wagner, D.: Benchmarking for Graph Clustering and Partitioning. In: Alhajj, R., Rokne, J. (eds.) Encyclopedia of Social Network Analysis and Mining, pp. 73–82. Springer, Heidelberg (2014)

5. Batsyn, M., Goldengorin, B., Maslov, E., Pardalos, P.: Improvements to MCS algorithm for the maximum clique problem. J. Comb. Optim. **27**(2), 397–416 (2014)
6. Battiti, R., Protasi, M.: Reactive local search for the maximum clique problem. Algorithmica **29**(4), 610–637 (2001)
7. Bourgeois, N., Escoffier, B., Paschos, V., van Rooij, J.M.: Fast algorithms for max independent set. Algorithmica **62**(1–2), 382–415 (2012)
8. Butenko, S., Pardalos, P., Sergienko, I., Shylo, V., Stetsyuk, P.: Finding maximum independent sets in graphs arising from coding theory. In: Proceedings of the ACM Symposium on Applied Computing (SAC 2002), pp. 542–546. ACM (2002)
9. Delling, D., Sanders, P., Schultes, D., Wagner, D.: Engineering route planning algorithms. In: Lerner, J., Wagner, D., Zweig, K.A. (eds.) Algorithmics of Large and Complex Networks. LNCS, vol. 5515, pp. 117–139. Springer, Heidelberg (2009)
10. Demetrescu, C., Goldberg, A.V., Johnson, D.S.: The Shortest Path Problem: 9th DIMACS Implementation Challenge, vol. 74. AMS (2009)
11. Feo, T.A., Resende, M.G.C., Smith, S.H.: A greedy randomized adaptive search procedure for maximum independent set. Oper. Res. **42**(5), 860–878 (1994)
12. Fomin, F.V., Kratsch, D.: Exact Exponential Algorithms. Springer, Heidelberg (2010)
13. Garey, M.R., Johnson, D.S.: Computers and intractability: a guide to the theory of np-completeness. In: Freeman, W.H. (1979)
14. Gemsa, A., Nöllenburg, M., Rutter, I.: Evaluation of labeling strategies for rotating maps. In: Gudmundsson, J., Katajainen, J. (eds.) SEA 2014. LNCS, vol. 8504, pp. 235–246. Springer, Heidelberg (2014)
15. Grosso, A., Locatelli, M., Della, F.C.: Combining swaps and node weights in an adaptive greedy approach for the maximum clique problem. J. Heuristics **10**(2), 135–152 (2004)
16. Grosso, A., Locatelli, M., Pullan, W.: Simple ingredients leading to very efficient heuristics for the maximum clique problem. J. Heuristics **14**(6), 587–612 (2008)
17. Hansen, P., Mladenović, N., Urošević, D.: Variable neighborhood search for the maximum clique. Discrete Appl. Math. **145**(1), 117–125 (2004)
18. Iwata, Y., Oka, K., Yoshida, Y.: Linear-time FPT algorithms via network flow. In: Proceedings of the 25th ACM-SIAM Symposium on Discrete Algorithms, SODA 2014, pp. 1749–1761. SIAM (2014)
19. Katayama, K., Hamamoto, A., Narihisa, H.: An effective local search for the maximum clique problem. Inform. Process. Lett. **95**(5), 503–511 (2005)
20. Kieritz, T., Luxen, D., Sanders, P., Vetter, C.: Distributed time-dependent contraction hierarchies. In: Festa, P. (ed.) SEA 2010. LNCS, vol. 6049, pp. 83–93. Springer, Heidelberg (2010)
21. Kunegis, J.: KONECT: The Koblenz network collection. In: Proceedings of the International Conference on World Wide Web Companion (WWW 13), pp. 1343–1350 (2013)
22. University of Milano Laboratory of Web Algorithms. Datasets
23. Lamm, S., Sanders, P., Schulz, C.: Graph partitioning for independent sets. In: Bampis, E. (ed.) SEA 2015. LNCS, vol. 9125, pp. 68–81. Springer, Heidelberg (2015)
24. Lamm, S., Sanders, P., Schulz, C., Strash, D., Werneck, R.F.: Finding near-optimal independent sets at scale. In: Proceedings of the 18th Workshop on Algorithm Engineering and Experiments (ALENEX 2016), pp. 138–150 (2016)
25. Liu, Y., Lu, J., Yang, H., Xiao, X., Wei, Z.: Towards maximum independent sets on massive graphs. Proc. VLDB Endow. **8**(13), 2122–2133 (2015)

26. Nemhauser, G.L., Trotter, L.E.: Vertex packings: Structural properties and algorithms. Math. Program. **8**(1), 232–248 (1975)
27. Pullan, W.J., Hoos, H.H.: Dynamic local search for the maximum clique. J. Arti. Int. Res. **25**, 159–185 (2006)
28. San Segundo, P., Matia, F., Rodriguez-Losada, D., Hernando, M.: An improved bit parallel exact maximum clique algorithm. Optim. Lett. **7**(3), 467–479 (2013)
29. Sander, P.V., Nehab, D., Chlamtac, E., Hoppe, H.: Efficient traversal of mesh edges using adjacency primitives. ACM Trans. Graph. **27**(5), 144:1–144:9 (2008)
30. San Segundo, P., Rodríguez-Losada, D., Jiménez, D.: An exact bit-parallel algorithm for the maximum clique problem. Comput. Oper. Res. **38**(2), 571–581 (2011)
31. Tarjan, R.E., Trojanowski, A.E.: Finding a maximum independent set. SIAM J. Comput. **6**(3), 537–546 (1977)
32. Tomita, E., Sutani, Y., Higashi, T., Takahashi, S., Wakatsuki, M.: A simple and faster branch-and-bound algorithm for finding a maximum clique. In: Rahman, M.S., Fujita, S. (eds.) WALCOM 2010. LNCS, vol. 5942, pp. 191–203. Springer, Heidelberg (2010)
33. Xiang, J., Guo, C., Aboulnaga, A.: Scalable maximum clique computation using mapreduce. In: Proceedings of the IEEE 29th International Conference on Data Engineering (ICDE 2013), pp. 74–85, April 2013
34. Xiao, M., Nagamochi, H.: Confining sets and avoiding bottleneck cases: A simple maximum independent set algorithm in degree-3 graphs. Theor. Comput. Sci. **469**, 92–104 (2013)

Computing Nonsimple Polygons
of Minimum Perimeter

Sándor P. Fekete[1]([✉]), Andreas Haas[1], Michael Hemmer[1], Michael Hoffmann[2],
Irina Kostitsyna[3], Dominik Krupke[1], Florian Maurer[1], Joseph S.B. Mitchell[4],
Arne Schmidt[1], Christiane Schmidt[5], and Julian Troegel[1]

[1] TU Braunschweig, Braunschweig, Germany
s.fekete@tu-bs.de
[2] ETH Zurich, Zurich, Switzerland
[3] TU Eindhoven, Eindhoven, The Netherlands
[4] Stony Brook University, Stony Brook, NY, USA
[5] Linköping University, Linköping, Sweden

Abstract. We provide exact and approximation methods for solving
a geometric relaxation of the Traveling Salesman Problem (TSP) that
occurs in curve reconstruction: for a given set of vertices in the plane, the
problem Minimum Perimeter Polygon (MPP) asks for a (not necessarily
simply connected) polygon with shortest possible boundary length. Even
though the closely related problem of finding a minimum cycle cover is
polynomially solvable by matching techniques, we prove how the topological structure of a polygon leads to NP-hardness of the MPP. On the
positive side, we show how to achieve a constant-factor approximation.

When trying to solve MPP instances to provable optimality by means
of integer programming, an additional difficulty compared to the TSP
is the fact that only a subset of subtour constraints is valid, depending
not on combinatorics, but on geometry. We overcome this difficulty by
establishing and exploiting additional geometric properties. This allows
us to reliably solve a wide range of benchmark instances with up to 600
vertices within reasonable time on a standard machine. We also show
that using a natural geometry-based sparsification yields results that are
on average within 0.5 % of the optimum.

Keywords: Traveling Salesman Problem (TSP) · Minimum Perimeter
Polygon (MPP) · Curve reconstruction · NP-hardness · Exact optimization · Integer programming · Computational geometry meets combinatorial optimization

1 Introduction

For a given set V of points in the plane, the Minimum Perimeter Polygon (MPP)
asks for a polygon P with vertex set V that has minimum possible boundary
length. An optimal solution may not be simply connected, so we are faced with
a geometric relaxation of the Traveling Salesman Problem (TSP).

© Springer International Publishing Switzerland 2016
A.V. Goldberg and A.S. Kulikov (Eds.): SEA 2016, LNCS 9685, pp. 134–149, 2016.
DOI: 10.1007/978-3-319-38851-9_10

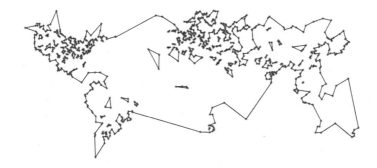

Fig. 1. A Minimum Perimeter Polygon for an instance with 960 vertices.

The TSP is one of the classic problems of Combinatorial Optimization. NP-hard even in special cases of geometric instances (such as grid graphs), it has served as one of the prototypical testgrounds for developing outstanding algorithmic approaches. These include constant-factor approximation methods (such as Christofides' 3/2 approximation [6] in the presence of triangle inequality, or Arora's [4] and Mitchell's [20] polynomial-time approximation schemes for geometric instances), as well as exact methods (such as Grötschel's optimal solution to a 120-city instance [14] or the award-winning work by Applegate et al. [2] for solving a 13509-city instance within 10 years of CPU time.) The well-established benchmark library TSPLIB [23] of TSP instances has become so widely accepted that it is used as a benchmark for a large variety of other optimization problems. See the books [15,18] for an overview of various aspects of the TSP and the books [3,7] for more details on exact optimization.

From a geometric point of view, the TSP asks for a shortest polygonal chain through a given set of vertices in the plane; as a consequence of triangle inequality, the result is always a simple polygon of minimum perimeter. Because of the fundamental role of polygons in geometry, this has made the study of TSP solutions interesting for a wide range of geometric applications. One such context is geometric shape reconstruction, where the objective is to re-compute the original curve from a given set of sample points; see Giesen [13], Althaus and Mehlhorn [1] or Dey et al. [9] for specific examples. However, this only makes sense when the original shape is known to be simply connected, i.e., bounded by a single closed curve. More generally, a shape may be multiply connected, with interior holes. Thus, computing a simple polygon may not yield the desired answer. Instead, the solution may be a Minimum Perimeter Polygon (MPP): for a set V of points in the plane, find a not necessarily simple polygon P with vertex set V, such that the boundary of P has smallest possible length[1]. See Fig. 1 for an optimal solution of an instance with 960 points; this also shows the possibly intricate structure of an MPP.

---

[1] Note that we exclude degenerate holes that consist of only one or two vertices.

While the problem MPP[2] asks for a cycle cover of the given set of vertices (as opposed to a single cycle required by the TSP), it is important to note that even the more general geometry of a polygon with holes imposes some topological constraints on the structure of boundary cycles; as a consequence, an optimal 2-factor (a minimum-weight cycle cover of the vertices, which can be computed in polynomial time) may not yield a feasible solution. Fekete et al. [11] gave a generic integer program for the MPP (and other related problems) that yields optimal solutions for instances up to 50 vertices. However, the main challenges were left unresolved. What is the complexity of computing an MPP? Is it possible to develop constant-factor approximation algorithms? And how can we compute provably optimal solutions for instances of relevant size?

Our Results

In this paper, we resolve the main open problems related to the MPP.

- We prove that the MPP is NP-hard. This shows that despite of the relationship to the polynomially solvable problem of finding a minimum 2-factor, dealing with the topological structure of the involved cycles is computationally difficult.
- We give a 3-approximation algorithm.
- We provide a general IP formulation with $O(n^2)$ variables to ensure a valid solution for the MPP.
- We describe families of cutting planes that significantly reduce the number of iterations needed to eliminate outer components and holes in holes, leading to a practically useful formulation.
- We present experimental results for the MPP, solving instances with up to 1000 points in the plane to provable optimality within 30 min of CPU time.
- We also consider a fast heuristic that is based on geometric structure, restricting the edge set to the Delaunay triangulation. Experiments on structured random point sets show that solutions are on average only about 0.5 % worse than the optimum, with vastly superior runtimes.

2   Complexity

Theorem 1. *The MPP problem is NP-hard.*

The proof is based on a reduction from the Minimum Vertex Cover problem for planar graphs. Details are omitted for lack of space; see the full version of the paper [12] for the detailed proof.

3   Approximation

In this section we show that the MPP can be approximated within a factor of 3. Note that we only sketch the general approach, skipping over some details for lack of space; a full proof is given in the full version of the paper [12].

---

[2] For simplicity, we will also refer to the problem of computing an MPP as "the MPP".

Theorem 2. *There exists a polynomial time 3-approximation for the MPP.*

Proof. We compute the convex hull, $CH(V)$, of the input set; this takes time $O(n \log h)$, where h is the number of vertices of the convex hull. Note that the perimeter, $|CH(V)|$, of the convex hull is a lower bound on the length of an optimal solution ($OPT \geq |CH(V)|$), since the outer boundary of any feasible solution polygon must enclose all points of V, and the convex hull is the minimum-perimeter enclosure of V.

Let $U \subseteq V$ be the input points interior to $CH(V)$. If $U = \varnothing$, then the optimal solution is given by the convex hull. If $|U| \leq 2$, we claim that an optimal solution is a simple (nonconvex) polygon, with no holes, on the set V, given by the TSP tour on V; since $|U| = 2$ is a constant, it is easy to compute the optimal solution in polynomial time, by trying all possible ways of inserting the points of U into the cycle of the points of V that lie on the boundary of the convex hull, $CH(V)$.

Thus, assume now that $|U| \geq 3$. We compute a minimum-weight 2-factor, denoted by $\gamma(U)$, on U, which is done in polynomial-time by standard methods [8]. Now, $\gamma(U)$ consists of a set of disjoint simple polygonal curves having vertex set U; the curves can be nested, with possibly many levels of nesting. We let F denote the directed *nesting forest* whose nodes are the cycles (connected components) of $\gamma(U)$ and whose directed edges indicate nesting (containment) of one cycle within another. Because an optimal solution consists of a 2-factor (an outer cycle, together with a set of cycles, one per hole of the optimal polygon), we know that $OPT \geq |\gamma(U)|$. (In an optimal solution, the nesting forest corresponding to the set of cycles covering all of V (not just the points U interior to $CH(V)$) is simply a single tree that is a star: a root node corresponding to the outer cycle, and a set of children adjacent to the root node, corresponding to the boundaries of the holes of the optimal polygon.) If the nesting forest F for our optimal 2-factor is a set of isolated nodes (i.e., there is no nesting among the cycles of the optimal 2-factor on U), then our algorithm outputs a polygon with holes whose outer boundary is the boundary of the convex hull, $CH(V)$, and whose holes are the (disjoint) polygons given by the cycles of $\gamma(U)$. In this case, the total weight of our solution is equal to $|CH(V)| + |\gamma(U)| \leq 2 \cdot OPT$.

Assume now that F has at least one nontrivial tree. We describe a two-phase process that transforms the set of cycles corresponding to F into a set of pairwise-disjoint cycles, each defining a simple polygon interior to $CH(V)$, with no nesting – the resulting simple polygons are disjoint, each having at least 3 vertices from $U \subset V$.

Phase 1 of the process transforms the cycles $\gamma(U)$ to a set of polygonal cycles that define *weakly simple* polygons whose interiors are pairwise disjoint. (A polygonal cycle β defines a *weakly simple* polygon P_β if P_β is a closed, simply connected set in the plane with a boundary, ∂P_β consisting of a finite union of line segments, whose traversal (e.g., while keeping the region P_β to one's left) is the (counterclockwise) cycle β (which can have line segments that are traversed twice, once in each direction).) The total length of the cycles at the end of phase 1 is at most 2 times the length of the original cycles, $\gamma(U)$. Then, phase 2 of the process transforms these weakly simple cycles into (strongly) simple cycles that define disjoint simple polygons interior to $CH(V)$. Phase 2 only does shortening operations on the weakly simple cycles; thus, the length of the resulting simple cycles at the end of phase 2 is at most 2 times the total length of $\gamma(U)$. Details of phase 1 and phase 2 processes are given in the full version of the paper. At the end of phase 2, we have a set of disjoint simple polygons within $CH(V)$, which serve as the holes of the output polygon, whose total perimeter length is at most $|CH(V)| + 2|\gamma(U)| \leq 3 \cdot OPT$.   □

4   IP Formulation

4.1   Cutting-Plane Approach

In the following we develop suitable Integer Programs (IPs) for solving the MPP to provable optimality. The basic idea is to use a binary variable $x_e \in \{0,1\}$ for any possible edge $e \in E$, with $x_e = 1$ corresponding to e being part of a solution P if and only if $x_e = 1$. This allows it to describe the objective function by $\min \sum_{e \in E} x_e c_e$, where c_e is the length of e. In addition, we impose a suitable set of linear constraints on these binary variables, such that they characterize precisely the set of polygons with vertex set V. The challenge is to pick a set of constraints that achieve this in a (relatively) efficient manner.

As it turns out (and is discussed in more detail in Sect. 5), there is a significant set of constraints that correspond to eliminating cycles within proper subsets $S \subset V$. Moreover, there is an exponential number of relevant subsets S, making it prohibitive to impose all of these constraints at once. The fundamental idea of a cutting-plane approach is that much fewer constraints are necessary for characterizing an optimal solution. To this end, only a relatively small subfamily of constraints is initially considered, leading to a relaxation. As long as solving the current relaxation yields a solution that is infeasible for the original problem, violated constraints are added in a piecemeal fashion, i.e., in *iterations*.

In the following, these constraints (which are initially omitted, violated by an optimal solution of the relaxation, then added to eliminate such infeasible solutions) are called *cutting planes* or simply *cuts*, as they remove solutions of a relaxation that are infeasible for the MPP.

4.2   Basic IP

We start with a basic IP that is enhanced with specific cuts, described in Sects. 5.2–5.4. We denote by E the set of all edges between two points of V, $\mathcal{C}$ a set of *invalid cycles* and $\delta(v)$ the set of all edges in E that are incident to $v \in V$. Then we optimize over the following objective function:

$$\min \sum_{e \in E} x_e c_e. \tag{1}$$

This is subject to the following constraints:

$$\forall v \in V : \sum_{e \in \delta(v)} x_e = 2, \tag{2}$$

$$\forall C \in \mathcal{C} : \sum_{e \in C} x_e \leq |C| - 1, \tag{3}$$

$$x_e \in \{0, 1\}. \tag{4}$$

For the TSP, $\mathcal{C}$ is simply the set of *all* subtours, making identification and separation straightforward. This is much harder for the MPP, where a subtour may end up being feasible by forming the boundary of a hole, but may also be required to connect with other cycles. Therefore, identifying valid inequalities requires more geometric analysis, such as the following. If we denote by CH the set of all convex hull points, then a cycle C is invalid if C contains:

1. at least one and at most $|CH| - 1$ convex hull points. (See Fig. 2(a))
2. all convex hull points but does not enclose all other points. (See Fig. 2(b))
3. no convex hull point but encloses other points. (See Fig. 2(c))

By $\mathcal{C}_i$ we denote the set of all invalid cycles with property i. Because there can be exponentially many invalid cycles, we add constraint (3) in separation steps.

For an invalid cycle with property 1, we use the equivalent cut constraint

$$\forall C \in \mathcal{C}_1 : \sum_{e \in \delta(C)} x_e \geq 2. \tag{5}$$

We use constraint (3) if $|C| \leq \frac{2n+1}{3}$ and constraint (5) otherwise, where $\delta(C)$ denotes the "cut" edges connecting a vertex $v \in C$ with a vertex $v' \notin C$. As argued by Pferschy and Stanek [22], this technique of *dynamic subtour constraints* (DSC) is useful, as it reduces the number of non-zero coefficients in the constraint matrix.

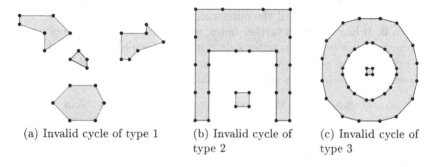

(a) Invalid cycle of type 1   (b) Invalid cycle of type 2   (c) Invalid cycle of type 3

Fig. 2. Examples of invalid cycles (red). Black cycles may be valid. (Color figure online)

4.3   Initial Edge Set

In order to quickly achieve an initial solution, we sparsify the $\Theta(n^2)$ input edges to the $O(n)$ edges of the Delaunay Triangulation, which naturally captures geometric nearest-neighbor properties. If a solution exists, this yields an upper bound. This technique has already been applied for the TSP by Jünger et al. [16]. In theory, this may not yield a feasible solution: a specifically designed example by Dillencourt shows that the Delaunay triangulation may be non-Hamiltonian [10]; this same example has no feasible solution for the MPP when restricted to Delaunay edges. We did not observe this behavior in practice.

CPLEX uses this initial solution as an upper bound, allowing it to quickly discard large solutions in a branch-and-bound manner. As described in Sect. 6, the resulting bounds are quite good for the MPP.

5   Separation Techniques

5.1   Pitfalls

When separating infeasible cycles, the Basic IP may get stuck in an exponential number of iterations, due to the following issues. (See Figs. 3–5 for illustrating examples.)

Problem 1: Multiple outer components containing convex hull points occur that (despite the powerful subtour constraints) do not get connected, because it is cheaper to, e.g., integrate subsets of the interior points. Such an instance can be seen in Fig. 3, where we have two equal components with holes. Since the two components are separated by a distance greater than the distance between their outer components and their interior points, the outer components start to include point subsets of the holes. This results in a potentially exponential number of iterations.

Problem 2: Outer components that do not contain convex hull points do not get integrated, because we are only allowed to apply a cycle cut on the outer component containing the convex hull points. An outer component that does not contain a convex hull point cannot be prohibited, as it may become a hole in later iterations. See Fig. 4 for an example in which an exponential number of iterations is needed until the outer components get connected.

Problem 3: If holes contain further holes, we are only allowed to apply a cycle cut on the outer hole. This outer hole can often cheaply be modified to fulfill the cycle cut but not resolve the holes in the hole. An example instance can be seen in Fig. 5, in which an exponential number of iterations is needed.

The second problem is the most important, as this problem frequently becomes critical on instances of size 100 and above. Holes in holes rarely occur on small instances but are problematic on instances of size >200. The first problem occurs only in a few instances.

In the following we describe three cuts that each solve one of the problems: The glue cut for the first problem in Sect. 5.2, the tail cut for the second problem in Sect. 5.3, and the HiH-Cut for the third problem in Sect. 5.4.

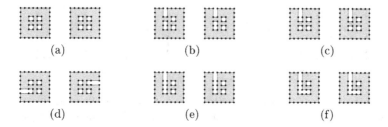

Fig. 3. (a)–(f) show consecutive iterations when trying to solve an instance using only constraint (5).

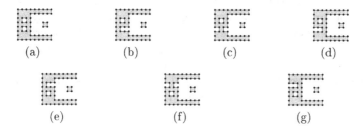

Fig. 4. (a)–(g) show consecutive iterations when trying to solve an instance using only constraint (3).

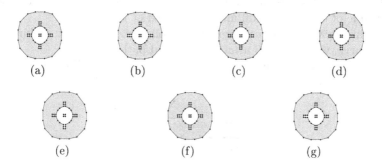

Fig. 5. (a)–(g) show consecutive iterations when trying to solve an instance using only constraint (3).

5.2   Glue Cuts

To separate invalid cycles of property 1 we use *glue cuts* (GC), based on a curve R_D from one unused convex hull edge to another (see Fig. 6). With $\mathcal{X}(R_D)$ denoting the set of edges crossing R_D, we can add the following constraint:

$$\sum_{e \in \mathcal{X}(R_D)} x_e \geq 2.$$

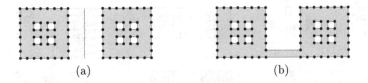

Fig. 6. Solving instance from Fig. 3 with a glue cut (red). (a) The red curve needs to be crossed at least twice; it is found using the Delaunay Triangulation (grey). (b) The first iteration after using the glue cut. (Color figure online)

Such curves can be found by considering a constrained Delaunay triangulation [5] of the current solution, performing a breadth-first-search starting from all unused convex hull edges of the triangulation. Two edges are adjacent if they share a triangle. Used edges are excluded, so our curve will not cross any used edge. As soon as two different search trees meet, we obtain a valid curve by using the middle points of the edges (see the red curve in Fig. 6).

For an example, see Fig. 6; as illustrated in Fig. 3, this instance is problematic in the Basic IP. This can we now be solved in one iteration.

5.3   Tail Cuts

An outer cycle C that does not contain any convex hull points cannot simply be excluded, as it may become a legal hole later. Such a cycle either has to be merged with others, or become a hole. For a hole, each curve from the hole to a point outside of the convex hull must be crossed at least once.

With this knowledge we can provide the following constraint, making use of a special curve, which we call a *tail* (see the red path in Fig. 7).

Let R_T be a valid tail and $\mathcal{X}(R_T)$ the edges crossing it. We can express the constraint in the following form:

$$\underbrace{\sum_{e \in \mathcal{X}(R_T) \setminus \delta(C)} x_e}_{\text{C gets surrounded}} + \underbrace{\sum_{e \in \delta(C)} x_e}_{\text{C merged}} \geq 1.$$

The tail is obtained in a similar fashion as the curves of the *Glue Cuts* by building a constrained Delaunay triangulation and doing a breadth-first search starting at the edges of the cycle. The starting points are not considered as part of the curve and thus the curve does not cross any edges of the current solution.

For an example, see Fig. 7; as illustrated in Fig. 4, this instance is problematic in the Basic IP. This can we now be solved in one iteration. Note that even though it is possible to cross the tail without making the cycle a hole, this is more expensive than simply merging it with other cycles.

5.4   Hole-in-Hole Cuts

The difficulty of eliminating holes in holes (Problem 3) is that they may end up as perfectly legal simple holes, if the outer cycle gets merged with the outer

(a)   (b)

Fig. 7. Solving the instance from Fig. 4 with a tail cut (red line). (a) The red curve needs to be crossed at least twice or two edges must leave the component. The red curve is found via the Delaunay Triangulation (grey). (b) The first iteration after using the tail cut. (Color figure online)

boundary. In that case, every curve from the hole to the convex hull *cannot* cross the used edges exactly two times (edges of the hole are ignored). One of the crossed edges has to be of the exterior cycle, while the other one cannot: otherwise would again leave the polygon. It also cannot be of an interior cycle, as it would have leave to leave that cycle again to reach the hole.

Therefore the inner cycle of a hole in hole either has to be merged, or all curves from it to the convex hull do not have exactly two used edge crossings. As it is impractical to argue over all curves, we only pick one curve P that currently crosses exactly two used edges (see the red curve in Fig. 8 with crossed edges in green).

Because we cannot express the inequality that P is not allowed to be crossed exactly two times as an linear programming constraint, we use the following weaker observation. If the cycle of the hole in hole becomes a simple hole, the crossing of P has to change. Let e_1 and e_2 be the two used edges that currently cross P and $\mathcal{X}(P)$ the set of all edges crossing P (including unused but no edges of H). We can express a change on P by

$$\underbrace{\sum_{e\in\mathcal{X}(P)\backslash\{e_1,e_2\}} x_e}_{\text{new crossing}} + \underbrace{(-x_{e_1} - x_{e_2})}_{e_1 \text{ or } e_2 \text{ vanishes}} \geq -1.$$

Together we obtain the following LP constraint for either H being merged or the crossing of P changing.

$$\underbrace{\sum_{e\in\delta(V_H,V\backslash V_H)} x_e}_{H \text{ merged}} + \underbrace{\sum_{e\in\mathcal{X}(P)\backslash\{e_1,e_2\}} x_e + (-x_{e_1} - x_{e_2})}_{\text{Crossing of } P \text{ changes}} \geq -1.$$

Again we use a breadth-first search on the constrained Delaunay triangulation starting from the edges of the hole in hole. Unlike the other two cuts we need to cross used edges. Thus, we get a shortest path search such that the optimal path primarily has a minimal number of used edges crossed and secondarily has a minimal number of all edges crossed.

For an example, see Fig. 8; as illustrated in Fig. 3, this instance is problematic in the Basic IP. This can now be solved in one iteration. The corresponding path

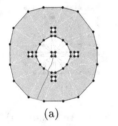
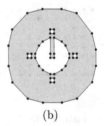

(a)   (b)

Fig. 8. Solving instance from Fig. 5 with hole in hole cut (red line). (a) The red line needs to be crossed at least two times or two edges must leave the component or one of the two existing edges (green) must be removed. The red line is built via Delaunay Triangulation. (b) The first iteration after using the hole in hole cut. (Color figure online)

is displayed in red and the two crossed edges are highlighted in green. Changing the crossing of the path is more expensive than simply connecting the hole in hole to the outer hole and thus the hole in hole gets merged.

6 Experiments

6.1 Implementation

Our implementation uses CPLEX to solve the relevant IPs. Important is also the geometric side of computation, for which we used the CGAL Arrangements package [24]. CGAL represents a planar subdivision using a doubly connected edge list (DCEL), which is ideal for detecting invalid boundary cycles.

6.2 Test Instances

While the TSPLIB is well recognized and offers a good mix of instances with different structure (ranging from grid-like instances over relatively uniform random distribution to highly clustered instances), it is rather sparse. Observing that the larger TSPLIB instances are all geographic in nature, we designed a generic approach that yields arbitrarily large and numerous clustered instances. This is based on illumination maps: A satellite image of a geographic region at night time displays uneven light distribution. The corresponding brightness values can be used as a random density function that can be used for sampling (see Fig. 12). To reduce noise, we cut off brightness values below a certain threshold, i.e., we set the probability of choosing the respective pixels to zero.

6.3 Results

All experiments were run on an *Intel Core i7-4770* CPU clocked at 3.40 GHz with 16 GB of RAM. We set a 30 min time limit to solve the instances. In Table 1, all results are displayed for every instance with more than 100 points that we

Table 1. The runtime in milliseconds of all variants on the instances of the TSPLib with more than 100 points, solved within 30 min. The number in the name of an instance indicates the number of points.

	BasicIP	+JS+DC+TC+ HIHC	+JS+TC+ HIHC	+JS+DC+ HIHC	+JS+DC+ TC	+DC+TC+ HIHC
eil101	-	575	445	-	527	1090
lin105	-	390	359	-	412	931
pr107	550	401	272	346	513	923
pr124	495	348	264	322	355	940
bier127	439	288	270	267	276	476
ch130	-	1758	1802	-	1594	2853
pr136	1505	964	1029	992	950	3001
gr137	-	1262	1361	-	1252	1724
pr144	6276	1028	2926	985	1030	2012
ch150	-	4938	5167	-	5867	7997
kroA150	-	3427	5615	-	3327	7474
kroB150	-	2993	2396	-	2943	5265
pr152	13285	2161	1619	10978	2151	19479
u159	13285	1424	1262	5339	1410	2513
rat195	106030	16188	19780	77216	16117	27580
d198	-	19329	155550	-	19398	41118
kroA200	-	26360	13093	-	26389	11844
kroB200	-	5492	6239	-	5525	15238
gr202	-	4975	7512	-	4304	9670
ts225	18902	7746	9750	7595	7603	60167
tsp225	91423	11600	9741	28756	11531	44297
pr226	-	8498	2800	-	7204	18848
gr229	-	5462	26478	-	10153	25674
gil262	-	23000	22146	-	-	72772
pr264	24690	6537	-	6719	6549	23641
a280	22023	3601	3857	3980	3619	12983
pr299	-	16251	355323	-	16173	85789
lin318	-	23863	1511219	-	24035	75312
linhp318	-	23107	1313680	-	23064	79352
rd400	-	111128	92995	-		302363
fl417	-	198013	-	-	215210	825808
gr431	-	56716	173609	-	78133	265416
pr439	-	46685	36592	-	48231	273873
pcb442	-	1356796	-	-	-	-
d493	-	359072	-	-	-	837229
att532	-	217679	256394	-	218665	817096
ali535	-	93771	427800	-	91828	323104
u574	-	371523	199114	-	-	1010276
rat575	-	417494	191198	-	580320	934988
p654	-	864066	-	-	-	-
d657	-	455378	253374	-	646148	1352747
gr666	-	366157	-	-	670818	-

solved within the time limit. The largest instance solved within 30 min is gr666 with 666 points, which took about 6 min. The largest instance solved out of the TSPLib so far is dsj1000 with 1000 points, solved in about 37 min. In addition, we generated 30 instances for each size, which were run with a time limit of 30 min.

We observe that even without using glue cuts and jumpstart, we are able to solve more than 50 % of the instances up to about 550 input points. Without the tail cuts, we hit a wall at 100 points, without the HiH-cut instances, at about 370 input points; see Fig. 9, which also shows the average runtime of all 30 instances for all variants. Instances exceeding the 30 min time limit are marked with a 30-minutes timestamp. The figure shows that using jumpstart shortens the runtime significantly; using the glue cut is almost as fast as the variant without the glue cut.

Figure 10 shows that medium-sized instances (up to about 450 points) can be solved in under 5 min. We also show that restricting the edge set to the Delaunay triangulation edges yields solutions that are about 0.5 % worse on average than the optimal solution. Generally the solution of the jumpstart gets very close to the optimal solution until about 530 points. After that, for some larger instances,

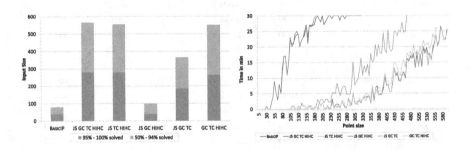

Fig. 9. (Left) Success rate for the different variants of using of the cuts, with 30 instances for each input size (*y*-axis). (Right) The average runtime of the different variants for all 30 instances. A non-solved instance is interpreted as 30 min runtime.

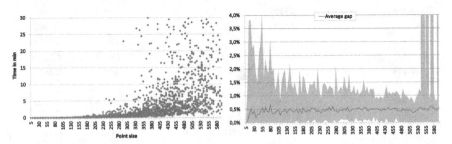

Fig. 10. (Left) The distribution of the runtime within 30 min for the case of using the jumpstart, glue cuts, tail cuts and HiH-cuts. (Right) The relative gap of the value on the edges of the Delaunay triangulation to the optimal value. The red area marks the range between the minimal and maximal gap.

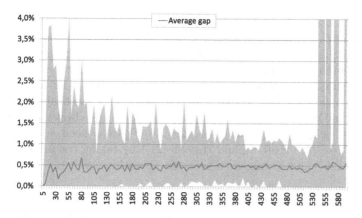

Fig. 11. The relative gap of the value on the edges of the Delaunay triangulation to the optimal value. The red area marks the range between the minimal and maximal gap. (Color figure online)

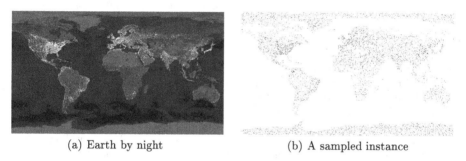

(a) Earth by night   (b) A sampled instance

Fig. 12. Using a brightness map as a density function for generating clustered point sets.

we get solutions on the edge set of the Delaunay triangulation that are up to 50 % worse than the optimal solution.

7 Conclusions

As discussed in the introduction, considering general instead of simple polygons corresponds to searching for a shortest cycle cover with a specific topological constraint: one outside cycle surrounds a set of disjoint and unnested inner cycles. Clearly, this is only one example of considering specific topological constraints. Our techniques and results should be applicable, after suitable adjustments, to other constraints on the topology of cycles. We gave a 3-approximation for the MPP; we expect that the MPP has a polynomial-time approximation scheme, base on PTAS techniques [4,20] for geometric TSP, and we will elaborate on this in a future version of the full paper.

There are also various practical aspects that can be explored further. It will be interesting to evaluate the practical performance of the theoretical approximation algorithm, not only from a practical perspective, but also to gain some insight on whether the approximation factor of 3 can be tightened. Pushing the limits of solvability can also be attempted, e.g., by using more advanced techniques from the TSP context. We can also consider sparsification techniques other than the Delaunay edges; e.g., the union between the best known tour and the k-nearest-neighbor edge set ($k \in \{2, 5, 10, 20\}$) has been applied for TSP by Land [17], or (see Padberg and Rinaldi [21]) by taking the union of k tours acquired by Lin's and Kernighan's heuristic algorithm [19].

Acknowledgements. We thank Stephan Friedrichs and Melanie Papenberg for helpful conversations. Parts of this work were carried out at the 30th Bellairs Winter Workshop on Computational Geometry (Barbados) in 2015. We thank the workshop participants and organizers, particularly Erik Demaine. Joseph Mitchell is partially supported by NSF (CCF-1526406). Irina Kostitsyna is supported by the Netherlands Organisation for Scientific Research (NWO) under project no. 639.023.208.

References

1. Althaus, E., Mehlhorn, K.: Traveling salesman-based curve reconstruction in polynomial time. SIAM J. Comput. **31**(1), 27–66 (2001)
2. Applegate, D.L., Bixby, R.E., Chvatal, V., Cook, W.J.: On the solution of traveling salesman problems. Documenta Mathematica – Journal der DeutschenMathematiker-Vereinigung, ICM, pp. 645–656 (1998)
3. Applegate, D.L., Bixby, R.E., Chvatal, V., Cook, W.J.: The Traveling Salesman Problem: A Computational Study. Princeton Series in Applied Mathematics. Princeton University Press, Princeton (2007)
4. Arora, S.: Polynomial time approximation schemes for Euclidean traveling salesman and other geometric problems. J. ACM **45**(5), 753–782 (1998)
5. Chew, L.P.: Constrained Delaunay triangulations. Algorithmica **4**(1–4), 97–108 (1989)
6. Christofides, N.: Worst-case analysis of a new heuristic for the Travelling Salesman Problem, Technical report 388, Graduate School of Industrial Administration, CMU (1976)
7. Cook, W.J.: In Pursuit of the Traveling Salesman: Mathematics at the Limits of Computation. Princeton University Press, Princeton (2012)
8. Cook, W.J., Cunningham, W.H., Pulleyblank, W.R., Schrijver, A.: Combinatorial Optimization. Wiley, New York (1998)
9. Dey, T.K., Mehlhorn, K., Ramos, E.A.: Curve reconstruction: connecting dots with good reason. Comput. Geom. **15**(4), 229–244 (2000)
10. Dillencourt, M.B.: A non-Hamiltonian, nondegenerate Delaunay triangulation. Inf. Process. Lett. **25**(3), 149–151 (1987)
11. Fekete, S.P., Friedrichs, S., Hemmer, M., Papenberg, M., Schmidt, A., Troegel, J.: Area- and boundary-optimal polygonalization of planar point sets. In: EuroCG 2015, pp. 133–136 (2015)
12. Fekete, S.P., Haas, A., Hemmer, M., Hoffmann, M., Kostitsyna, I., Krupke, D., Maurer, F., Mitchell, J.S.B., Schmidt, A., Schmidt, C., Troegel, J.: Computing nonsimple polygons of minimum perimeter. CoRR, abs/1603.07077 (2016)

13. Giesen, J.: Curve reconstruction, the traveling salesman problem and Menger's theorem on length. In: Proceedings of 15th Annual Symposium on Computational Geometry (SoCG), pp. 207–216 (1999)
14. Grötschel, M.: On the symmetric travelling salesman problem: solution of a 120-city problem. Math. Program. Study **12**, 61–77 (1980)
15. Gutin, G., Punnen, A.P.: The Traveling Salesman Problem and Its Variations. Springer, New York (2007)
16. Jünger, M., Reinelt, G., Rinaldi, G.: The traveling salesman problem. In: Handbooks in Operations Research and Management Science, vol. 7, pp. 225–330 (1995)
17. Land, A.: The solution of some 100-city Travelling Salesman Problems, Technical report, London School of Economics (1979)
18. Lawler, E.L., Lenstra, J.K., Rinnooy-Kan, A.H.G., Shmoys, D.B.: The Traveling Salesman Problem: A Guided Tour of Combinatorial Optimization. Wiley, Chichester (1985)
19. Lin, S., Kernighan, B.W.: An effective heuristic algorithm for the traveling-salesman problem. Oper. Res. **21**(2), 498–516 (1973)
20. Mitchell, J.S.B.: Guillotine subdivisions approximate polygonal subdivisions: a simple polynomial-time approximation scheme for geometric TSP, k-MST, and related problems. SIAM J. Comput. **28**(4), 1298–1309 (1999)
21. Padberg, M., Rinaldi, G.: A branch-and-cut algorithm for the resolution of large-scale symmetric traveling salesman problems. SIAM Rev. **33**(1), 60–100 (1991)
22. Pferschy, U., Stanek, R.: Generating subtour constraints for the TSP from pure integer solutions. Department of Statistics and Operations Research, University of Graz, Technical report (2014)
23. Reinelt, G.: TSPlib - a traveling salesman problem library. ORSA J. Comput. **3**(4), 376–384 (1991)
24. Wein, R., Berberich, E., Fogel, E., Halperin, D., Hemmer, M., Salzman, O., Zukerman, B.: 2D arrangements. In: CGAL User and Reference Manual, 4.3rd edn. CGAL Editorial Board (2014)

Sparse Subgraphs for 2-Connectivity
in Directed Graphs

Loukas Georgiadis[1], Giuseppe F. Italiano[2], Aikaterini Karanasiou[1],
Charis Papadopoulos[1(✉)], and Nikos Parotsidis[2]

[1] University of Ioannina, Ioannina, Greece
{loukas,akaranas,charis}@cs.uoi.gr
[2] Università di Roma "Tor Vergata", Rome, Italy
{giuseppe.italiano,nikos.parotsidis}@uniroma2.it

Abstract. Let G be a strongly connected directed graph. We consider the problem of computing the smallest strongly connected spanning subgraph of G that maintains the pairwise 2-vertex-connectivity of G, i.e., the 2-vertex-connected blocks of G (2VC-B). We provide linear-time approximation algorithms for this problem that achieve an approximation ratio of 6. Based on these algorithms, we show how to approximate, in linear time, within a factor of 6 the smallest strongly connected spanning subgraph of G that maintains respectively: both the 2-vertex-connected blocks and the 2-vertex-connected components of G (2VC-B-C); all the 2-connectivity relations of G (2C), i.e., both the 2-vertex- and the 2-edge-connected components and blocks. Moreover, we provide heuristics that improve the size of the computed subgraphs in practice, and conduct a thorough experimental study to assess their merits in practical scenarios.

1 Introduction

Let $G = (V, E)$ be a directed graph (digraph), with m edges and n vertices. G is *strongly connected* if there is a directed path from each vertex to every other vertex. The *strongly connected components* of G are its maximal strongly connected subgraphs. A vertex (resp., an edge) of G is a *strong articulation point* (resp., a *strong bridge*) if its removal increases the number of strongly connected components. A digraph G is *2-vertex-connected* if it has at least three vertices and no strong articulation points; G is *2-edge-connected* if it has no strong bridges. The *2-vertex-* (resp., *2-edge-*) *connected components* of G are its maximal 2-vertex- (resp., 2-edge-) connected subgraphs. Let v and w be two distinct vertices: v and w are *2-vertex-connected* (resp., *2-edge-connected*), denoted by $v \leftrightarrow_{2v} w$ (resp., $v \leftrightarrow_{2e} w$), if there are two internally vertex-disjoint (resp., two edge-disjoint) directed paths from v to w and two internally vertex-disjoint (resp., two edge-disjoint) directed paths from w to v (a path from v

---

G.F. Italiano and N. Parotsidis—Partially supported by MIUR under Project AMANDA.

A.V. Goldberg and A.S. Kulikov (Eds.): SEA 2016, LNCS 9685, pp. 150–166, 2016.
DOI: 10.1007/978-3-319-38851-9_11

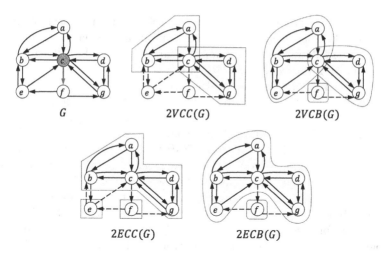

Fig. 1. A strongly connected digraph G with a strong bridge (c, f) and a strong articulation point c shown in red (better viewed in color), the 2-vertex-connected components and blocks of G, and the 2-edge-connected components and blocks of G. Vertex f forms a trivial 2-edge-connected and 2-vertex-connected block. (Color figure online)

to w and a path from w to v need not be either vertex- or edge- disjoint). A 2-*vertex-connected block* (resp., 2-*edge-connected block*) of a digraph $G = (V, E)$ is a maximal subset $B \subseteq V$ such that $u \leftrightarrow_{2v} v$ (resp., $u \leftrightarrow_{2e} v$) for all $u, v \in B$. Note that, as a (degenerate) special case, a 2-vertex- (resp., 2-edge-) connected block might consist of a singleton vertex only: we denote this as a *trivial* 2-*vertex-* (resp., 2-*edge-*) *connected block*. In the following, we will consider only non-trivial 2-vertex- and 2-edge- connected blocks. Since there is no danger of ambiguity, we will call them simply 2-vertex- and 2-edge-connected blocks.

Differently from undirected graphs, in digraphs 2-vertex and 2-edge connectivity have a much richer and more complicated structure, and indeed 2-connectivity problems on directed graphs appear to be more difficult than their undirected counterparts. In particular, in digraphs 2-vertex- (resp., 2-edge-) connected blocks can be different from the 2-vertex- (resp., 2-edge-) connected components, i.e., two vertices may be 2-vertex- (resp., 2-edge-) connected but lie in different 2-vertex- (resp., 2-edge-) connected components (see Fig. 1). This is not the case for undirected graphs. Moreover, for undirected graphs it has been known for over 40 years how to compute the 2-edge- and 2-vertex- connected components in linear time [25]. In the case of digraphs, however, it was shown only recently how to compute the 2-edge- and 2-vertex- connected blocks in linear time [11,12], and the best current bound for computing the 2-edge- and the 2-vertex- connected components in digraphs is not even linear, but it is $O(n^2)$ [16].

In this paper we investigate problems where we wish to find a smallest spanning subgraph of G (i.e., with minimum number of edges) that maintains certain 2-connectivity requirements in addition to strong connectivity. Problems of this nature are fundamental in network design, and have several practical applications [24]. Specifically, we consider computing a smallest strongly connected spanning subgraph of a digraph G that maintains the following properties: the pairwise 2-vertex-connectivity of G, i.e., the 2-vertex-connected blocks of G (2VC-B); the 2-vertex-connected components of G (2VC-C); both the 2-vertex-connected blocks and components of G (2VC-B-C). This complements our previous study of the edge-connectivity versions of these problems [13], that we refer to as 2EC-C (maintaining 2-edge-connected components), 2EC-B (maintaining 2-edge-connected blocks), and 2EC-B-C (maintaining 2-edge-connected blocks and components). Finally, we also consider computing a smallest spanning subgraph of G that maintains all the 2-connectivity relations of G (2C), that is, simultaneously the 2-vertex-connected and the 2-edge-connected components and blocks. Note that all these problems are NP-hard [9,13], so one can only settle for efficient approximation algorithms. Computing small spanning subgraphs is of particular importance when dealing with large-scale graphs, say graphs having hundreds of million to billion edges. In this framework, one big challenge is to design linear-time algorithms, since algorithms with higher running times might be practically infeasible on today's architectures.

Related Work. Computing a smallest k-vertex-(resp., k-edge-) connected spanning subgraph of a given k-vertex- (resp. k-edge-) connected digraph is NP-hard for any $k \geq 1$ (and for $k \geq 2$ for undirected graphs) [9]. The case for $k = 1$ is to compute a smallest strongly connected spanning subgraph (SCSS) of a given digraph. This problem was originally studied by Khuller et al. [20], who provided a polynomial-time algorithm with an approximation guarantee of 1.64. This was improved to 1.61 by the same authors [21]. Later on, Vetta announced a further improvement to $3/2$ [27], and Zhao et al. [28] presented a faster linear-time algorithm at the expense of a larger $5/3$-approximation factor. For the smallest k-edge-connected spanning subgraph (kECSS), Laehanukit et al. [23] gave a randomized $(1+1/k)$-approximation algorithm. For the smallest k-vertex-connected spanning subgraph (kVCSS), Cheriyan and Thurimella [4], gave a $(1 + 1/k)$-approximation algorithm that runs in $O(km^2)$ time. For $k = 2$, the running time of Cheriyan and Thurimella's algorithm was improved to $O(m\sqrt{n} + n^2)$, based on a linear-time 3-approximation for 2VCSS [10]. We also note that there has been extensive work on more general settings where one wishes to approximate minimum-cost subgraphs that satisfy certain connectivity requirements. See, e.g., [6], and the survey [22]. The previous results on kECSS and kVCSS immediately imply an approximation ratio smaller than 2 for 2EC-C and 2VC-C [13,19]. While there has been substantial progress for 2EC-C and 2VC-C, problems 2EC-B and 2VC-B (i.e., computing sparse subgraphs with the same pairwise 2-edge or 2-vertex connectivity) seem substantially harder. Jaberi [18] was the first to consider several optimization problems related to 2EC-B and 2VC-B and proposed approximation algorithms. The approximation ratio in his algorithms, however, is linear in the number of strong bridges for 2EC-B and in the number of strong articulation points for 2VC-B, and hence $O(n)$ in the worst case. In [13], linear-time 4-approximation algorithms for 2EC-B and 2EC-B-C were presented.

It seems thus natural to ask whether one can design linear-time algorithms which achieve small approximation guarantees for 2VC-B, 2VC-B-C and 2C.

Our Results. In this paper we address this question by presenting practical approximation algorithms for the 2VC-B, 2VC-B-C and 2C problems. We stress that the approach in this paper is substantially different from [13], since vertex connectivity is typically more involved than edge connectivity and requires several novel ideas and non-trivial techniques. In particular, differently from [13], our starting point in this paper is the recent framework for strong connectivity and 2-connectivity problems in digraphs [14], combined with the notions of *divergent spanning trees* and *low-high orders* [15] (defined below). Building on this new framework, we can obtain *sparse certificates* also for the 2-vertex-connected blocks. In our context, a sparse certificate of a strongly connected digraph G is a strongly connected spanning subgraph $C(G)$ of G with $O(n)$ edges that maintains the 2-vertex-connected blocks of G. We show that our constructions achieve a 6-approximation for 2VC-B in linear time. Then, we extend our algorithms so that they compute a 6-approximation for 2VC-B-C and 2C. These algorithms also run in linear time once the 2-vertex and the 2-edge-connected components of G are available; if not, the current best running time for computing them is $O(n^2)$ [16]. Then we provide efficient implementations of these algorithms that run very fast in practice. We also present several heuristics that improve the quality (i.e., the number of edges) of the computed spanning subgraphs. Finally, we assess how all these algorithms perform in practical scenarios by conducting a thorough experimental study, and report its main findings.

2   Preliminaries

A *flow graph* is a digraph such that every vertex is reachable from a distinguished start vertex. Let $G = (V, E)$ be a strongly connected digraph. For any vertex $s \in V$, we denote by $G(s) = (V, E, s)$ the corresponding flow graph with start vertex s; all vertices in V are reachable from s since G is strongly connected. The *dominator relation* in $G(s)$ is defined as follows: A vertex u is a *dominator* of a vertex w (u *dominates* w) if every path from s to w contains u; u is a *proper dominator* of w if u dominates w and $u \neq w$. The dominator relation in $G(s)$ can be represented by a rooted tree, the *dominator tree* $D(s)$, such that u dominates w if and only if u is an ancestor of w in $D(s)$. If $w \neq s$, we denote by $d(w)$ the parent of w in $D(s)$. The dominator tree of a flow graph can be computed in linear time, see, e.g., [2,3]. An edge (u, w) is a *bridge* in $G(s)$ if all paths from s to w include (u, w).[1] Italiano et al. [17] gave linear-time algorithms for computing all the strong bridges and all the strong articulation points of a digraph G. Their algorithms use the dominators and the bridges of flow graphs $G(s)$ and $G^R(s)$, where s is an arbitrary start vertex and G^R is the digraph that results from G after reversing edge directions. A spanning tree T of a flow graph $G(s)$ is a tree

---

[1] Throughout, we use consistently the term *bridge* to refer to a bridge of a flow graph $G(s)$ and the term *strong bridge* to refer to a strong bridge in the original graph G.

with root s that contains a path from s to v for all vertices v. Two spanning trees T_1 and T_2 rooted at s are *edge-disjoint* if they have no edge in common. A flow graph $G(s)$ has two such spanning trees if and only if it has no bridges [26]. Two spanning trees are *maximally edge-disjoint* if the only edges they have in common are the bridges of $G(s)$. Two (maximally) edge-disjoint spanning trees can be computed in linear-time by an algorithm of Tarjan [26], using the disjoint set union data structure of Gabow and Tarjan [8]. Two spanning trees T_1 and T_2 rooted at s are *divergent* if for all vertices v, the paths from s to v in T_1 and T_2 share only the dominators of v. A *low-high order* δ on $G(s)$ is a preorder of the dominator tree $D(s)$ such for all $v \neq s$, $(d(v), v) \in E$ or there are two edges $(u, v) \in E$, $(w, v) \in E$ such that u is less than v ($u <_\delta v$), v is less than w ($v <_\delta w$), and w is not a descendant of v in $D(s)$. Every flow graph $G(s)$ has a pair of maximally edge-disjoint divergent spanning trees and a low-high order, both computable in linear-time [15].

Let T be a dfs tree of a digraph G rooted at s. For a vertex u, we denote by $loop(u)$ the set of all descendants x of u in T such that there is a path from x to u in G containing only descendants of u in T. Since any two vertices in $loop(u)$ reach each other, $loop(u)$ induces a strongly connected subgraph of G. Furthermore, *loops* define a laminar family (i.e., for any two vertices u and v, we have $loop(u) \cap loop(v) = \emptyset$, or $loop(v) \subseteq loop(u)$, or $loop(u) \subseteq loop(v)$). The *loop nesting tree* L of a strongly connected digraph G with respect to T, is the tree in which the parent of any vertex $v \neq s$ is the nearest proper ancestor u of v such that $v \in loop(u)$. The loop nesting tree can be computed in linear time [3, 26].

3   Approximation Algorithms and Heuristics for 2VC-B

Let $G = (V, E)$ be the input strongly connected digraph. In problem 2VC-B, we wish to compute a strongly connected spanning subgraph G' of G that has the same 2-vertex-connected blocks of G, with as few edges as possible. We consider the following approach. Start with the empty graph $G' = (V, \emptyset)$, and add as few edges as possible until G' is guaranteed to have the same 2-vertex-connected blocks as G. We consider three linear-time algorithms that apply this approach. The first two are based on the sparse certificates for 2-vertex-connected blocks from [12, 14], which use divergent spanning trees. The third is a new algorithm that selects the edges of G' with the help of low-high orders.

Divergent Spanning Trees. We can compute a sparse certificate $C(G)$ for the 2-vertex-connected blocks of a strongly connected digraph G using the algorithm of [12], which is based on a linear-time construction of two divergent spanning trees of a flow graph [15]. We refer to this algorithm as DST-B. Let s be an arbitrarily chosen start vertex in G. Recall that we denote by $G(s)$ the flow graph with start vertex s, by $G^R(s)$ the flow graph obtained from $G(s)$ after reversing edge directions, and by $D(s)$ and $D^R(s)$ the dominator trees of $G(s)$ and $G^R(s)$ respectively. Also, let $C(v)$ and $C^R(v)$ be the set of children of v in $D(s)$ and $D^R(s)$ respectively. For each vertex r, let $C^k(r)$ denote the level k

descendants of r, where $C^0(r) = \{r\}$, $C^1(r) = C(r)$, and so on. For each vertex $r \neq s$ that is not a leaf in $D(s)$ we build the *auxiliary graph* $G_r = (V_r, E_r)$ of r as follows. The vertex set of G_r is $V_r = \cup_{k=0}^3 C^k(r)$ and it is partitioned into a set of *ordinary* vertices $V_r^o = C^1(r) \cup C^2(r)$ and a set of *auxiliary* vertices $V_r^a = C^0(r) \cup C^3(r)$. The auxiliary graph G_r results from G by contracting the vertices in $V \setminus V_r$ as follows. All vertices that are not descendants of r in $D(s)$ are contracted into r. For each vertex $w \in C^3(r)$, we contract all descendants of w in $D(s)$ into w. We use the same definition for the auxiliary graph G_s of s, with the only difference that we let s be an ordinary vertex. In order to bound the size of all auxiliary graphs, we eliminate parallel edges during those contractions. We call an edge $e \in E_r \setminus E$ a *shortcut* edge of G_r. That is, a shortcut edge is formed by the contraction of a part of G into an auxiliary vertex of G_r. Thus, a shortcut edge is not an original edge of G but corresponds to at least one original edge, and is adjacent to at least one auxiliary vertex.

Algorithm DST-B selects the edges that are inserted into $C(G)$ in three phases. During the construction, the algorithm may choose a shortcut edge or a reverse edge to be inserted into $C(G)$. In this case we insert the associated original edge instead. Also, an edge may be selected multiple times, so we remove multiple occurrences of such edges in a postprocessing step. In the first phase, we insert into $C(G)$ the edges of two maximally edge-disjoint divergent spanning trees, $T_1(G(s))$ and $T_2(G(s))$ of $G(s)$. In the second phase we process the auxiliary graphs of $G(s)$ that we refer to as the *first-level auxiliary graphs*. For each such auxiliary graph $H = G_r$, we compute two maximally edge-disjoint divergent spanning trees $T_1(H^R(r))$ and $T_2(H^R(r))$ of the corresponding reverse flow graph $H^R(r)$ with start vertex r. We insert into $C(G)$ the edges of these two spanning trees. It can be proved that, at the end of this phase, $C(G)$ induces a strongly connected spanning subgraph of G. Finally, in the last phase we process the *second-level auxiliary graphs*, which are the auxiliary graphs of H^R for all first-level auxiliary graphs H. Let H_q^R be a second-level auxiliary graph of H^R. For every strongly connected component S of $H_q^R \setminus q$, we choose an arbitrary vertex $v \in S$ and compute a spanning tree of S and a spanning tree of S^R, and insert their edges into $C(G)$.

This construction inserts $O(n)$ edges into $C(G)$, and therefore achieves a constant approximation ratio for 2VC-B. However, due to the use of auxiliary vertices and two levels of auxiliary graphs, we do not have a good bound for this constant. (The first-level auxiliary graphs have at most $4n$ vertices and $4m + n$ edges in total [12].) We propose a modification of DST-B, that we call DST-B modified: For each auxiliary graph, we do not select in $C(G)$ the edges of its two divergent spanning trees that have only auxiliary descendants. Also, for every second-level auxiliary graph, during the computation of its strongly connected components we include the chosen edges that already form a strongly connected component.

More precisely, algorithm DST-B modified works as follows. In the first two phases, we try reuse as many edges as possible when we build the divergent spanning trees of $G(s)$ and of its auxiliary graphs. In the third phase of the construction we need to solve the smallest SCSS problem for each strongly connected component S in the second-level auxiliary graphs H_q after the deletion of the root vertex q. We do this by running a modified version of the linear-time 5/3-approximation algorithm of Zhao et al. [28]. The algorithm of Zhao et al. a SCSS of a strongly connected graph by performing a depth-first search traversal of the input graph. During the dfs traversal, any cycle that is detected is contracted into a single vertex. We modify this approach so that we can avoid inserting new edges into the sparse certificate as follows. Since we only care about the ordinary vertices in S, we can construct a subgraph of S that contains edges already added in $C(G)$. We compute the strongly connected components of this subgraph and contract them. Then we apply the algorithm of Zhao et al. on the contracted graph of S. Furthermore, during the dfs traversal we give priority to edges already added in $C(G)$. We can apply a similar idea in the second phase of the construction as follows. The algorithms of [15] for computing two divergent spanning trees of a flow graph use the edges of a dfs spanning tree, together with at most $n-1$ other edges. Hence, we can modify the dfs traversal so that we give priority to edges already added in $C(G)$.

Divergent Spanning Trees and Loop Nesting Trees. An alternative linear-time algorithm to compute a sparse certificate $C(G)$ for the 2-vertex-connected blocks can be obtained via loop nesting trees, as described in [14]. As in algorithm DST-B, we compute two maximally edge-disjoint divergent spanning trees T_1 and T_2 of $G(s)$, and insert their edges into $C(G)$. But instead of computing auxiliary graphs, we compute a loop nesting tree L of $G(s)$ and insert into $C(G)$ the edges that define L. These are the edges of a dfs tree of $G(s)$, and at most $n-1$ additional edges that are required to define the loops of $G(s)$. (See [15,26] for the details.) Then, we repeat the same process in the reverse direction, i.e., for $G^R(s)$. As shown in [14], a spanning subgraph having the same dominator trees and loop nesting trees (in both directions) as the digraph G, has the same 2-edge- and 2-vertex-connected blocks as G. We refer to this algorithm as DLN-B.

Theorem 1. *Algorithm DLN-B achieves an approximation ratio of 6, in linear time, for problem 2VC-B.*

Proof. Consider first the "forward" pass of the algorithm. It adds at most $2(n-1)$ edges for the two divergent spanning trees, and at most $2(n-1)$ edges that define a loop nesting tree of $G(s)$. By [15,26], both these constructions use the edges of a dfs tree of $G(s)$ and some additional edges. Hence, we can use the same dfs tree to compute the divergent spanning trees and the loop nesting tree. This gives a total of at most $3(n-1)$ edges. Similarly, the "reverse" pass computes at most $3(n-1)$ edges, so algorithm DLN-B selects at most $6(n-1)$ edges. Since the resulting subgraph must be strongly connected, any valid solution to problem 2VC-B has at least n edges, so DLN-B achieves a 6-approximation. By [15,26], both the computation of a pair of divergent spanning trees and of a loop nesting tree can be done in linear time, hence DLN-B also runs in linear time.   □

Low-High Orders and Loop Nesting Trees. Now we introduce a new linear-time construction of a sparse certificate, via low-high orders, that we refer to as

LHL-B. The algorithm consists of two phases. In the first phase, we insert into $C(G)$ the edges that define the loop nesting trees L and L^R of $G(s)$ and $G^R(s)$, respectively, as in algorithm DLN-B. In the second phase, we insert enough edges so that $C(G)$ (resp., $C^R(G)$) maintains a low-high order of $G(s)$ (resp., $G^R((s))$). Let δ be a low-high order on $G(s)$. Subgraph $C(G)$ satisfies the low-high order δ if, for each vertex $v \neq s$, one of the following holds: (a) there are two edges (u, v) and (w, v) in $C(G)$ such that $u <_\delta v$, $v <_\delta w$, and w is not a descendant of v in $D(s)$; (b) $(d(v), v)$ is a strong bridge of G and is contained in $C(G)$; or (c) $(d(v), v)$ is an edge of G that is contained in $C(G)$, and there is another edge (u, v) in $C(G)$ such that $u <_\delta v$ and $u \neq d(v)$.

Theorem 2. *Algorithm LHL-B is correct and achieves an approximation ratio of 6 for problem 2VC-B, in linear time.*

Proof. By construction, the sparse certificate $C(G)$ computed by LHL-B satisfies a low-high order δ of $G(s)$. This implies that $C(G)$ contains two divergent spanning trees T_1 and T_2 of $G(s)$ [15]. Moreover, cases (b) and (c) of the construction ensure that T_1 and T_2 are maximally edge-disjoint. This is because when case (a) does not apply for a vertex v, then $C(G)$ contains $(d(v), v)$. Also, $d(v)$ is the only vertex u that satisfies $u <_\delta v$ if and only if $(d(v), v)$ is a strong bridge. Hence, $C(G)$ indeed contains two maximally edge-disjoint divergent spanning trees of $G(s)$. Similarly, $C(G)$ also contains two maximally edge-disjoint divergent spanning trees of $G^R(s)$. So the correctness of LHL-B follows from the fact that DLN-B is correct.

Next we bound the approximation ratio of LHL-B. The edges selected to maintain a loop nesting tree L of $G(s)$ contain at least one entering edge for each vertex $v \neq s$. This means that it remains to include at most one edge for each vertex $v \neq s$ in order to satisfy a low-high order of $G(s)$. The symmetric arguments holds for the reverse direction as well, so $C(G)$ contains at most $6(n-1)$ edges, which gives an approximation ratio of 6.   □

We note that both DLN-B and LHN-B also maintain the 2-edge-connected blocks of the input digraph. We use this fact in Sect. 4, where we compute a sparse subgraph that maintains all 2-connectivity relations. We can improve the solution computed by the above algorithms by using the following filter.

Two Vertex-Disjoint Paths Test. We test if $G' \setminus (x, y)$ contains two vertex-disjoint paths from x to y. If this is the case, then we remove edge (x, y); otherwise, we keep the edge (x, y) in G' and proceed with the next edge. For doing so, we define the modified graph G'' of G' after vertex-splitting (see, e.g., [1]): for each vertex v, replace v by two vertices v^+ and v^-, and add the edge (v^-, v^+). Then, we replace each edge (u, w) in G' by (u^+, v^-) in G'', so v^- has the edges entering v and v^+ has the edges leaving v. Now we can test if G' still has two vertex-disjoint paths from x to y after deleting (x, y) by running two iterations of the Ford-Fulkerson augmenting paths algorithm [7] for finding two edge-disjoint paths on G'' by treating x^+ as the source and y^- as the sink. Note that we need to compute G'' once for all such tests. If an edge (x, y) is deleted from G', then we also delete (x^+, y^-) from G''. Since G' has $O(n)$ edges, this test takes $O(n)$ time per edge, so the total running time is $O(n^2)$. We refer to this filter as 2VDP. In our implementations we applied 2VDP on the outcome of DLN-B in order to assess our algorithms with a solution close to minimum. For the 2VC-B problem the algorithm obtained after applying such a filter is called 2VDP-B. In order to improve the running time of 2VDP in practice, we apply a speed-up heuristic for *trivial edges* (x, y): if x belongs to a 2-vertex-connected block and has outdegree two or y belongs to a 2-vertex-connected block and has indegree two, then (x, y) must be included in the solution.

4   Approximation Algorithms and Heuristics for 2C

To get an approximate solution for problem 2C, we combine our algorithms for 2VC-B with algorithms that approximate 2VCSS [4,10]. We also take advantage of the fact that every 2-vertex-connected component is contained in a 2-edge-connected component. This property suggests the following approach for 2C. First, we compute the 2-vertex-connected components of G and solve the 2VCSS problem independently for each such component. Then, we apply one of the algorithms DLN-B or LHL-B for 2VC-B on G. Since the sparse certificate from DLN-B or LHL-B also maintain the 2-edge-connected blocks, it remains to include edges that maintain the 2-edge-connected components of G. We can find these edges in a *condensed graph* $\check{G}$ defined as follows. Digraph $\check{G}$ is formed from G by contracting each 2-vertex-connected component of G into a single supervertex. Note that any two 2-vertex-connected components may have at most one vertex in common: if two such components share a vertex, they are contracted into the same supervertex. The resulting digraph $\check{G}$ is a multigraph since the contractions can create loops and parallel edges. For any vertex v of G, we denote by $\check{v}$ the supervertex of $\check{G}$ that contains v. Every edge $(\check{u}, \check{v})$ of $\check{G}$ is associated with the corresponding original edge (u, v) of G. Now we describe the main steps of our algorithm for 2C:

1. Compute the 2-vertex-connected components. Solve independently the 2VCSS problem for each such component, using the linear-time algorithm of [10].
2. Form the condensed multigraph $\check{G}$, and compute its 2-edge-connected components. Solve independently the 2ECSS problem for each such component, using edge-disjoint spanning trees [13].
3. Execute the DNL-B or LHL-B algorithms on the original graph G and compute a sparse certificate for the 2-edge- and the 2-vertex-connected blocks.

The solution to the 2C problem consists of the edges selected in each step of the algorithm. Note that in Step 2, we should allow 2-edge-connected components of size two because such a component may correspond to the union of 2-vertex-connected components of the original graph. We consider two versions of our algorithm, DLN-2C and LHL-2C, depending on the algorithm for the 2VC-B problem used in Step 3.

Theorem 3. *Algorithms DLN-2C and LHL-2C compute a 6-approximation for problem 2C. Moreover, if the 2-edge- and the 2-vertex- connected components of G are available, then the algorithms run in linear time.*

Proof. Let n_v be the number of vertices of G that belong to some 2-vertex-connected component of G. Also, let $\check{n}$ be the number of vertices in $\check{G}$, and let $\check{n}_e$ be the number of vertices of $\check{G}$ that belong to some 2-edge-connected component of $\check{G}$. By the analysis in the proof of Theorem 4, the algorithm for 2VC-B-C selects less than $6(n + n_v)$ edges. For the 2ECSS problems, we can compute a 2-approximate solution in linear-time as in [13], using edge-disjoint spanning trees [5,26]. Let $\check{C}$ be a 2-edge-connected component of $\check{G}$. We select an arbitrary vertex $\check{v} \in \check{C}$ as a root and compute two edge-disjoint spanning trees in the flow graph $\check{C}(\check{v})$ and two edge-disjoint spanning trees in the reverse flow graph $\check{C}^R(\check{v})$. Thus, we select less than $4\check{n}_e$ edges. Hence, the subgraph computed by the algorithm has less than $6(n + n_c + \check{n}_e)$ edges.

Now consider any solution to 2C. It has to include $2n_c + 2\check{n}_e$ edges in order to maintain the 2-vertex and the 2-edge-connected components of G. Moreover, since the resulting subgraph must be strongly connected, there must be at least one edge entering each of the $\check{n} - \check{n}_e$ vertices of $\check{G}$ that do not belong in a 2-edge-connected component of $\check{G}$. Thus, the optimal solution has at least $2n_c + \check{n}_e + \check{n}$ edges. Note that $\check{n}_c + \check{n} \geq n$, so the optimal solution has at least $n + n_c + \check{n}_e$ edges and the approximation ratio of 6 follows.

Finally, we show that all three steps of the algorithms DLN-2C and LHL-2C run in linear time given the 2-edge- and the 2-vertex- connected components of G. This is immediate for Steps 1 and 3. In Step 2, we do not need to compute the 2-edge-connected components of $\check{G}$ from scratch, but we can form them from the 2-edge-connected components of G using contractions. Let C be a 2-edge-connected component of G. We contract each 2-vertex-connected component of G contained in C into a single supervertex. Then, the resulting digraph $\check{C}$ is a 2-edge-connected component of $\check{G}$.   □

If we wish to improve the quality of the computed solution G', we can apply the 2VDP filter, and the analogous 2-edge-disjoint paths filter 2EDP, as follows. In Step 1, we run the 2VDP filter for the edges computed by the linear-time algorithm of [10]. This produces a minimal solution for 2VCSS in each 2-vertex-connected component of G. Similarly, in Step 2, we run the 2EDP filter for the edges of the edge-disjoint spanning trees computed in each 2-edge-connected component of $\check{G}$. This produces a minimal solution for 2ECSS in each 2-edge-connected component of $\check{G}$. Finally, we run the 2VDP filter on the whole G', but only consider the edges added in Step 3 of our algorithm, since the edges from Steps 1 and 2 are needed to maintain the 2-vertex- and the 2-edge-connected components. We implemented this algorithm, using DLN-B for Step 3, and refer to it as 2VDP-2C.

Approximation Algorithms and Heuristics for 2VC-B-C. Executing Steps 1 and 3 of the above algorithm described for 2C, is enough to produce a certificate for the 2VC-B-C problem. If we use DLN-B or LHL-B for Step 3,

then we obtain a 6-approximate solution for 2VC-B-C. We call the corresponding algorithms DLN-B-C and LHL-B-C, respectively.

Theorem 4. *There is a polynomial-time algorithm for 2VC-B-C that achieves
an approximation ratio of 6. Moreover, if the 2-vertex-connected components of
G are available, then the algorithm runs in linear time.*

Proof. A result in [10] shows that, given a 2-vertex-connected digraph with ν
vertices, we can compute in linear time a 2-vertex-connected spanning subgraph
that has less than 6ν edges. Hence, if n_c is the number of vertices that belong
in a 2-vertex-connected component of G, then applying this algorithm to each
2-vertex-connected component selects less than $6n_c$ edges. Finally, we apply the
construction of a sparse certificate for the 2-vertex-connected blocks which selects
at most $6(n-1)$ edges by Theorems 1 or 2. Hence, the subgraph computed by
the algorithm has less than $6(n+n_c)$. One the other hand, any solution to 2VC-
B-C has to include at least $2n_c$ edges for the 2-vertex-connected components of
G, and at least $n - n_c$ edges in order to obtain a strongly connected subgraph.
Thus, the optimal solution has at least $n + n_c$ edges, so the approximation ratio
of 6 follows.   □

As in the 2VC-B and 2C problems, we can improve the quality of the computed solution by applying the 2VDP filter for the edges that connect different
2-vertex-connected components. We implemented this algorithm, using DLN-B
for Step 3, and refer to it as 2VDP-B-C.

5   Experimental Analysis

We implemented the algorithms previously described: 5 for 2VC-B, 3 for 2VC-
B-C, and 3 for 2C, as summarized in Table 1. All implementations were written in C++ and compiled with g++ v.4.4.7 with flag -O3. We performed our
experiments on a GNU/Linux machine, with Red Hat Enterprise Server v6.6: a
PowerEdge T420 server 64-bit NUMA with two Intel Xeon E5-2430 v2 processors and 16 GB of RAM RDIMM memory. Each processor has 6 cores sharing a
15 MB L3 cache, and each core has a 2 MB private L2 cache and 2.50 GHz speed.
In our experiments we did not use any parallelization, and each algorithm ran
on a single core. We report CPU times measured with the getrusage function.
All our running times were averaged over ten different runs.

For the experimental evaluation we use the datasets shown in Table 2. We
measure the quality of the solution computed by algorithm A on problem $\mathcal{P}$
by a *quality ratio* defined as $q(A, \mathcal{P}) = \delta_{avg}^A / \delta_{avg}^{\mathcal{P}}$, where δ_{avg}^A is the average
vertex indegree of the subgraph computed by A and $\delta_{avg}^{\mathcal{P}}$ is a lower bound on
the average vertex indegree of the optimal solution for $\mathcal{P}$. Specifically, for 2VC-B
and 2VC-B-C we define $\delta_{avg}^B = (n + k)/n$, where n is the total number of vertices
of the input digraph and k is the number of vertices that belong in (nontrivial)

Table 1. The algorithms considered in our experimental study. The worst-case bounds refer to a digraph with n vertices and m edges. Running times indicated by † assume that the 2-vertex-connected components of the input digraph are available; running times indicated by ‡ assume that also the 2-edge-connected components are available.

Algorithm	Problem	Technique	Time
DST-B	2VC-B	Original sparse certificate from [12] based on divergent spanning trees	$O(m+n)$
DST-B modified	2VC-B	Modified sparse certificate from [12]	$O(m+n)$
DLN-B	2VC-B	Sparse certificate from [14] based on divergent spanning trees and loop nesting trees	$O(m+n)$
LHL-B	2VC-B	New sparse certificate based on low-high orders and loop nesting trees	$O(m+n)$
2VDP-B	2VC-B	2VDP filter applied on the digraph produced by DLN-B	$O(n^2)$
DLN-B-C	2VC-B-C	DST-B combined with the linear-time 2VCSS algorithm of [10]	$O(m+n)^†$
LHL-B-C	2VC-B-C	LHL-B combined with the linear-time 2VCSS algorithm of [10]	$O(m+n)^†$
2VDP-B-C	2VC-B-C	2VDP filter applied on the digraph produced by DLN-B-C	$O(n^2)$
DLN-2C	2C	DLN-B-C combined with the linear-time 2ECSS algorithm using edge-disjoint spanning trees	$O(m+n)^‡$
LHL-2C	2C	LHL-B-C combined with the linear-time 2ECSS algorithm using edge-disjoint spanning trees	$O(m+n)^‡$
2VDP-2C	2C	2VDP and 2EDP filters applied on the digraph produced by DLN-2C	$O(n^2)$

2-vertex-connected blocks[2]. We set a similar lower bound δ_{avg}^C for 2C, with the only difference that k is the number of vertices that belong in (nontrivial) 2-edge-connected blocks, since every 2-vertex-connected component or block is contained in a 2-edge-connected block. Note that the quality ratio is an upper bound of the actual approximation ratio. The smaller the values of $q(A, \mathcal{P})$ (i.e., the closer to 1), the better is the approximation obtained by algorithm A for problem $\mathcal{P}$.

We now report the results of our experiments with all the algorithms considered for problems 2VC-B and 2C. For the 2VC-B problem, the quality ratio of the spanning subgraphs computed by the different algorithms is shown in

---

[2] This follows from the fact that in the sparse subgraph the k vertices in blocks must have indegree at least two, while the remaining $n - k$ vertices must have indegree at least one, since we seek for a strongly connected spanning subgraph.

Table 2. Real-world graphs sorted by file size of their largest SCC; n is the number of vertices, m the number of edges, and δ_{avg} is the average vertex indegree; s^* is the number of strong articulation points; δ_{avg}^B and δ_{avg}^C are lower bounds on the average vertex indegree of an optimal solution to 2VC-B and 2C, respectively.

Dataset	n	m	File size	δ_{avg}	s^*	δ_{avg}^B	δ_{avg}^C	Type
Rome99	3353	8859	100 KB	2.64	789	1.76	1.76	road network
P2p-Gnutella25	5153	17695	203 KB	3.43	1840	1.60	1.60	peer2peer
P2p-Gnutella31	14149	50916	621 KB	3.59	5357	1.56	1.56	peer2peer
Web-NotreDame	53968	296228	3,9 MB	5.48	9629	1.50	1.50	web graph
Soc-Epinions1	32223	443506	5,3 MB	13.76	8194	1.56	1.56	social network
USA-road-NY	264346	733846	11 MB	2.77	46476	1.80	1.80	road network
USA-road-BAY	321270	800172	12 MB	2.49	84627	1.69	1.69	road network
USA-road-COL	435666	1057066	16 MB	2.42	120142	1.68	1.68	road network
Amazon0302	241761	1131217	16 MB	4.67	69616	1.74	1.74	prod. co-purchase
WikiTalk	111881	1477893	18 MB	13.20	14801	1.45	1.45	social network
Web-Stanford	150532	1576314	22 MB	10.47	14801	1.62	1.58	web graph
Amazon0601	395234	3301092	49 MB	8.35	69387	1.82	1.82	prod. co-purchase
Web-Google	434818	3419124	50 MB	7.86	89838	1.59	1.58	web graph
Web-Berkstan	334857	4523232	68 MB	13.50	53666	1.56	1.51	web graph

Table 3 (left) and Fig. 3 (top), while their running times are given and plotted in Table 4 (left) and Fig. 2 (left), respectively. Similarly, for the 2VC-B-C and 2C problems, the quality ratio of the spanning subgraphs computed by the different algorithms is shown in Table 3 (right) and Fig. 3 (bottom), while their running times are given and plotted in Table 4 (right) and Fig. 2 (right), respectively.

We observe that all our algorithms perform well in terms of the quality of the solution they compute. Indeed, the quality ratio is less than 2.5 for all algorithms and inputs. Our modified version of DST-B performs consistently better than the original version. Also in all cases, LHL-B computed a higher quality solution than DLN-B. For most inputs, DST-B modified computes a sparser graph than LHL-B, which is somewhat surprising given the fact that we do not have a good bound for the (constant) approximation ratio of DST-B modified. On the other hand, LHL-B is faster than DST-B modified by a factor of 4.15 on average and has the additional benefit of maintaining both the 2-vertex and the 2-edge-connected blocks. The 2VDP filter provides substantial improvements of the solution, since all algorithms that apply this heuristic have consistently better quality ratios (1.38 on average and always less than 1.87). However, this is paid with much higher running times, as those algorithms can be even 5 orders of magnitude slower than the other algorithms.

From the analysis of our experimental data, all algorithms achieve consistently better approximations for road networks than for most of the other graphs in our data set. This can be explained by taking into account the macroscopic structure of road networks, which is rather different from other networks. Indeed,

Table 3. Quality ratio $q(A, \mathcal{P})$ of the solutions computed for 2VC-B, 2VC-B-C and 2C.

Dataset	DST-B	DST-B modified	DLN-B	LHL-B	2VDP-B	DLN-B-C	LHL-B-C	2VDP-B-C	DLN-2C	LHL-2C	2VDP-2C
Rome99	1.384	1.363	1.432	1.388	1.170	1.462	1.459	1.199	1.462	1.459	1.198
P2p-Gnutella25	1.726	1.602	1.713	1.568	1.234	1.712	1.568	1.234	1.712	1.568	1.234
P2p-Gnutella31	1.717	1.647	1.732	1.602	1.273	1.732	1.573	1.273	1.732	1.573	1.273
Web-NotreDame	2.072	2.067	2.108	2.085	1.588	2.232	2.149	1.628	2.250	2.180	1.638
Soc-Epinions1	2.082	1.964	2.213	2.027	1.475	2.474	2.411	1.572	2.474	2.411	1.573
USA-road-NY	1.255	1.251	1.371	1.357	1.168	1.376	1.374	1.175	1.376	1.374	1.175
USA-road-BAY	1.315	1.311	1.374	1.365	1.242	1.375	1.379	1.246	1.375	1.379	1.246
USA-road-COL	1.308	1.307	1.354	1.348	1.249	1.357	1.357	1.252	1.357	1.357	1.252
Amazon0302	1.918	1.791	1.849	1.719	1.245	2.020	1.928	1.386	2.032	1.944	1.399
WikiTalk	2.145	2.126	2.281	2.190	1.796	2.454	2.441	1.863	2.454	2.441	1.863
Web-Stanford	2.115	2.019	2.130	2.078	1.572	2.287	2.257	1.622	2.238	2.209	1.584
Amazon0601	1.926	1.793	1.959	1.747	1.196	2.241	2.155	1.278	2.242	2.157	1.279
Web-Google	2.052	2.004	2.083	2.051	1.485	2.306	2.335	1.585	2.338	2.372	1.602
Web-Berkstan	2.302	2.233	2.290	2.275	1.692	2.472	2.492	1.767	2.410	2.431	1.717

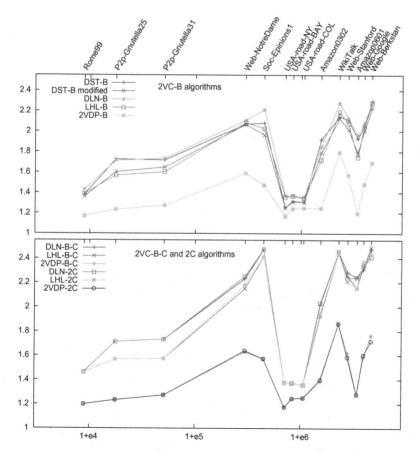

Fig. 2. The plotted quality ratios taken by Table 3.

Table 4. Running times in seconds of the algorithms for 2VC-B, 2VC-B-C and 2C.

Dataset	DST-B	DST-B modified	DLN-B	LHL-B	2VDP-B	DLN-B-C	LHL-B-C	2VDP-B-C	DLN-2C	LHL-2C	2VDP-2C
Rome99	0.014	0.018	0.004	0.005	0.264	0.032	0.034	0.122	0.034	0.036	0.122
P2p-Gnutella25	0.027	0.032	0.008	0.007	1.587	0.042	0.042	0.729	0.051	0.053	0.725
P2p-Gnutella31	0.070	0.094	0.024	0.027	13.325	0.119	0.119	5.613	0.143	0.149	5.422
Web-NotreDame	0.335	0.486	0.059	0.080	97.355	0.491	0.521	27.091	0.573	0.600	27.746
Soc-Epinions1	0.258	0.309	0.089	0.110	92.812	0.606	0.621	54.559	0.602	0.664	54.548
USA-road-NY	1.095	1.402	0.261	0.360	2546.484	2.227	2.337	991.092	2.153	2.415	995.913
USA-road-BAY	1.659	2.152	0.316	0.435	4089.389	2.153	2.298	1429.443	2.296	2.476	1447.318
USA-road-COL	2.439	3.050	0.438	0.603	7739.256	3.770	3.969	3093.258	3.938	4.228	3064.297
Amazon0302	2.101	2.410	0.517	0.675	3503.910	4.708	5.017	2244.856	5.135	5.509	2094.263
WikiTalk	1.777	2.125	0.355	0.473	1158.855	2.179	2.133	943.690	2.203	2.513	924.810
Web-Stanford	1.756	2.395	0.429	0.564	1174.984	2.037	2.313	279.236	2.561	2.487	317.115
Amazon0601	3.532	3.924	1.363	1.605	15349.126	9.793	10.038	8065.680	11.669	11.397	8696.212
Web-Google	4.837	5.467	1.533	1.968	26299.714	9.789	10.172	5095.600	11.535	12.979	5128.337
Web-Berkstan	3.239	5.261	0.690	0.869	6301.410	4.670	4.872	1595.033	5.178	5.601	1546.041

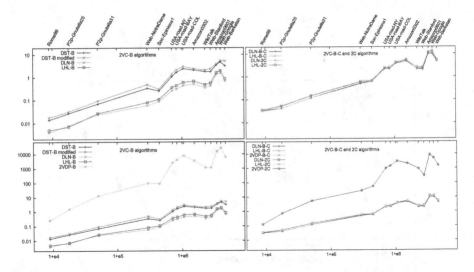

Fig. 3. Running times in seconds with respect to the number of edges (in log-log scale) taken by Table 4. The upper plots get a close-up view of the fastest algorithms by not considering 2VDP-B, 2VDP-B-C and 2VDP-2C.

road networks are very close to be "undirected": i.e., whenever there is an edge (x, y), there is also the reverse edge (y, x) (except for one-way roads). Roughly speaking, road networks mainly consist of the union of 2-vertex-connected components, joined together by strong bridges, and their 2-vertex-connected blocks coincide with their 2-vertex-connected components. In this setting, a sparse strongly connected subgraph of the condensed graph will preserve both blocks and components. On the other hand, such a gain on the solution for the road networks is balanced at the cost of their additional running time.

In addition, our experiments highlight interesting tradeoffs between practical performance and quality of the obtained solutions. In particular, the fastest algorithms for the 2VC-B problem are the ones based on loop-nesting trees (DLN-B and LHL-B), with LHL-B achieving consistently better solutions than DLN-B.

References

1. Ahuja, R.K., Magnanti, T.L., Orlin, J.B.: Network Flows: Theory, Algorithms, and Applications. Prentice-Hall Inc., Upper Saddle River (1993)
2. Alstrup, S., Harel, D., Lauridsen, P.W., Thorup, M.: Dominators in linear time. SIAM J. Comput. **28**(6), 2117–2132 (1999)
3. Buchsbaum, A.L., Georgiadis, L., Kaplan, H., Rogers, A., Tarjan, R.E., Westbrook, J.R.: Linear-time algorithms for dominators and other path-evaluation problems. SIAM J. Comput. **38**(4), 1533–1573 (2008)
4. Cheriyan, J., Thurimella, R.: Approximating minimum-size k-connected spanning subgraphs via matching. SIAM J. Comput. **30**(2), 528–560 (2000)
5. Edmonds, J.: Edge-disjoint branchings. In: Rustin, B. (ed.) Combinatorial Algorithms, pp. 91–96. Academic Press, New York (1972)
6. Fakcharoenphol, J., Laekhanukit, B.: An $o(\log^2 k)$-approximation algorithm for the k-vertex connected spanning subgraph problem. In: Proceedings of the 40th ACM Symposium on Theory of Computing, STOC 2008, pp. 153–158, New York, NY, USA, ACM (2008)
7. Ford, L.R., Fulkerson, D.R.: Maximal flow through a network. Can. J. Math. **8**, 399–404 (1956)
8. Gabow, H.N., Tarjan, R.E.: A linear-time algorithm for a special case of disjoint set union. J. Comput. Syst. Sci. **30**(2), 209–221 (1985)
9. Garey, M.R., Johnson, D.S.: Computers and Intractability: A Guide to the Theory of NP-Completeness. W. H. Freeman & Co., New York (1979)
10. Georgiadis, L.: Approximating the smallest 2-vertex connected spanning subgraph of a directed graph. In: Demetrescu, C., Halldórsson, M.M. (eds.) ESA 2011. LNCS, vol. 6942, pp. 13–24. Springer, Heidelberg (2011)
11. Georgiadis, L., Italiano, G.F., Laura, L., Parotsidis, N.: 2-Edge connectivity in directed graphs. SODA 2015, pp. 1988–2005 (2015)
12. Georgiadis, L., Italiano, G.F., Laura, L., Parotsidis, N.: 2-Vertex connectivity in directed graphs. In: Halldórsson, M.M., Iwama, K., Kobayashi, N., Speckmann, B. (eds.) ICALP 2015. LNCS, vol. 9134, pp. 605–616. Springer, Heidelberg (2015)
13. Georgiadis, L., Italiano, G.F., Papadopoulos, C., Parotsidis, N.: Approximating the smallest spanning subgraph for 2-edge-connectivity in directed graphs. In: Bansal, N., Finocchi, I. (eds.) Algorithms - ESA 2015. LNCS, vol. 9294, pp. 582–594. Springer, Heidelberg (2015). doi:10.1007/978-3-662-48350-3_49
14. Georgiadis, L., Italiano, G.F., Parotsidis, N.: A new framework for strong connectivity and 2-connectivity in directed graphs. CoRR, November 2015. arXiv:1511.02913
15. Georgiadis, L., Tarjan, R.E.: Dominator tree certification and divergent spanning trees. ACM Trans. Algorithms **12**(1), 11:1–11:42 (2015)
16. Henzinger, M., Krinninger, S., Loitzenbauer, V.: Finding 2-edge and 2-vertex strongly connected components in quadratic time. In: Halldórsson, M.M., Iwama, K., Kobayashi, N., Speckmann, B. (eds.) ICALP 2015. LNCS, vol. 9134, pp. 713–724. Springer, Heidelberg (2015)

17. Italiano, G.F., Laura, L., Santaroni, F.: Finding strong bridges and strong articulation points in linear time. Theor. Comput. Sci. **447**, 74–84 (2012)
18. Jaberi, R.: Computing the 2-blocks of directed graphs. RAIRO-Theor. Inf. Appl. **49**(2), 93–119 (2015)
19. Jaberi, R.: On computing the 2-vertex-connected components of directed graphs. Discrete Applied Mathematics, (2015, to appear)
20. Khuller, S., Raghavachari, B., Young, N.E.: Approximating the minimum equivalent digraph. SIAM J. Comput. **24**(4), 859–872 (1995). Announced at SODA 1994, 177–186
21. Khuller, S., Raghavachari, B., Young, N.E.: On strongly connected digraphs with bounded cycle length. Discrete Appl. Math. **69**(3), 281–289 (1996)
22. Kortsarz, G., Nutov, Z.: Approximating minimum cost connectivity problems. In: Gonzalez, T.F. (ed.) Approximation Algorithms and Metaheuristics. Chapman & Hall/CRC, Boca Raton (2007)
23. Laekhanukit, B., Oveis Gharan, S., Singh, M.: A rounding by sampling approach to the minimum size k-arc connected subgraph problem. In: Czumaj, A., Mehlhorn, K., Pitts, A., Wattenhofer, R. (eds.) ICALP 2012, Part I. LNCS, vol. 7391, pp. 606–616. Springer, Heidelberg (2012)
24. Nagamochi, H., Ibaraki, T.: Algorithmic Aspects of Graph Connectivity, 1st edn. Cambridge University Press, New York (2008)
25. Tarjan, R.E.: Depth-first search and linear graph algorithms. SIAM J. Comput. **1**(2), 146–160 (1972)
26. Tarjan, R.E.: Edge-disjoint spanning trees and depth-first search. Acta Informatica **6**(2), 171–185 (1976)
27. Vetta, A.: Approximating the minimum strongly connected subgraph via a matching lower bound. In: SODA, pp. 417–426 (2001)
28. Zhao, L., Nagamochi, H., Ibaraki, T.: A linear time 5/3-approximation for the minimum strongly-connected spanning subgraph problem. Inf. Process. Lett. **86**(2), 63–70 (2003)

Worst-Case-Efficient Dynamic Arrays in Practice

Jyrki Katajainen$^{(\boxtimes)}$

Department of Computer Science, University of Copenhagen,
Universitetsparken 5, 2100 Copenhagen East, Denmark
jyrki@di.ku.dk

Abstract. The basic operations of a dynamic array are **operator**[ ],
push_back, and pop_back. This study is an examination of variations of
dynamic arrays that support these operations at $O(1)$ worst-case cost.
In the literature, many solutions have been proposed, but little information is available on their mutual superiority. Most library implementations only guarantee $O(1)$ amortized cost per operation. Four variations
with good worst-case performance were benchmarked: (1) resizable array
relying on doubling, halving, and incremental copying; (2) level-wise-
allocated pile; (3) sliced array with fixed-capacity slices; and (4) block-
wise-allocated pile. Let $|\mathcal{V}|$ denote the size of the values of type $\mathcal{V}$ and $|\mathcal{V}*|$
the size of the pointers to values of type $\mathcal{V}$, both measured in bytes. For
an array of n values and a slice of S values, the space requirements of the
considered variations were at most $12|\mathcal{V}|n+O(|\mathcal{V}*|)$, $2|\mathcal{V}|n+O(|\mathcal{V}*|\lg n)$,
$|\mathcal{V}|(n + S) + O(|\mathcal{V}*|n/S)$, and $|\mathcal{V}|n + O((|\mathcal{V}| + |\mathcal{V}*| + |\mathcal{V}**|)\sqrt{n})$ bytes,
respectively. A sliced array that uses a few per cent of extra space turned
out to be a reasonable solution in practice. In general, for worst-case-
efficient variations, the operations were measurably slower than those for
the C++ standard-library implementation. Moreover, slicing can make the
structures fragile, so measures to make them more robust are proposed.

1 Introduction

A one-dimensional array is a fundamental data structure that is needed in
most applications. Its dynamic variant allows growing and shrinkage at one end.
This paper studies practical implementations of dynamic arrays. Several variations programmed in C++ [22] for the CPH STL [6] (**namespace cphstl**) are
described and experimentally compared against each other and to the implementation shipped with the **g++** compiler (**namespace std**). The class template **std::vector** [4, Clause 23.3.6] is a dynamic array that allows random
access to its values using indices and iterators. The main aim of this study was
to avoid some of the drawbacks known for most existing implementations of
std::vector:

- Support **operator**[ ], push_back, and pop_back at $O(1)$ worst-case cost
  (i.e. instead of $O(1)$ amortized cost per push_back).
- Ensure that the memory overhead is never more than a few per cent (instead
  of 100 % or more).

© Springer International Publishing Switzerland 2016
A.V. Goldberg and A.S. Kulikov (Eds.): SEA 2016, LNCS 9685, pp. 167–183, 2016.
DOI: 10.1007/978-3-319-38851-9_12

- Make manual space management by the function shrink_to_fit unnecessary (i.e. fit the amount of allocated space to the number of elements stored).
- Do not move values because of dynamization (i.e. keep references, pointers, and iterators to the values valid if possible).

Array. Let x be a variable that names a cell storing a value of type $\mathcal{V}$ and let p be a variable that names a cell storing an address. More specifically, the *address* of a value is a pointer to the cell where the value is stored. In the programming languages like C [13] and C++ [22], the type of p is $\mathcal{V}*$. These concepts are bound together by the address-of and contents-of operators:

$\mathcal{V}*$ operator&(): A call of the address-of operator &x returns the address of the cell named by x.

$\mathcal{V}\&$ operator*(): A call of the contents-of operator *p returns a reference to the value stored at the cell pointed to by p.

Let $\mathbb{N}$ be an alias for the type of counters and indices. An *array* A stores a sequence of values of the same type $\mathcal{V}$ and supports the operations:

construction: Create an array of the given size by allocating space from the static storage, the stack, or the heap. In the case of the heap, the memory allocation must be done by calling malloc or **operator new**[].

destruction: If an array is allocated from the static storage or the stack, it will be destroyed automatically when the end of its enclosing scope is reached. But, if an array is allocated from the heap, its space must be explicitly released by calling free or **operator delete**[] after the last use.

operator $\mathcal{V}*$(): Convert the name of an array to a pointer to its first value as, for example, in the assignment $\mathcal{V}*$ p = A.

$\mathcal{V}\&$ Operator[](${\mathbb{N}}$ i): For an index i, a call of the subscripting operator A[i] returns *(A + i), i.e. a reference to the value stored at the cell pointed to by pointer A + i.

The important features of an array are (1) that its size is fixed at construction time and (2) that its values are stored in a contiguous memory segment. Hence, the subscripting operator can be supported at constant cost by simple arithmetic, e.g. by going from the beginning of the array $i \cdot |\mathcal{V}|$ bytes forward, where $|\mathcal{V}|$ denotes the size of a value of type $\mathcal{V}$ in bytes.

Dynamic Array. A *dynamic array* can grow and shrink at one end after its construction. The class template std::vector [4, Clause 23.3.6] is parameterized with two type parameters:

$\mathcal{V}$: the type of the values stored and

$\mathcal{A}$: the type of the allocator used to allocate space and construct a value in that place, and to destroy a value and deallocate the reserved space.

The configuration of a dynamic array is specified by two quantities: *size*, i.e. the number of values stored, and *capacity*, i.e. the number of cells allocated for storing the values. Additionally, `std::vector` supports iterators that are generalizations of pointers. In particular, iterator operation `begin` makes the conversion operator from the name of an array to the address of its first value superfluous. Let $\mathcal{I}$ be the type of the iterators. Compared to an array, the most important new operations are the following:

$\mathcal{I}$ `begin()` **const:** Return an iterator pointing at the first value of A.
$\mathcal{I}$ `end()` **const:** Return an iterator pointing at the non-existing past-the-end value of A. If A is empty, then `A.begin()` $==$`A.end()`.
$\mathbb{N}$ `size()` **const:** Get the number of values stored in A.
`void resize(`$\mathbb{N}$ `n):` Set the number of values stored in A to n.
$\mathbb{N}$ `capacity()` **const:** Get the capacity of A.
`void reserve(`$\mathbb{N}$ `N):` Set the capacity of A to N.
`void push-back(`$\mathcal{V}$`& const x):` Append a copy of x at the end of A.
`void pop-back():` Destroy the last value of A. Precondition: A is not empty.

Often, `begin`, `end`, `size`, and `capacity` are easy to realize at $O(1)$ worst-case cost; `resize` at $O(|n - n'|)$ worst-case cost, n being the old size and n' the new size; and `reserve` at $O(n)$ worst-case cost. In fact, there should be support for a larger set of operations (move-based `push_back`, copy/move construction, copy/move assignment, `swap`, `clear`), but we will not discuss this boilerplate code here. An interested reader may consult the source code for details (see "Software Availability" at the end of the paper).

The following question-answer (**Q-A**) pair captures our vision.

Q: What is the best way of implementing a dynamic array in a software library?

A: Provide a set of kernels that can be easily extended to a full implementation with necessary convenience functions, and let the user of the library select the kernel that suits best for her or his needs.

To realize this vision, the bridge design pattern [23, Sect. 14.4] has been used when implementing container classes. Each container class provides a large set of members, which make the use convenient, but only a small kernel is used in the implementation of these members. By changing the kernel, which is yet another type parameter, a user can tailor the container to his exact needs, either related to safety or performance. As to the safety features, we refer to [11] (referential integrity) and [22, Sect. 13.6] (exception safety). In this paper we focus on the space efficiency of the kernels and the time efficiency of the operations `operator[]`, `push_back`, and `pop_back`. In the worst-case set-up, the space and time efficiency have not been examined thoroughly in the past (cf. [11, Ex. 2]).

Amortized Solution. The standard way of dynamizing an array is to use doubling and halving (see, e.g. [5, Sect. 17.4]). The values are stored in a contiguous memory segment, but when it becomes full, a new, two times larger segment is allocated and all values are moved to there; finally the old segment is released. When the current segment is only one quarter full, a new segment that is half the size of the old one is allocated and all values are moved to the new segment, and then the old segment is released. Both push_back and pop_back have a linear cost in the worst case, but their amortized cost is $O(1)$ since at least $n/2$ elements must be added or $n/4$ elements must be removed before a reorganization occurs again. Thus, we can charge the $O(n)$ reorganization cost to these modifying operations and achieve a constant amortized cost per operation. If the data structure stores n values, the capacity of the current segment can be as large as $4n$ and during the reorganization another segment of size $2n$ must be allocated before the old can be released. Thus, in the worst-case scenario, the amount of space reserved for values can be as high as $6n$. Naturally, other space-time trade-offs could be obtained by applying the reorganizations more frequently.

Worst-Case-Efficient Solutions. One way of deamortizing the above solution is to let, during a reorganization, two memory segments coexist, call them X and Y, and to move the values from X to Y incrementally in connection with the forthcoming modifying operations. Imaginarily, the moves happen instantly. However, if the index of the accessed value is smaller than the size of X, the value can be found from there. In connection with every push_back, if possible, one value from the end of X is moved to Y at the same relative position and the new incoming value is placed at the end of Y. In connection with every pop_back, if possible, two values are moved from the end of X to Y at the same relative positions and the value at the end of Y is popped out. This is repeated until X becomes empty, after which it can be released and Y can take its place. Such an incremental reorganization starts whenever only one segment X exists, and it is either full (then the size of Y will be twice the size of X) or it is one quarter full (then the size of Y will be half the size of X).

This solution—which we call a *resizable array*—is part of computing folklore; we use it as a baseline for other worst-case-efficient implementations. Because the two segments coexist in memory, in the worst-case scenario, the amount of extra space used can be even larger than that needed in the amortized case. Namely, if X is one quarter full, it can take $(1/8)n$ pop_back operations before X will be released. Therefore, just before X is released, the amount of space allocated for it is about $8n$ and the amount of space allocated for Y is about $4n$. Based on this discussion, we can conclude that, in the worst case, the amount of space allocated for values is upper bounded by $12n$ and the leading constant in this bound cannot be improved without changing the reorganization strategy.

As to the space consumption, the folklore solution is far from optimal. Namely, Brodnik et al. [3] proved that, when memory is to be allocated block-wise, for a dynamic array of size n, the space bound $n + \Omega(\sqrt{n})$ is optimal, $n + O(\sqrt{n})$ is achievable, and at the same time the operations **operator**[ ], push_back and pop_back can be supported at $O(1)$ worst-case cost.

Test Set-up. In our experiments we considered the following implementations:

std::vector: This was the standard-library implementation that shipped with our g++ compiler (version 4.8.4). It stored the values in one segment, push_back relied on doubling, and **pop_back** was a noop—memory was released only at the time of destruction. Compared to the other alternatives, this version only supported **push_back** at $O(1)$ amortized cost.

cphstl::resizable_array: This solution relied on doubling, halving, and incremental copying as described above.

cphstl::pile: This version implemented the level-wise-allocated pile described in [9]. The data was split into a logarithmic number of contiguous segments, values were not moved due to reorganizations, and the three operations of interest were all supported at $O(1)$ worst-case cost.

cphstl::sliced_array: This version imitated the standard-library implementation of a double-ended queue. It was like a page table where the directory was implemented as a resizable array and the pages (memory segments) were arrays of fixed capacity (512 values).

cphstl::space_efficient_array: This version was as the block-wise-allocated pile described in [9], but the implementation was simplified by seeing it as a pile of hashed array trees [20]. This version matched the space and time bounds proved to be optimal in [3].

These implementations were benchmarked on a laptop computer that had the following hardware and software specifications at the time of experimentation:

processor: Intel® Core™ i5-2520M CPU @ 2.50GHz × 4
word size: 64 bits
L₁ instruction cache: 32 KB, 64 B per line, 8-way associative
L₁ data cache: 32 KB, 64 B per line, 8-way associative
L₂ cache: 256 KB, 64 B per line, 8-way associative
L₃ cache: 3.1 MB, 64 B per line, 12-way associative
main memory: 3.8 GB, 8 KB per page
operating system: Ubuntu 14.04 LTS
Linux kernel: 3.13.0-83-generic
compiler: g++ version 4.8.4
compiler options: -O3 -std=c++11 -Wall -DNDEBUG -msse4.2 -mabm

In each test, an array of integers of type **int** was used as input. The average running time, the number of value moves, and the amount of space were the performance indicators considered. In the experiments, only four problem sizes were considered: 2^{10}, 2^{15}, 2^{20}, and 2^{25}. For a problem of size n, each experiment was repeated $2^{26}/n$ (or $2^{27}/n$ times) and the mean was reported.

2 Motivating Example: Reverse

Consider the function **reverse** which reverses the order of values in a sequence. According to the C++ standard [4, Clause 25.3.10], its interface is as follows:

```
template <typename I>
void reverse(I, I);
```

The iterators of type $\mathcal{I}$ are assumed to be bidirectional or stronger. This interface forces the algorithm to perform the permutation in-place. For this problem, for an input of size n, $\lfloor (3/2)n \rfloor$ is known to be a lower bound for the number of value moves performed (see, for example, [21, Theorem 11.1]). To surpass this lower bound, we use a more natural interface:

```
template <typename S>
void reverse(S&);
```

Now the input is a reference to a sequence of type $\mathcal{S}$. In Fig. 1, we provide two programs that carry out the reversal. The swap-based implementation is the one used in most standard-library implementations. However, the move-based implementation is more interesting. It heavily relies on the fact that the underlying sequence (1) is space efficient and (2) does not perform any value moves because of reorganizations. If this is the case, values are just moved once from one sequence to another and at the end the handles to these sequences are swapped.

```
template <typename I>
void reverse(I f, I l) {
  while (true) {
    if (f == l or f == --l) {
      return;
    }
    else {
      std::swap(*f, *l);
      ++f;
    }
  }
}

template <typename S>
void reverse(S& s) {
  reverse(s.begin(), s.end());
}
```

```
template <typename S, typename T>
void reverse_copy(S& in, T& out) {
  auto n = in.size();
  while (n != 0) {
    --n;
    out.push_back(std::move(in[n]));
    in.pop_back();
  }
}

template <typename S>
void reverse(S& s) {
  S t;
  reverse_copy(s, t);
  s.swap(t);
}
```

Fig. 1. Swap-based **reverse** (left) and move-based **reverse** (right)

A *sliced array* maintains a resizable array of pointers to contiguous memory segments, each of the same size. Only the last segment may be partially full. When cphstl::sliced_array is used in the move-based algorithm, one slice will be non-full from both sequences. When a slice is processed in the input, it can be released and reused in the output. Of course, both algorithms could also be run using std::vector. For the swap-based algorithm, there is no space penalty since the algorithm is fully in-place, but for std::vector the move-based algorithm will use much more space since the space is released first at the time of destruction.

Table 1. Characteristics of the two reversal algorithms; n denotes the size of the input and S the size of a slice used by `cpshtl::sliced_array`; $-$ means that `std::vector` does not give any space guarantee; the running times were measured for $n = 2^{25}$

Reverse	Array	Moves	Time/n [ns]	Values	Pointers
Swap-based	Vector	$1.5n$	0.88	$-$	$O(1)$
Swap-based	Sliced	$1.5n$	2.25	$n + S$	$O(n/S)$
Move-based	Vector	$2n$	3.83	$-$	$O(1)$
Move-based	Sliced	$1n$	5.17	$n + 2S$	$O(n/S)$

The characteristics of the algorithms for `std::vector` and `cphstl::sliced_array` are summarized in Table 1. These simple experiments show the following: (1) When move assignments are expensive, one should consider using the move-based reversal algorithm; (2) For `std::vector`, the subscripting operator is fast; (3) Reorganizations that move data behind the scenes may harm the performance.

3   Space Efficiency

In principle, a dynamic array that is asymptotically optimal with respect to the amount of extra space used is conceptually simple. However, it seems that the research articles (see, e.g. [3,7,9,11,19]), where such structures have been proposed, have failed to disseminate this simplicity to the textbook authors since such a data structure is seldom described in a textbook. Let us make yet another attempt to capture the essence of such a structure.

Hashed Array Tree. Assume that the maximum capacity of the array is fixed beforehand; let it be N. A *hashed array tree*, introduced by Sitarski [20], is a sliced array where each slice is set to be of size $O(\sqrt{N})$. To make the subscripting operator fast, it is advantageous to let the size be a power of two. Also, the directory will be of size $O(\sqrt{N})$ (i.e. this extra space is solely used for pointers) and there will be at most one non-full memory segment of size $O(\sqrt{N})$ (i.e. this extra space is used for data). From a sliced array this structure inherits the property that the values are never moved because of dynamization. If wanted, the structure could be made fully dynamic by quadrupling and quartering the current capacity whenever necessary [14], but after this the performance guarantees would be amortized, not worst-case.

Pile of Arrays. This data structure was introduced in [9] where it was called a *level-wise-allocated pile*; we call it simply `cphstl::pile`. It took its inspiration from the binary heap of Williams [24]. Instead of using a single memory segment for storing the values, the data is split into a logarithmic number of contiguous memory segments, which increase exponentially in size and of which only the last may be partially full. In a sense, this is like a binary heap, but each level of

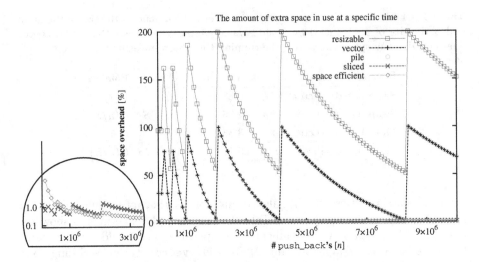

Fig. 2. The amount of extra space in use after n push_back operations for different array implementations; inside the half circle the curves for the two space-efficient alternatives are zoomed out

this heap is a separate array. A directory is needed for storing pointers to the allocated memory segments. Since the size of this directory is only logarithmic, the space for it can often be allocated statically. In a fully dynamic solution the directory is implemented as a resizable array. When there are n values, the size of last non-full memory segment is at most n, so this is an upper bound for the amount of extra space needed for values. In order to realize the subscripting operator at $O(1)$ worst-case cost, it must be assumed that the whole-number logarithm of a positive integer can be computed at $O(1)$ worst-case cost.

Pile of Hashed Array Trees. In [9], this data structure was called a *block-wise-allocated pile*; here we call it cphstl::space_efficient_array. At each level of a pile, the maximum capacity is fixed. Hence, by implementing each level as a hashed array tree, we get a dynamic array that needs extra space for at most $O(\sqrt{n})$ pointers and at most $O(\sqrt{n})$ values, n being the number of values stored.

Space Test. To understand the space efficiency of different array implementations in practice, we performed a *space test* where we executed n push_back operations and measured the amount of memory in use at the end. We repeated this for several values of n. The obtained results are shown in Fig. 2.

More precisely, we measured the memory overhead (i.e. the amount of space used minus the amount of space used by the input) in per cents. The numbers varied between one per mill and 200 per cent, the latter meaning that the amount of memory reserved was large enough to store $3n$ values. The measurements were carried out by using an allocator that counted the number of bytes allocated; it delegated its actual work to std::allocator. During its lifetime, a data

contiguous array

```
V* index_to_address(N i) const {
  return A + i;
}
```

resizable array

```
V* index_to_address(N i) const {
  if (i < X_size) {
    return X + i;
  }
  return Y + i;
}
```

pile

```
N whole_number_logarithm(N x) {
  asm("bsr %0, %0\n"
    : "=r"(x)
    : "0"(x)
  );
  return x;
}
```

```
V* index_to_address(N i) const {
  if (i < 2) {
    return directory[0] + i;
  }
  N h = whole_number_logarithm(i);
  return directory[h] + i − (1 ≪ h);
}
```

sliced array

```
V* index_to_address(N i) const {
  return directory[i ≫ shift] + (i & mask);
}
```

space-efficient array

```
V* index_to_address(N i) const {
  if (i < 2) {
    return directory[0].index_to_address(i);
  }
  N h = whole_number_logarithm(i);
  N Δ = i − (1 ≪ h);
  return directory[h].index_to_address(Δ);
}
```

Fig. 3. Implementation of the `index_to_address` function needed by `operator[]`, for different arrays; the meaning of the class variables should be clear from the context

structure could use several allocators. All these allocators had the same base and it was this base that was responsible for collecting and reporting the final counts.

In theory, there is a significant difference between the extra space of $O(\sqrt{n})$ and $O(n)$ values and/or pointers, but, as seen from the curves in Fig. 2, the space overhead of n/c pointers, for a large integer c, and much fewer values may be equally good in practice. For both space-efficient alternatives, the observed space overhead was less than 4 %, often even less. For the implementations based on doubling, the space overhead could be as high as 100 %. In the space test, `std::vector` and `cphstl::pile` had exactly the same space overhead for all values of n. Even in this simple test, for a resizable array, the space overhead could be as high as 200 %.

4   Subscripting Operator

The key feature of an array is that it supports random access to its values at constant worst-case cost. Moreover, this operation should be fast because it is employed so frequently. In all our implementations, the subscripting operator was implemented in an identical way:

```
V& operator[](N i) {
  return *index_to_address(i);
}
```

As the name suggests, the function index_to_address converts the given index to a pointer to the position where the desired value resizes. In Fig. 3, implementations of this function are shown for different arrays.

Our preliminary experiments revealed that, for a pile and its space-efficient variant, the whole-number-logarithm function needed by the index_to_address function had to be implemented using inline assembly code. Otherwise, the subscripting operator would have been unacceptably slow.

Sorting Tests. After code tuning, we performed two simple tests that used different kinds of arrays in sorting. These benchmarks exercised the subscripting operator extensively. In the *introsort test*, we called the standard-library std::sort routine (introsort [16]) for a sequence of n values. The purpose of this test was to determine the efficiency of sequential access. In the *heapsort test*, we called the standard-library std::partial_sort routine (heapsort [24]) for a sequence of n values. Here the purpose was to determine the efficiency of random access. In these sorting tests, we measured the overall running time for different values of n, and we report the average running time per $n \lg n$. In each test, the input was a random permutation of integers $\langle 0, 1, \ldots, n-1 \rangle$.

The results for introsort are given in Table 2 and those for heapsort in Table 3. It was expected that more complicated code would have its consequences for the running times. Compared to std::vector, integer sorting becomes a constant factor slower with these worst-case-efficient arrays. For a pile and its space-efficient variant, the cost of computing the whole-number logarithm in connection with each access is noticeable, even though we implemented it in assembly language. For all arrays, random access (trusted by heapsort) was significantly slower than sequential access (mostly used by introsort).

Table 2. Results of the introsort tests; running time per $n \lg n$ [ns]

n	Vector	Resizable	Pile	Sliced	Space efficient
2^{10}	3.56	6.18	9.31	8.35	12.0
2^{15}	3.56	5.96	8.99	8.05	11.6
2^{20}	3.48	5.84	8.80	7.91	11.3
2^{25}	3.48	5.79	8.67	7.80	11.2

5 Iterator Operators

An *iterator* is a generalization of a pointer that specifies a position when traversing a sequence (for an introduction to iterators and iterator categories, see, e.g. [21, Chapter 10]). Let $\mathcal{I}$ be the type of the iterators under consideration and let $\mathbb{Z}$ be the type specifying a distance between two positions. In this review we concentrate on three operations that have direct counterparts for pointers.

Table 3. Results of the heapsort tests; running time per $n \lg n$ [ns]

n	Vector	Resizable	Pile	Sliced	Space efficient
2^{10}	4.83	8.89	17.1	12.5	20.3
2^{15}	4.94	8.47	16.6	12.3	19.8
2^{20}	7.18	10.7	17.8	15.7	21.8
2^{25}	23.5	27.7	33.3	37.0	39.8

$\mathcal{V}$& **operator*( ) const:** The *deferencing* operator has the same semantics as the contents-of operator for pointers, i.e. it returns a reference to the value stored at the current position.

$\mathcal{I}$& **operator++( ):** The *pre-increment* operator has the same semantics as the corresponding pointer operator, i.e. it returns a reference to an iterator that points to the successor of the value stored at the current position.

$\mathcal{I}$& **operator+=($\mathbb{Z}$ i):** The *addition-assignment* operator is used to move the iterator to the position that refers to the value that is i positions forward (or backward if i is negative) from the current position.

Traditionally, the iterator support is provided by implementing two iterator classes, one for mutable iterators and another for immutable iterators, inside every container class in the library in question (see, e.g. the implementations provided in [18]). This leads to a lot of redundant code. Austern [1] proposed an improvement were the mutable and **const** versions were implemented in one generic class. We have gone one step further [8]: We provide one generic iterator class template that can be used to get both iterator variants for any container that supports the subscript operator and the function **size**.

Rank Iterators. In the class template **cphstl::rank_iterator**, we use three concepts: (1) A *rank* is an integer which specifies the number of values that precede a value in the given sequence; (2) An *owner* is the sequence where the referred value resizes; (3) A *sentinel* is a rank of a value whose position is unspecified. A *rank iterator* is implemented as a (pointer, rank) pair where the pointer refers to the owner of the encapsulated value and the rank is the index of that value within the owner. A sentinel is used for defensive-programming purposes to perform bounds checking.

```
static N constexpr sentinel = std::
  numeric_limits<N>::max();

V& operator*() const {
  return (*owner_p)[rank];
}

rank_iterator& operator++() {
  ++rank;
  if (rank == (*owner_p).size()) {
    rank = sentinel;
  }
  return *this;
}
```

```
rank_iterator& operator+=(Z n) {
  Z new_place = rank;
  if (rank == sentinel) {
    new_place = (*owner_p).size();
  }
  new_place += n;
  if (new_place < 0) {
    rank = sentinel;
    return *this;
  }
  rank = N(new_place);
  if (rank >= (*owner_p).size()) {
    rank = sentinel;
  }
  return *this;
}
```

Fig. 4. Implementation of the basic iterator operations for rank iterators; owner_p and rank are the class variables denoting a pointer to the owner and the rank, respectively

For a sequence of type S, the types of its iterators are as follows:

```
using iterator = cphstl::rank_iterator<S>;
using const_iterator = cphstl::rank_iterator<S const>;
```

These classes provide the full functionality of a random-access iterator. The implementations of the three important member functions are given in Fig. 4.

Iterator Tests. When analysing the efficiency of rank iterators, we used two tests. In the *sequential-access iterator test*, we initialized an array of size n by visiting each position once. This iterator test exercised derefencing (operator*) and successor (operator++) operators. In the *random-access iterator test*, we also initialized an array of size n by visiting each position once, but there was a gap of 617 values between consecutive visits. This iterator test exercised deferencing (operator*) and addition-assignment (operator+=) operators. All other calculations were done using integers (e.g. no iterator comparisons were done).

In our preliminary experiments, we compared the performance of std::vector and cphstl::contiguous_array, of which the latter used our rank iterators. For these data structures the iterator operations were equally fast, so our generic rank iterator has only little, if any, overhead.

The results of the iterator tests are given in Tables 4 and 5. As to the cost of slicing, on an average, even for $\lceil n/512 \rceil$ slices, the time overhead is about a factor of two. We consider this to be good taking into account that for cphstl::sliced_array the space overhead is never extremely high.

Table 4. Results of the sequential-access iterator tests; running time per n [ns]

n	Vector	Resizable	Pile	Sliced	Space efficient
2^{15} 2^{20} 2^{25}	0.82	1.50	3.15	1.80	3.99

Table 5. Results of the random-access iterator tests; running time per n [ns]

n	Vector	Resizable	Pile	Sliced	Space efficient
2^{10}	1.54	1.90	3.44	2.72	5.91
2^{15}	2.54	2.55	3.20	2.94	5.66
2^{20}	10.9	10.9	11.2	11.3	11.4
2^{25}	14.4	14.4	14.6	17.2	16.7

6 Modifying Operations

Modification Tests. In the *growth test*, we executed n push_back operations repeatedly. In the *shrinkage test*, we created a sequence of size n and then measured the running time used by n repeated pop_back operations. As before, we measured the overall running time and report the average running time per operation for different values of n. The obtained results are shown in Tables 6 and 7.

Compared to an amortized solution that kept the difference between the capacity and size within a permitted range (not discussed earlier), for a resizable array relying on doubling, halving, and incremental copying, the average cost of push_back increased a bit since we could not rely on copying of values in chunks. Also, when we release memory, pop_back is no more free of cost. On the other hand, cphstl::pile and cphstl::sliced_array do not move any values, so they are faster than cphstl::resizable_array. For cphstl::space_efficient_array, the relatively large running times are a consequence of complicated code.

Table 6. Results of the growth tests; running time per n [ns]

n	Vector	Resizable	Pile	Sliced	Space efficient
2^{10}	4.23	5.18	5.65	4.65	10.3
2^{15}	3.52	6.39	5.16	4.63	7.35
2^{20}	4.78	8.48	5.12	4.60	6.92
2^{25}	4.15	8.42	4.55	4.58	6.75

Table 7. Results of the shrinkage tests; running time per n [ns]

n	Vector	Resizable	Pile	Sliced	Space efficient
2^{10}	0.0	3.62	3.08	2.56	8.15
2^{15}	0.0	2.99	2.15	2.60	5.55
2^{20}	0.0	2.86	2.27	2.41	5.17
2^{25}	0.0	2.91	2.11	2.43	5.07

7 Robustness

When our kernels are used to build a container with the same functionality as std::vector, we cannot be standard compliant in one respect [4, Clause 23.3.6]: The values are no more stored in a contiguous memory segment. In this section we consider situations where slicing and slice boundaries can make the structures fragile. We also describe measures that will make the structures more robust.

Break-Down Tests. In our first malicious experiment, we created many small arrays and studied at which point the driver crashed. Recall that our test computer had 3.8 GB of main memory. The actual experiment was as follows:

1. Create a new empty array (elements of type int, four bytes each).
2. Insert 2^{20} elements into this array using push_back.
3. Remove $2^{20} - 1$ elements from this array using pop_back.
4. Repeat this until we get an out-of-memory signal.

That is, how many single-element arrays one can have simultaneously in memory, if the arrays have been bigger at some earlier point in time?

The results obtained varied a bit depending on the memory usage of the other processes run on the test computer, but the numbers on Table 8 speak for themselves. In this kind of application environment, the approach of not releasing allocated memory can have disastrous consequences. To improve the situation with the sliced array, the slices could be made smaller or the first slice could be implemented as a resizable array.

Gap-Crossing Tests. Because of slicing, the worst-case running of one individual push_back and pop_back depends on the efficiency of memory management. In the theoretical analysis, we assumed that the allocator operations allocate that allocates a memory segment and deallocate that releases it have the worst-case

Table 8. Results of the break-down tests; number of repetitions before receiving an out-of-memory signal

Vector	Resizable	Pile	Sliced	Space efficient
804	33 554 432	16 777 216	1 048 448	8 388 473

Table 9. Results of the gap-crossing tests; average running time per (pop_back, push_back) pair [ns]; number of identified gaps in brackets

n	Vector	Resizable	Pile	Sliced	Space efficient
2^{10}	4.48 [11]	3.39 [11]	9.48 [10]	46.7 [2]	31.6 [62]
2^{15}	4.53 [16]	3.85 [16]	8.16 [15]	47.9 [64]	24.5 [382]
2^{20}	4.31 [21]	3.64 [21]	7.60 [20]	49.6 [2 048]	29.6 [2 046]
2^{25}	4.30 [26]	3.46 [26]	118 [25]	49.1 [65 536]	24.4 [12 286]

cost of $O(1)$, independent of the size of the processed segment. By running the instruction-cost micro-benchmark from Bentley's book [2, Appendix 3], it was possible to verify that this assumption did not hold in our test environment.

To see whether the memory-management costs are visible when crossing the gaps between the slices, we carried out one more experiment:

1. Identify where the segment boundaries are.
2. Execute a sequence of push_back operations, but after crossing a gap, execute many additional pairs of pop_back and push_back operations.
3. Report the average running time per (pop_back, push_back) pair.

The obtained results (Table 9) should be compared to those for push_back (Table 6) and pop_back (Table 7) obtained under non-malicious conditions. Of the tested arrays, a resizable array was the most robust since it deamortized the cost of allocations and deallocations over a sequence of modifying operations, and each of these operations touched at most three elements every time. As the opposite, for the largest instance, a pile became very slow because it was forced to allocate and deallocate big chunks of memory repeatedly. The approach used in a resizable array could be used to make the other structures more robust, too. Instead of releasing a segment immediately after it becomes empty, some delay could be introduced so that allocations followed by deallocations were avoided.

8   Discussion

To summarize, a theoretician may think that a solution guaranteeing the worst-case cost of $O(1)$ per operation and the memory overhead of $O(\sqrt{n})$ would be preferable since both bounds are optimal. However, based on the results of our experiments, we have to conclude that, when both the time and space efficiency are important, a sliced array is a good solution. Our implementation supports all the basic operations at $O(1)$ worst-case cost, since we used a worst-case-efficient resizable array to implement the directory, and the observed memory overhead was less than 2 % when n was large, although asymptotically, when the slice size is S, extra space may be needed for S values and $O(n/S)$ pointers. In general, the cutting of the data into slices did not make the operations much slower; in a sequential scan it was not a problem to skip over $\lfloor n/S \rfloor$ slice boundaries.

One reason for inefficiency seems to be the complexity of the formula used for computing the address of the cell where the requested value is. On the other hand, when implementating an industry-strength kernel, special measures must be taken to avoid bad behaviour in situations where subsequent operations are forced to jump back and forth over slice boundaries.

Software Availability

The programs discussed and benchmarked are available via the home page of the CPH STL (www.cphstl.dk) in the form of a technical report and a `tar` file.

Acknowledgements. This work builds on the work of many students who implemented the prototypes of the programs discussed in this paper. From the version-control system of the CPH STL, I could extract the following names—I thank them all: Tina A. G. Andersen, Filip Bruman, Marc Framvig-Antonsen, Ulrik Schou Jörgensen, Mads D. Kristensen [14], Daniel P. Larsen, Andreas Milton Maniotis [8], Bjarke Buur Mortensen [9,10,15], Michael Neidhardt [17], Jan Presz, Wojciech Sikora-Kobylinski, Bo Simonsen [11,12,17], Jens Peter Svensson, Mikkel Thomsen, Claus Ullerlund, Bue Vedel-Larsen, and Christian Wolfgang.

References

1. Austern, M.: Defining iterators and const iterators. C/C++ User's J. **19**(1), 74–79 (2001)
2. Bentley, J.: Programming Pearls, 2nd edn. Addison Wesley Longman Inc., Reading (2000)
3. Brodnik, A., Carlsson, S., Demaine, E.D., Munro, J.I., Sedgewick, R.D.: Resizable arrays in optimal time and space. In: Dehne, F., Gupta, A., Sack, J.-R., Tamassia, R. (eds.) WADS 1999. LNCS, vol. 1663, pp. 37–48. Springer, Heidelberg (1999)
4. The C++ Standards Committee: Standard for Programming Language C++. Working Draft N4296, ISO/IEC (2014)
5. Cormen, T.H., Leiserson, C.E., Rivest, R.L., Stein, C.: Introduction to Algorithms, 3rd edn. The MIT Press, Cambridge (2009)
6. The CPH STL: Department of Computer Science, University of Copenhagen (2000–2016). http://cphstl.dk/
7. Goodrich, M.T., Kloss II, J.G.: Tiered vectors: efficient dynamic arrays for rank-based sequences. In: Dehne, F., Gupta, A., Sack, J.-R., Tamassia, R. (eds.) WADS 1999. LNCS, vol. 1663, pp. 205–216. Springer, Heidelberg (1999)
8. Katajainen, J., Maniotis, A.M.: Conceptual frameworks for constructing iterators for compound data structures–electronic appendix I: component-iterator and rank-iterator classes. CPH STL Report 2012-3, Department of Computer Science, University of Copenhagen, Copenhagen (2012)
9. Katajainen, J., Mortensen, B.B.: Experiences with the design and implementation of space-efficient deques. In: Brodal, G.S., Frigioni, D., Marchetti-Spaccamela, A. (eds.) WAE 2001. LNCS, vol. 2141, pp. 39–50. Springer, Heidelberg (2001)
10. Katajainen, J., Mortensen, B.B.: Experiences with the design and implementation of space-efficient deques. CPH STL Report 2001-7, Department of Computer Science, University of Copenhagen, Copenhagen (2001)

11. Katajainen, J., Simonsen, B.: Adaptable component frameworks: using `vector` from the C++ standard library as an example. In: Jansson, P., Schupp, S. (eds.) 2009 ACM SIGPLAN Workshop on Generic Programming, pp. 13–24. ACM, New York (2009)

12. Katajainen, J., Simonsen, B.: Vector framework: electronic appendix. CPH STL Report 2009–4, Department of Computer Science, University of Copenhagen, Copenhagen (2009)

13. Kernighan, B.W., Ritchie, D.M.: The C Programming Language, 2nd edn. Prentice Hall PTR, Englewood Cliffs (1988)

14. Kristensen, M.D.: Vector implementation for the CPH STL. CPH STL Report 2004–2, Department of Computer Science, University of Copenhagen, Copenhagen (2004)

15. Mortensen, B.B.: The deque class in the Copenhagen STL: first attempt. CPH STL Report 2001–4, Department of Computer Science, University of Copenhagen, Copenhagen (2001)

16. Musser, D.R.: Introspective sorting and selection algorithms. Software Pract. Exper. **27**(8), 983–993 (1997)

17. Neidhardt, M., Simonsen, B.: Extending the CPH STL with LEDA APIs. CPH STL Report 2009–8, Department of Computer Science, University of Copenhagen, Copenhagen (2009)

18. Plauger, P.J., Stepanov, A.A., Lee, M., Musser, D.R.: The C++ Standard Template Library. Prentice Hall PTR, Upper Saddle River (2001)

19. Raman, R., Raman, V., Rao, S.S.: Succinct dynamic data structures. In: Dehne, F., Sack, J.-R., Tamassia, R. (eds.) WADS 2001. LNCS, vol. 2125, pp. 426–437. Springer, Heidelberg (2001)

20. Sitarski, E.: Algorithm alley: HATs: hashed array trees: fast variable-length arrays. Dr. Dobb's J. **21**(11) (1996). http://www.drdobbs.com/database/algorithm-alley/184409965

21. Stepanov, A.A., Rose, D.E.: From Mathematics to Generic Programming. Pearson Education Inc., Upper Saddle River (2015)

22. Stroustrup, B.: The C++ Programming Language, 4th edn. Pearson Education Inc., Upper Saddle River (2013)

23. Vandervoorde, D., Josuttis, N.M.: C++ Templates: The Complete Guide. Pearson Education Inc., Boston (2003)

24. Williams, J.W.J.: Algorithm 232: heapsort. Commun. ACM **7**(6), 347–348 (1964)

On the Solution of Circulant Weighing Matrices Problems Using Algorithm Portfolios on Multi-core Processors

Ilias S. Kotsireas[1], Panos M. Pardalos[2], Konstantinos E. Parsopoulos[3(✉)], and Dimitris Souravlias[3]

[1] Department of Physics and Computer Science,
Wilfrid Laurier University, Waterloo, ON, Canada
`ikotsire@wlu.ca`
[2] Department of Industrial and Systems Engineering,
University of Florida, Gainesville, FL, USA
`pardalos@ufl.edu`
[3] Department of Computer Science and Engineering,
University of Ioannina, Ioannina, Greece
`{kostasp,dsouravl}@cse.uoi.gr`

Abstract. Research on the existence of specific classes of combinatorial matrices such as the Circulant Weighing Matrices (CWMs) lies in the core of diverse theoretical and computational efforts. Modern metaheuristics have proved to be valuable tools for solving such problems. Recently, parallel Algorithm Portfolios (APs) composed of established search algorithms and sophisticated resource allocation procedures offered significant improvements in terms of time efficiency and solution quality. The present work aims at shedding further light on the latent quality of parallel APs on solving CWM problems. For this purpose, new AP configurations are considered along with specialized procedures that can enhance their performance. Experimental evaluation is conducted on a computationally restrictive, yet widely accessible, multi-core processor computational environment. Statistical analysis is used to reveal performance trends and extract useful conclusions.

Keywords: Algorithm Portfolios · Circulant Weighing Matrices · Computational optimization · Multi-core processors

1 Introduction

Combinatorial matrices are involved in various significant applications ranging from statistical experimentation to coding theory and quantum information processing [3,8,23]. Special types of combinatorial matrices have been extensively investigated. Circulant Weighing Matrices (CWMs) constitute an important class in this framework. The existence of finite or infinite classes of CWMs has been the core subject in several theoretical works [2,4,9,10,12].

© Springer International Publishing Switzerland 2016
A.V. Goldberg and A.S. Kulikov (Eds.): SEA 2016, LNCS 9685, pp. 184–200, 2016.
DOI: 10.1007/978-3-319-38851-9_13

Metaheuristics have proved to be very useful in cases where theoretical approaches have not provided adequate insight. The application of metaheuristics requires the transformation of the CWM existence problem to a combinatorial optimization task [6,7,18,19,21,22]. Recently, prevailing metaheuristics have been used in the Algorithm Portfolios (APs) framework [16,17] for solving CWM problems in parallel computational environments [28]. Sophisticated resource allocation schemes based on market trading procedures were used in those approaches, achieving high standards of performance. The provided results suggested that APs can remarkably enhance the time performance and quality of solution of their constituent algorithms in CWM problems [28]. Also, they verified the domination of trajectory-based approaches against population-based stochastic algorithms.

The present work aims at extending the previous studies by offering further insight regarding the performance of interactive and non-interactive parallel APs. Based on previous findings, the established Tabu Search (TS) algorithm and the previously unused Iterated Local Search (ILS) approach compose the considered APs. Additionally, a sequence-comparison scheme that prevents TS from revisiting classes of equivalent sequences is introduced.

The experimental evaluation of the APs is conducted on a low-specification parallel hardware, i.e., a common multi-core processor, in contrast to the abundant grid-computing environment of previous studies [28]. The overall performance of the APs is investigated in terms of time efficiency and solution quality on two representative CWM problems. Additionally, the impact of the number of concurrently running algorithms on the overall performance is investigated. Diverse homogeneous and heterogeneous APs with various parameter configurations are also considered.

The rest of the paper is structured as follows: Sect. 2 formulates the CWM problem as a combinatorial optimization task. The employed individual algorithms and APs are described in Sect. 3. Experimental analysis is reported in Sect. 4, and the paper concludes in Sect. 5.

2 Circulant Weighing Matrices

Circulant Weighing Matrices (CWMs) [4] refer to a special type of combinatorial matrices. A square $n \times n$ matrix W defined as,

$$W = (w_{ij}), \qquad w_{ij} \in \{-1, 0, 1\}, \qquad i, j = 1, \ldots, n,$$

is a CWM of *order* n and *weight* k^2, denoted as $CW(n, k^2)$, if it satisfies the condition,

$$W\,W^\top = k\,I_n,$$

where I_n is the identity matrix of size n, and $W^\top$ is the transpose of W. Thus, a CWM is primarily a *weighing matrix*. Additionally, each row of a CWM, except the first one, is obtained through a right cyclic shift of its preceding row. Hence, the complete matrix can be fully defined by its first row. A significant amount of research has been devoted to theoretical and experimental investigations on the existence of CWMs of various orders and weights [1,5,10,29].

Metaheuristics have been employed in cases where theoretical efforts have been fruitless. In these cases, the problem is solved as a permutation optimization one, aiming at the detection of the first row of the considered CWM type. The underlying objective function is based on the concept of *Periodic Autocorrelation Function* (PAF) [20]. The defining row of a CWM is a ternary sequence,

$$x = (x_1, x_2, \ldots, x_n) \in \{-1, 0, +1\}^n,$$

of length n, and its PAF values are defined as,

$$PAF_x(s) = \sum_{i=1}^{n} x_i \, x_{i+s}, \qquad s = 1, 2, \ldots, \left\lceil \frac{n}{2} \right\rceil. \tag{1}$$

CWMs with zero PAF values have special research interest [19,20,28]. In addition, it has been proved that admissible sequences have exactly k^2 non-zero components, with $k(k+1)/2$ components being equal to $+1$ and $k(k-1)/2$ components assuming the -1 value.

Let $S_{(n,k)}$ be the search space that contains all admissible ternary sequences that define CWMs of order n and weight k^2. Then, the objective function of the corresponding combinatorial optimization problem is defined as,

$$\min_{x \in S_{(n,k)}} \; f(x) = \sum_{s=1}^{\left\lceil \frac{n}{2} \right\rceil} |PAF_x(s)| = \sum_{s=1}^{\left\lceil \frac{n}{2} \right\rceil} \left| \sum_{i=1}^{n} x_i \, x_{i+s} \right|, \tag{2}$$

where $i + s$ is taken modulo n when $i + s > n$. Obviously, the global minimizer of this optimization problem is a sequence with zero PAF values for all s. Experimental evidence has shown that the difficulty of a $CW(n, k^2)$ problem increases with the order n (length of sequence) and, particularly, with the weight k^2.

3   Employed Algorithms

In the following paragraphs, we briefly describe the employed individual algorithms as well as the considered APs. For presentation purposes, we assume that the considered optimization problem is given in the general form,

$$\min_{x \in S} \; f(x),$$

where S in the corresponding search space.

3.1   Iterated Local Search

Iterated Local Search (ILS) defines a simple and straightforward framework for probing complex search spaces. Its main requirement is the use of a suitable

Table 1. Pseudocode of the ILS algorithms.

Iterated Local Search (ILS)
1 : $x_{\text{ini}} \leftarrow GetInitialSequence(S)$
2 : $x^* \leftarrow LocalSearch(x_{\text{ini}})$
3 : $S^* \leftarrow \{x^*\}$
4 :   **While** (not stopping) **Do**
5 :   **If** $(rand() < \rho)$ **Then**
6 : $x_{\text{ini}} \leftarrow GetInitialSequence(S^*)$
7 :   **Else**
8 : $x_{\text{ini}} \leftarrow GetInitialSequence(S)$
9 :   **End If**
10 : $x^* \leftarrow LocalSearch(x_{\text{ini}})$
11 : $S^* \leftarrow S^* \cup \{x^*\}$
12 :   **End While**
13 : $x_{\text{best}} \leftarrow \arg\min_{x^* \in S^*} f(x^*)$
14 :   **Report** x_{best}

local search procedure for the problem at hand. The local search is initiated to a randomly selected sequence x_{ini} and generates a trajectory that eventually reaches the nearest local minimizer x^*. This is achieved by iteratively selecting downhill moves within the close neighborhood of the current sequence.

In discrete spaces such as the ones in the studied CWM problems, the close neighborhood of a sequence is defined as the finite set of sequences with the smallest possible distance from it. Typically, Hamming distance is used for this purpose. The local search procedure usually scans the whole neighborhood of the current sequence and makes the move with the highest improvement (*neighborhood-best* strategy). Alternatively, it can make a move to the first improving sequence found in the neighborhood (*first-best* strategy). The detected local minimizer is archived in a set S^*. Then, a new trajectory is started from a new initial sequence [24].

In its simplest form, ILS generates new trajectories by randomly sampling new initial sequences in the search space according to a (typically Uniform) distribution. This is the well-known *Random Restarts* approach. The most common stopping criteria are the detection of a prespecified number of local minimizers or a maximum computational budget in terms of running time or function evaluations. Although random restarts were shown to be sufficient in various problems, relevant research suggests that efficiency can be increased if already detected local minimizers from the set S^* are exploited during the search [24]. Typically, this refers to the generation of new initial sequences by perturbing previously detected local minimizers.

The two initialization approaches can also be combined. Naturally, this scheme introduces new parameters to the algorithm. Specifically, the user needs

Table 2. Pseudocode of the TS algorithms.

Tabu Search (TS)
1 : $TL \leftarrow \emptyset$
2 : $\boldsymbol{x} \leftarrow GetInitialSequence(S)$
3 : $UpdateTabuList(TL, \boldsymbol{x})$
4 : $\boldsymbol{x}_{\text{best}} \leftarrow \boldsymbol{x}$
5 :   **While** (not stopping) **Do**
6 : $\boldsymbol{x}' \leftarrow ProbeNeighborhood(N(\boldsymbol{x}), TL)$
7 : $UpdateTabuList(TL, \boldsymbol{x}')$
8 :   **If** $(f(\boldsymbol{x}') < f(\boldsymbol{x}_{\text{best}}))$ **Then**
9 : $\boldsymbol{x}_{\text{best}} \leftarrow \boldsymbol{x}'$
10 :   **End If**
11 :   **If** (trajectory termination) **Then**
12 : $\boldsymbol{x} \leftarrow NewInitialSequence(S, \boldsymbol{x}_{\text{best}}, \rho)$
13 :   **End If**
14 :   **End While**
15 :   **Report** $\boldsymbol{x}_{\text{best}}$

to specify a probability $\rho \in [0, 1]$ of using perturbation-based restarts as well as the criteria for selecting the local minimizers from the set S^*.

The ILS algorithm is given in pseudocode in Table 1. Each call of $rand()$ returns a real-valued random number in the range $[0, 1]$, while the function $GetInitialSequence()$ implements the sampling procedures for the search space S and the set S^*. For a comprehensive presentation of ILS the reader is referred to [24].

3.2 Tabu Search

Tabu Search (TS) is among the most popular and well-studied metaheuristics. Since its formal introduction in [13,14], TS has been applied on numerous problems spanning various fields of discrete optimization [11,15,26]. The basic motivation for the development of TS originated from the necessity of search algorithms to overcome local minimizers. This was achieved by equipping the algorithms with descent and hill-climbing capabilities.

In descent mode, the local search procedure of TS follows the baseline of the ILS approach described in the previous section. After the detection of a local minimizer, the algorithm begins ascending by reversing from downhill to uphill moves in the neighborhood $N(\boldsymbol{x})$ of the current sequence $\boldsymbol{x}$. This continues until a local maximizer is reached. Subsequently, a new descent phase takes place etc.

In order to avoid retracing the same trajectories, a memory structure that stores the most recent moves and prevents the algorithm from revisiting them is used in TS. In practice, the memory comprises a finite list structure, also called *tabu list* (TL), where the most recently visited sequence replaces the oldest one.

The use of memory cannot fully prevent TS from getting trapped in misleading trajectories that drive the search away from global minimizers. In such cases, it is beneficial to restart the algorithm on a new sequence if the current trajectory has not improved the best solution for a prespecified number of iterations or elapsed time.

Similarly to ILS, new initial sequences can be generated either randomly within the whole search space S or through perturbations of already detected local minimizers. The latter approach can be effective in problems where local minimizers are closely clustered.

A simple form of the TS algorithm is reported in Table 2, where the parameter $\rho \in [0,1]$ defines the probability of restarting the algorithm on a perturbation of the best-so-far solution x_{best}. Other crucial parameters are the size of the tabu list, s_{tabu}, as well as the number of non-improving steps, T_{nis}, before restarting a trajectory. Further details on TS and its applications can be found in [11,15,26].

3.3   Algorithm Portfolios

Algorithm Portfolios (APs) [17] define schemes composed of multiple individual algorithms that share the available computational budget. An AP may consist of multiple copies of one algorithm with the same or different parameters (*homogeneous AP*) or different algorithms (*heterogeneous AP*). All the algorithms run concurrently in either one or multiple processors (CPUs). If a single CPU is used, the algorithms' execution is alternated according to a time assignment schedule. In multi-core or parallel systems, the algorithms share the hardware resources, i.e., number of available CPUs [16].

Relevant studies have shown that proper resource allocation between the constituent algorithms can render APs more efficient than the standalone algorithms, both in serial [17] and parallel [16] computational environments. Moreover, information exchange between the algorithms (*interactive APs*) can be highly beneficial [25]. Motivated from these studies, a new parallel AP with a sophisticated time budget allocation scheme that is based on market-trading procedures was proposed in [27] and successfully applied on the CWM problems in [28]. The specific AP comprised various search algorithms. Among them, TS was shown to be the most effective one [28].

The previous studies offered useful insight on the performance of APs on CWM problems, leaving prosperous ground for further investigation. The AP in [28] was based on the special trading-based time allocation rather than the plain interactive AP model. The experimental results offered clear evidence that trajectory-based approaches were dominant in terms of solution quality. Moreover, the highly-effective TS algorithm was considered only with the neighborhood best strategy, which is an effective but also computationally demanding approach.

Another important issue in parallel APs is the effect of the number of algorithms and, consequently, the number of nodes that are concurrently used. The experiments in [28] were conducted on a computer cluster where a large number of processors were available. However, it would be interesting to evaluate the APs also on the widely accessible multi-core processors, which typically offer only a small number of CPUs to the user. For instance, modern Intel© i7 processors consist of 4 actual cores that offer 8 CPUs by using hyper-threading technology[1]. Each CPU can concurrently run multiple algorithms in different computation threads at the cost of slower execution, since the algorithms are alternatively executed. Given a prespecified running-time budget, it is compelling to investigate whether it is preferable to use small number of algorithms (not exceeding the number of available CPUs) in order to attain faster execution or use higher numbers of algorithms (thereby promoting exploration) with slower execution.

Another interesting issue that emanates from previous TS applications is related to the criteria of accepting a new sequence through comparisons with the ones stored in TL. The typical comparison has been based solely on the Hamming distance between the compared sequences, i.e., a pairwise comparison of their corresponding components. Thus, a new sequence was accepted only if it had non-zero distance from all stored sequences in TL. Although this approach adheres to the typical rules applied in various TS applications, it can become inefficient in CWM problems.

The reason lies on the specific properties of CWM matrices. Specifically, a given sequence x defines the same CWM with all right-hand shifted sequences produced from it. In simple words, the sequence x defines a whole class of sequences that produce the same CWM. These equivalent sequences have non-zero Hamming distances between them. Thus, the comparison criterion in previous TS approaches cannot prevent the acceptance of a sequence that is equivalent with one already included in TL. Tabu lists of large size as in [28] can ameliorate this deficiency but they impose additional computational burden. For this reason, it is preferable to modify the comparison procedure such that a candidate new sequence is accepted only if it differs from all sequences in TL as well as from all their right-hand shifts.

The present work attempts to shed light on the aforementioned issues. The employed parallel APs are outlined in Table 3. The number of nodes, m, refers to the number of threads required by the AP and can exceed the number of available CPUs. The parallel AP is based on a standard master-slave parallelization model, where the master (node 0) is devoted to book-keeping and information-sharing between the algorithms. Both homogeneous and heterogeneous APs consisting of the TS and ILS algorithms are studied. The simple Random Restart variant of ILS was used, along with the local search described in the previous sections. Further details for the algorithms are given in the following section.

---

[1] http://www.intel.com/content/www/us/en/processors/core/core-i7-processor. html.

Table 3. Pseudocode of the parallel Algorithm Portfolio approach with m nodes.

Algorithm Portfolio (m nodes)
Master Node ($i = 0$)

1 :	**Initialize** $(m-1)$ slave nodes and assign an algorithm to each one.
2 :	$x_{\text{best}} \leftarrow GetInitialSequence(S)$
3 :	$SendSequence(i, x_{\text{best}})$, $i = 1, \ldots, m-1$
3 :	**While** (nodes still running) **Do**
4 :	$\quad GetMessage(i)$
5 :	$\quad$ **If** (node i improved x_{best}) **Then**
6 :	$\quad\quad UpdateBest(x_{\text{best}})$
7 :	$\quad$ **Else If** (node i requests best sequence update) **Then**
8 :	$\quad\quad SendSequence(i, x_{\text{best}})$
9 :	$\quad$ **End If**
10 :	**End While**
11 :	**Report** x_{best}

Slave Nodes ($i = 1, \ldots, m-1$)

1 :	**Initialize** assigned algorithm.
2 :	$ReceiveSequence(0, x_{\text{best}})$
3 :	**While** (allocated time not exceeded) **Do**
4 :	$\quad$ **Execute** one iteration of the algorithm.
5 :	$\quad$ **If** (new x_{best} found) **Then**
6 :	$\quad\quad SendSequence(0, x_{\text{best}})$
7 :	$\quad$ **Else If** (best sequence update is needed) **Then**
8 :	$\quad\quad RequestSequenceUpdate(0)$
9 :	$\quad$ **End If**
10 :	**End While**
11 :	**Finalize** node

4   Experimental Analysis

The experimental analysis consisted of two phases. In the first phase, all algorithms were applied on the representative 33-dimensional $CW(33, 25)$ problem, in order to statistically analyze their performance. The specific problem was selected due to its guaranteed solution existence, moderate size, high weight ($k^2 = 25$), and reasonable convergence times of the algorithms. The second phase consisted of the application of the best-performing algorithms on the more challenging 48-dimensional $CW(48, 36)$ problem. This is a well-studied problem that was used as benchmark in previous studies [28]. The number of sequence components that assume each value of the set $\{-1, 0, +1\}$ for both problems is reported in Table 4.

Table 4. Details of the considered representative problems.

Problem	Length	Weight	Dim.	+1	−1	0
$CW(33,25)$	33	25	33	15	10	8
$CW(48,36)$	48	36	48	21	15	12

Table 5. Parameter values for the employed algorithms.

Param.	Description	Value(s)
m	number of nodes (threads)	$\{8, 16, 64\}$
nss	neighborhood search strategy	$\{$neighb. best (nb), first best (fb)$\}$
s_{tabu}	tabu list size	$\{5, 10\}$
T_{nis}	non-improving iterations before restart	$\{100, 1000\}$
ρ	probability of perturbing best solution	$\{0.00, 0.01\}$
p_{type}	algorithm parameters' type	$\{$fixed (f), random (r)$\}$
T_{max}	maximum running time	$300\,s$

We considered APs composed solely of TS or ILS algorithms, henceforth denoted as "TS" or "ILS", respectively. Also, we considered mixed APs embracing both algorithms, henceforth denoted as "MIX". Initially, an extensive experimental study was conducted for all combinations (full factorial design) of the parameter values reported in Table 5. Specifically, for each portfolio type (TS, ILS, or MIX), we considered the cases of $m = 8$, 16, and 64 threads running on a single processor with 8 CPUs available (note that the number of slave nodes is $m - 1$). In TS-based APs all slave nodes were occupied by TS algorithms, while in ILS-based APs they were devoted to ILS. In MIX APs, the TS algorithm was assigned to the odd-indexed nodes $(1, 3, \ldots)$ and ILS algorithms were running on even-indexed nodes $(2, 4, \ldots)$.

All experiments were conducted on a single-processor Intel© i7-4770 3.40 GHz machine with 8 GB DDR3 RAM, providing 8 available CPUs under Ubuntu Linux 14.04. There was no suppression of the operating system procedures during the runs. For the parallelization, the Open MPI libraries were used with the GCC 4.8.4 compiler. All source codes were developed in the C programming language.

In the TS and ILS algorithms, both the neighborhood-best (nb) and first-best (fb) strategies were considered for neighborhood search. New trajectories were either initialized on random perturbations of the best-so-far solution with probability $\rho = 0.01$ or solely on random new points (denoted as $\rho = 0.00$). In the first case, mild perturbations of the best solution were used, consisting of 1 up to 3 distinct random swaps of the sequence's components.

The TS algorithms require some additional parameters. The tabu list size s_{tabu} in our experiments was set to rather small values, namely 5 and 10. These values are significantly smaller than in previous studies where it was set equal

Table 6. Results for the 3 best-performing approaches per AP type and number of nodes, as well as for the 5 overall best APs for the $CW(33, 25)$ problem. The "*" symbol denotes randomized-parameters APs and, if followed by a number, e.g., "*s", it denotes that the upper bound of the corresponding randomized parameter is s.

TS-based APs

m	nss	s_{tabu}	T_{nis}	ρ	p_{type}	Suc.(%)	Time	Loc. Min.
8	*	5	*100	0.00	r	100.0	24.6(26.4)	14080.9
8	fb	5	100	0.00	f	100.0	26.8(28.2)	17535.1
8	*	10	*100	0.00	r	100.0	32.9(34.4)	10488.2
16	*	5	*100	*0.01	r	100.0	25.7(27.4)	7574.2
16	*	5	*100	0.00	r	100.0	31.6(26.5)	8937.5
16	fb	5	100	0.00	f	100.0	38.4(30.2)	12494.7
64	fb	5	100	0.00	f	100.0	25.0(25.5)	2032.3
64	fb	5	100	0.01	f	100.0	25.5(23.8)	2100.9
64	*	5	100	0.00	r	100.0	29.0(31.8)	2063.8

ILS-based APs

m	nss	s_{tabu}	T_{nis}	ρ	p_{type}	Suc.(%)	Time	Loc. Min.
8	nb	-	-	0.00	f	100.0	11.0(14.6)	32512.7
8	nb	-	-	0.01	f	100.0	9.6(11.7)	28486.7
8	fb	-	-	0.00	f	100.0	6.6(4.8)	42418.4
16	fb	-	-	0.01	f	100.0	2.8(4.3)	8762.9
16	nb	-	-	0.00	f	100.0	8.5(9.0)	12413.9
16	nb	-	-	0.01	f	100.0	12.2(11.7)	17587.3
64	fb	-	-	0.00	f	100.0	4.2(4.5)	3447.3
64	fb	-	-	0.01	f	100.0	4.3(5.3)	3388.9
64	*	-	-	0.00	r	100.0	7.7(9.9)	3927.3

MIX APs

m	nss	s_{tabu}	T_{nis}	ρ	p_{type}	Suc.(%)	Time	Loc. Min.
8	fb	5	100	0.00	f	100.0	10.3(12.1)	31847.9
8	fb	10	100	0.01	f	100.0	9.9(10.2)	29267.2
8	*	5	*100	*0.01	r	100.0	9.8(12.2)	20060.1
16	fb	5	100	0.01	f	100.0	7.6(11.5)	12508.8
16	fb	10	1000	0.01	f	100.0	8.3(11.8)	12792.8
16	fb	10	100	0.01	f	100.0	7.6(7.5)	11695.5
64	*	10	*1000	0.00	r	100.0	9.9(15.8)	2681.8
64	fb	10	1000	0.00	f	100.0	8.0(6.1)	3318.4
64	fb	5	100	0.00	f	100.0	9.1(7.9)	3967.3

OVERALL BEST APs

Alg.	m	nss	s_{tabu}	T_{nis}	ρ	p_{type}	Suc.(%)	Time	Loc. Min.
ILS	16	fb	-	-	0.01	f	100.0	2.8(4.3)	8762.9
ILS	64	fb	-	-	0.01	f	100.0	4.3(5.3)	3388.9
ILS	64	fb	-	-	0.00	f	100.0	4.2(4.5)	3447.3
MIX	16	fb	5	100	0.01	f	100.0	7.6(11.5)	12508.8
MIX	16	fb	10	1000	0.01	f	100.0	8.3(11.8)	12792.8

to the length of the sequences (order of the CWM) [28]. The reduction was implied by the new scheme for comparisons between the current sequence and the stored ones in TL, as described in Sect. 3.3. The tolerance T_{nis} of non-improving moves before restart was set to 100 and 1000. Larger values of T_{nis} result in longer trajectories and, hence, better local exploration around recently visited minimizers. Smaller values promote global exploration because the algorithm is restarted more frequently. All combinations of the corresponding parameters were considered for each AP type.

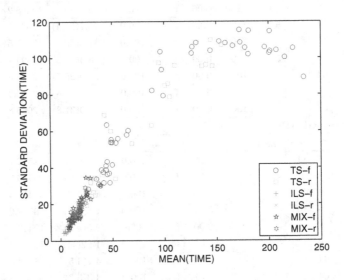

Fig. 1. Mean vs standard deviation of time required per AP type (TS, ILS, MIX) for fixed (-f) and random (-r) parameters.

In addition to the fixed-parameters APs, randomized-parameters variants were also studied. In these cases, the algorithms in the AP were allowed to randomly select new parameter values, among the available ones in Table 5, for each new trajectory. Thus, there was a total number of 162 different APs in our experiments. Each AP was independently applied on the $CW(33, 25)$ problem for a maximum time of $T_{max} = 300\,s$. Since the algorithms in the APs involve stochastic decisions, 25 independent experiments were conducted per AP for statistical analysis purposes.

For each individual combination of type (TS, ILS, MIX) and number of nodes (8, 16, 64), the performances of the corresponding APs were pairwisely tested using the Wilcoxon ranksum test with 0.05 significance level in order to identify statistically significant differences. The comparisons were primarily based on the successes of the APs in detecting globally optimal solutions and, secondarily, on the required running times. Then the APs were sorted according to the achieved scores.

The three best-performing APs per case are reported in Table 6 along with their parameters. For each reported AP, the percentage of success in detecting a global minimizer, the mean and standard deviation (in parenthesis) of the required time in seconds, as well as the mean number of visited local minimizers per slave node, averaged over the 25 experiments are reported. For the MIX APs, the parameters s_{tabu} and T_{nis} refer to their constituent TS algorithms.

In a second round of comparisons, all the 162 APs were statistically compared against each other, aiming at finding the overall best-performing approaches. The Wilcoxon ranksum test with 0.05 significance level was used also in this case, and the APs were sorted according to their scores. The five most promising APs are reported in the lower part of Table 6. Furthermore, Fig. 1 illustrates the mean value versus the standard deviation of the time required per AP type (TS, ILS, MIX) for fixed (-f) and random (-r) parameters. Figure 2 shows the average time required per AP type (TS, ILS, MIX) for 8, 16, and 64 nodes.

Table 6 offers interesting evidence for each AP type. First, we can notice that the best TS-based APs required higher average running times and visited less local minimizers than ILS-based and MIX APs. This is also observed in Fig. 1 where TS-based APs occupy the upper-right part of the figure. The observed time-performance profiles are reasonable, since TS spends a fraction of its computational budget for procedures related to checking and updating the tabu list, as well as for hill-climbing. Nevertheless, TS was highly effective in detecting the global minimizer. Also, we can see that the small TL size, $s_{\text{tabu}} = 5$, was dominant in the best-performing TS-based APs because larger tabu lists require additional comparisons and, consequently, reduce convergence speed. This is also in line with the dominant $T_{\text{nis}} = 100$ parameter, which promotes shorter trajectories.

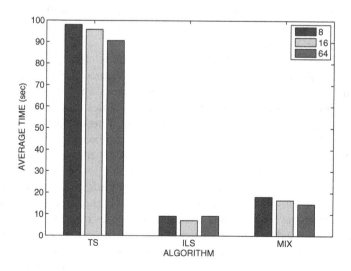

Fig. 2. Average time required per AP type (TS, ILS, MIX) for 8, 16, and 64 nodes.

Moreover, Table 6 reveals a trend of (almost linear) reduction of the number of visited minimizers with the number of nodes. This may seem counter-intuitive, because additional nodes offer higher numbers of concurrent trajectories (although at slower speed in the multi-core processor). However, it can be explained by the increase of the exploration capabilities of the algorithms, which lead to faster detection of a global minimizer. The rapid convergence is also reflected to the declining average execution times in Fig. 2. Thus, the number of concurrent trajectories seems to be highly beneficial in TS-based APs despite the possible slowdown in the AP's execution.

Another interesting observation is that TS-based APs with fixed parameters (p_{type} = "f") profited from the first-best neighborhood search (nss = "fb"). This is related to the previous comment on the significant time consumed for comparing the complete neighborhood of the current sequence with the whole tabu list. Obviously, making a move directly after a sequence of adequate quality is detected in the current neighborhood, can spare significant amount of execution time without reducing effectiveness.

Finally, the results show that APs with randomized parameters can perform equally well with the ones with fixed parameters, especially when random initialization is preferred against perturbations of the best solution. The latter can be a consequence of the inherent ability of TS to visit neighboring sequences through hill-climbing.

The ILS-based APs achieved the lowest average convergence times as depicted in Fig. 2. This is reasonable, since ILS exploits its computational budget solely in descent moves towards the nearest local minimizer. However, there is an interesting effect of the number of nodes on the average running time. As we can see in Fig. 2, doubling the number of nodes from 8 to 16 results in improved average time, but further increase to 64 nodes produces negative effects on performance since the trajectories are significantly slowed down. This verifies the existence of a trade-off between the number of concurrent trajectories and time efficiency for the ILS-based APs on multi-core processors.

The three best-performing ILS-based APs show a clear preference to fixed parameters configuration, as we notice in Table 6. Since ILS does not have an inherent mechanism for searching neighboring minimizers, there is a balanced preference between completely random new trajectories and the use of perturbations of the best solution. Also, the neighborhood-best approach seems to be more beneficial, i.e., ILS prefers to conduct steepest descent to the nearest minimizer. The number of visited local minimizers with respect to the employed nodes shows similar trends with the TS-based APs.

The performance profile of MIX APs combines performance aspects of both TS and ILS. The effect of the number of nodes on the average running time appears to follow the same trends with TS, since the running time of the AP is primarily consumed by the TS algorithms. On the other hand, the number of visited minimizers is comparable to the ILS-based APs. However, there are some peculiarities on the parameters of the three best-performing approaches, as revealed in Table 6.

Specifically, we observe that the first-best approach was dominant, obviously because it enhances the TS algorithms. However, APs with higher TL sizes and longer trajectories appear more frequently among the best ones, especially for higher number of nodes. Thus, there seems to be an interesting division of labor in MIX APs, where TS algorithms offer the AP's exploitation ability and ILS undertake the exploration task. Furthermore, perturbation-based initialization appears to be very competitive to the pure randomized one.

Overall, the synergism between TS and ILS in the MIX approaches seems to equip the APs with combined dynamics. The extra cost due to the tabu list-related procedures is counterbalanced by the first-best neighborhood search. This way, the spared computation time allows for longer trajectories and larger tabu lists.

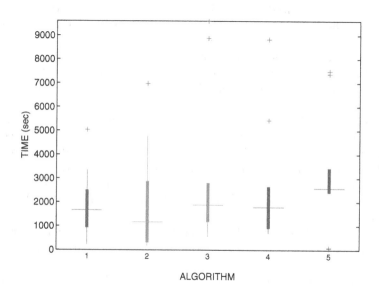

Fig. 3. Boxplots of running time of the best-performing APs on the $CW(48, 6)$ problem. Green color stands for 64-nodes APs, while red color stands for the 16-nodes APs. (Color figure online)

In a second round of comparisons, where each AP was compared against all the rest, three ILS and two MIX approaches were distinguished. As it is reported in Table 6, all these APs were based on fixed parameters and high number of nodes. Interestingly, a clear domination of the perturbation-based initialization of new trajectories is noticed.

The five distinguished approaches were further assessed on the more challenging $CW(48, 36)$ problem, which has been used as a benchmark in previous works. The maximum time per experiment for this case was $T_{\max} = 10800\,\text{s}$ (3 h). In all experiments and algorithms, a global minimizer was detected. The boxplots of the running time of the APs over 25 experiments is illustrated in Fig. 3, where the APs appear in the same order as in Table 6.

Wilcoxon ranksum tests with 0.05 significance level showed no significant differences in terms of running time among them. However, their average times are favorably compared to those in previous studies [28], although strict comparisons would be questionable due to the completely different hardware and experimental configurations used in the present work. Nevertheless, it is a clear indication of the potential of the presented APs in solving CWM problems.

5   Conclusions

The present work enriched our insight on the performance of parallel APs on CWM problems. Enhanced TS- and ILS-based APs were used. Also, mixed APs composed by both algorithms were considered. Experimentation was focused on the widely accessible multi-core processor computational environment. Two representative test problems were used in order to investigate the performance of the APs as well as the influence of the requested number of nodes (threads), which defines also the number of the AP's algorithms, on the time efficiency and solution quality.

A rich variety of both homogeneous and heterogeneous APs were considered under various parameter settings, offering useful conclusions. ILS-based APs were significantly faster than TS-based ones, and they showed different response when the number of nodes was increased. Fixed parameters were shown to dominate randomized ones. Also, the effect of the time-consuming neighborhood-best strategy was counterbalanced by smaller tabu lists in TL-based APs. Shorter trajectories were clearly preferred in TS-based APs. Nevertheless, the best-performing mixed APs assumed also longer trajectories for the TS constituent algorithms, since running time was sparred by the ILS constituent algorithms of the AP.

Future work will consider further refinements of the AP as well as more extensive investigations of the identified trade-offs among their different properties.

Acknowledgements. Research is partially supported by the Paul and Heidi Brown Preeminent Professorship in Industrial & Systems Engineering, University of Florida.

References

1. Ang, M., Arasu, K., Ma, S., Strassler, Y.: Study of proper circulant weighing matrices with weigh 9. Discrete Math. **308**, 2802–2809 (2008)
2. Arasu, K., Dillon, J., Jungnickel, D., Pott, A.: The solution of the waterloo problem. J. Comb. Theor. Ser. A **71**, 316–331 (1995)
3. Arasu, K., Gulliver, T.: Self-dual codes over fp and weighing matrices. IEEE Trans. Inf. Theor. **47**(5), 2051–2055 (2001)
4. Arasu, K., Gutman, A.: Circulant weighing matrices. Cryptogr. Commun. **2**, 155–171 (2010)

5. Arasu, K., Leung, K., Ma, S., Nabavi, A., Ray-Chaudhuri, D.: Determination of all possible orders of weight 16 circulant weighing matrices. Finite Fields Appl. **12**, 498–538 (2006)
6. Chiarandini, M., Kotsireas, I., Koukouvinos, C., Paquete, L.: Heuristic algorithms for hadamard matrices with two circulant cores. Theoret. Comput. Sci. **407**(1–3), 274–277 (2008)
7. Cousineau, J., Kotsireas, I., Koukouvinos, C.: Genetic algorithms for orthogonal designs. Australas. J. Comb. **35**, 263–272 (2006)
8. van Dam, W.: Quantum algorithms for weighing matrices and quadratic residues. Algorithmica **34**, 413–428 (2002)
9. Eades, P.: On the existence of orthogonal designs. Ph.D. thesis, Australian National University, Canberra (1997)
10. Eades, P., Hain, R.: On circulant weighing matrices. Ars Comb. **2**, 265–284 (1976)
11. Gendreau, M., Potvin, J.Y.: Tabu search. In: Gendreau, M., Potvin, J.Y. (eds.) Handbook of Metaheuristics, pp. 41–59. Springer, New York (2010)
12. Geramita, A., Sebery, J.: Orthogonical Designs: Quadratic Forms and Hadamard Matrices. Lecture Notes in Pure and Applied Mathematics. Marcel Dekker, Inc., New York (1979)
13. Glover, F.: Tabu search - part I. ORSA J. Comput. **1**, 190–206 (1989)
14. Glover, F.: Tabu search - part II. ORSA J. Comput. **2**, 4–32 (1990)
15. Glover, F., Laguna, M.: Tabu Search. Kluwer Academic Publishers, Norwell (1997)
16. Gomes, C.P., Selman, B.: Algorithm portfolio design: theory vs. practice. In: Proceedings of Thirteenth Conference on Uncertainty in Artificial Intelligence, pp. 190–197 (1997)
17. Huberman, B.A., Lukose, R.M., Hogg, T.: An economics approach to hard computational problems. Science **27**, 51–53 (1997)
18. Kotsireas, I.S., Parsopoulos, K.E., Piperagkas, G.S., Vrahatis, M.N.: Ant-based approaches for solving autocorrelation problems. In: Dorigo, M., Birattari, M., Blum, C., Christensen, A.L., Engelbrecht, A.P., Groß, R., Stützle, T. (eds.) ANTS 2012. LNCS, vol. 7461, pp. 220–227. Springer, Heidelberg (2012)
19. Kotsireas, I.: Algorithms and metaheuristics for combinatorial matrices. In: Pardalos, P., Du, D.Z., Graham, R.L. (eds.) Handbook of Combinatorial Optimization, pp. 283–309. Springer, New York (2013)
20. Kotsireas, I., Koukouvinos, C., Pardalos, P., Shylo, O.: Periodic complementary binary sequences and combinatorial optimization algorithms. J. Comb. Optim. **20**(1), 63–75 (2010)
21. Kotsireas, I., Koukouvinos, C., Pardalos, P., Simos, D.: Competent genetic algorithms for weighing matrices. J. Comb. Optim. **24**(4), 508–525 (2012)
22. Kotsireas, I.S., Parsopoulos, K.E., Piperagkas, G.S., Vrahatis, M.N.: Ant-based approaches for solving autocorrelation problems. In: Dorigo, M., Birattari, M., Blum, C., Christensen, A.L., Engelbrecht, A.P., Groß, R., Stützle, T. (eds.) ANTS 2012. LNCS, vol. 7461, pp. 220–227. Springer, Heidelberg (2012)
23. Koukouvinos, C., Seberry, J.: Weighing matrices and their applications. J. Stat. Plan. Infer. **62**(1), 91–101 (1997)
24. Lourenço, H.R., Martin, O.C., Stützle, T.: Iterated local search: framework and applications. In: Gendreau, M., Potvin, J.Y. (eds.) Handbook of Metaheuristics, pp. 363–397. Springer, New York (2010)
25. Peng, F., Tang, K., Chen, G., Yao, X.: Population-based algorithm portfolios for numerical optimization. IEEE Trans. Evol. Comp. **14**(5), 782–800 (2010)
26. Pham, D., Karaboga, D.: Intelligent Optimisation Techniques: Genetic Algorithms, Tabu Search, Simulated Annealing and Neural Networks. Springer, London (2000)

27. Souravlias, D., Parsopoulos, K.E., Alba, E.: Parallel algorithm portfolio with market trading-based time allocation. In: Proceedings International Conference on Operations Research 2014 (OR 2014) (2014)
28. Souravlias, D., Parsopoulos, K.E., Kotsireas, I.S.: Circulant weighing matrices: a demanding challenge for parallel optimization metaheuristics. Optim. Lett. (2015)
29. Strassler, Y.: The classification of circulant weighing matrices of weight 9. Ph.D. thesis, Bar-Ilan University (1997)

Engineering Hybrid DenseZDDs

Taito Lee[1]($^{(\boxtimes)}$), Shuhei Denzumi[2]($^{(\boxtimes)}$), and Kunihiko Sadakane[2]($^{(\boxtimes)}$)

[1] Department of Creative Informatics, Graduate School of Information Science
and Technology, The University of Tokyo, Tokyo, Japan
ri.taito@ci.i.u-tokyo.ac.jp
[2] Department of Mathematical Informatics, Graduate School of Information Science
and Technology, The University of Tokyo, Tokyo, Japan
{denzumi,sada}@mist.i.u-tokyo.ac.jp

Abstract. ZDDs (Zero-suppressed Binary Decision Diagrams) [Minato 93] have been proposed to store set families compactly. Though more compact than other representations, they still use large amount of memory to support dynamic operations such as taking union and intersection of set families. DenseZDDs and Hybrid DenseZDDs [Denzumi et al. 2014] have been proposed to compress the size of static and dynamic ZDDs, respectively. There exist however no implementations of Hybrid DenseZDDs and their practical performance is unknown. This paper engineers a practical implementation of Hybrid DenseZDDs. Because of our new compression algorithm, our new Hybrid DenseZDDs run in reasonable time using little working memory. Experimental results on the frequent itemset mining problem show that our algorithm uses 33 % of memory compared with a standard ZDD at the cost of 40 % increase in running time.

Keywords: Zero-suppressed binary decision diagram · Succinct data structure

1 Introduction

ZDDs (Zero-suppressed Binary Decision Diagrams) [7] are compact representations of set families (sets of subsets). ZDDs are variants of BDDs (Binary Decision Diagrams) [2] which are used to compactly represent Boolean functions. In this paper we focus on ZDDs; however our techniques can be also applied to BDDs.

In the area of data mining or combinatorial optimization, it is useful to store all candidates of solutions because we want to change the objective function in multi-criterion optimization problems. It is however costly to store the candidates because there is an exponential number of them in some cases. ZDDs can be used to store such data compactly.

Let S be a set and $\mathcal{F}$ be a set family of S, that is, $\mathcal{F} \subset 2^S$. The ZDD representing $\mathcal{F}$ is a directed acyclic graph in which each source-sink path corresponds to an element of $\mathcal{F}$ (its precise definition is given in Sect. 2). By sharing

© Springer International Publishing Switzerland 2016
A.V. Goldberg and A.S. Kulikov (Eds.): SEA 2016, LNCS 9685, pp. 201–216, 2016.
DOI: 10.1007/978-3-319-38851-9_14

nodes corresponding to the same subsets of elements of $\mathcal{F}$, The number of nodes are greatly reduced. More interestingly, because of its canonical form, ZDDs support efficient set operations such as union and intersection of set families without extracting shared nodes. It is therefore possible to perform set operations on set families of exponential size in polynomial time, if the number of nodes of ZDDs is polynomial. There are many applications of ZDDs [8,10].

Though ZDDs are more compact than other representations, they still require a large amount of memory. For example, an implementation of ZDD [9] uses about 28 bytes for each node. Therefore huge main memory is required to represent large set families, which motivates to compress ZDDs. Implementations of ZDDs can be classified into three types [3].

- Dynamic: The ZDD can be modified. New nodes can be added to the ZDD.
- Static: The ZDD cannot be modified. Only query operations are supported.
- Freeze-dried: All the information of the ZDD is stored, but it cannot be used before restoration.

There is a freeze-dried representation of BDDs/ZDDs which compresses a node into 1–2 bits [4]. A static representation called *DenseZDD* has been proposed [3], which uses 1/5 to 1/10 of memory as compared with a dynamic ZDD [9]. In their paper, Hybrid DenseZDDs are also proposed to support dynamic operations by combining static DenseZDDs and ordinary dynamic ZDDs. However they gave only an algorithm to check existence of a node in a Hybrid DenseZDD and they did not implement it. A direct implementation will not reduce the working memory because to convert a dynamic ZDD into a static one, it will use the amount of memory proportional to that of a dynamic ZDD. It is therefore not clear if the working memory of Hybrid DenseZDDs is smaller than ordinary ZDDs.

In this paper, we give the first implementation of Hybrid DenseZDDs. Our new compression algorithm enables us to reduce the working memory. As a result, Hybrid DenseZDDs use much smaller memory than ordinary ones with little increase in running time. Our main result is the following theorem:

Theorem 1. *Let m and n be the number of DZ-nodes and Z-nodes before compression, respectively, and we refer to the number of DZ-nodes and Z-nodes after compression as m' and n', respectively. The number of nodes including dummy nodes in the new zero-edge tree is denoted as u. Our algorithm re-compress takes $\mathcal{O}(m + n + n' \log n' + u)$ time, and uses $3m + o(m) + \mathcal{O}(n'(\log m' + \log n'))$ bits.*

Our algorithm is superior to the existing one [3] theoretically and practically. In theory, the working space of our algorithm is roughly the size of the compressed representation of a ZDD plus $\mathcal{O}(n \log n)$, while that of [3] is $\mathcal{O}((m + n) \log(m + n))$. In practice, experimental results on the N-queen, the frequent itemset mining, and logic function minimization problems are given to show effectiveness of our algorithm.

2   Preliminaries

2.1   Zero-Suppressed Binary Decision Diagrams (ZDDs)

A *zero-suppressed binary decision diagram* (a ZDD) [7] is a variant of a binary decision diagram [2], suitable for manipulating set families. Let $S = \{1, 2, \ldots, k\}$ be a set and $\mathcal{F}$ be a set family of S. The ZDD representing $\mathcal{F}$ is a directed acyclic graph G having two terminal nodes. Each nonterminal node v has an integer label $index(v) \in \{1, \ldots, k\}$ called the *index* of v. For any nonterminal node v, $index(v)$ is larger than the indices of its children. Each nonterminal node v always has two children, denoted by $zero(v)$ and $one(v)$ and called the *0-child* and *1-child*, respectively. The edges from nonterminals to their 0-child (1-child resp.) are called *0-edges* (*1-edges* resp.). The two terminal nodes are labeled **0** and **1** and called *0-terminal node* and *1-terminal node*, respectively. We define $triple(v) = \langle index(v), zero(v), one(v) \rangle$, called the *attribute triple* of v. We define the *size* of the graph, denoted by $|G|$, as the number of its nonterminals. Figure 1 shows an example of a ZDD.

We define the *join* of families $F_1, F_2 \subset S$ as $F_1 \sqcup F_2 = \{ S_1 \cup S_2 \mid S_1 \in F_1, S_2 \in F_2 \}$. Then the set family $\mathcal{F}(v)$ represented by a node $v \in G$ is defined as follows.

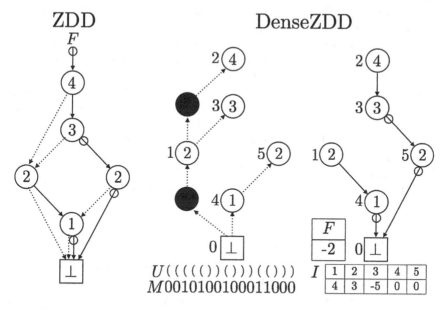

Fig. 1. (Left) The ZDD representing $\{\{4, 3, 2\}, \{4, 3, 1\}, \{4, 3\}, \{4, 2, 1\}, \{2, 1\}, \emptyset\}$ and the corresponding DenseZDD. Terminal nodes are denoted by squares, and nonterminal nodes are denoted by circles. The 0-edges are denoted by dotted arrows, and the 1-edges are denoted by solid arrows. (Middle) The zero-edge tree of the corresponding DenseZDD. Black nodes are dummy nodes. (Right) The one-child array I and the tree represented by it.

Table 1. Basic operations supported by ZDD.

$index(v)$	Returns the index of node v
$zero(v)$	Returns the 0-child of node v
$one(v)$	Returns the 1-child of node v
$getnode(i, v_0, v_1)$	Generates (or makes a reference to) a node v with index i and two child nodes $v_0 = zero(v)$ and $v_1 = one(v)$.
	If such a node does not exist, return the 0-terminal node

Definition 1 (set family represented by ZDD). *The set family $F(v)$ represented by a node v of a ZDD G is defined as follows: (1) If v is the 1-terminal node, $F(v) = \{\emptyset\}$, (1) If v is the 0-terminal node, $F(v) = \emptyset$, (3) If v is a nonterminal node, $F(v) = (\{index(v)\} \sqcup F(one(v))) \cup F(zero(v))$.*

Paths from a node v to the 1-terminal node **1** and elements of $F(v)$ have one-to-one correspondence. Namely, let $v_1, v_2, v_3, \ldots, v_k$ be the nodes on the path from v to **1** such that their 1-edges are on the path. Then the set $\{index(v_1), index(v_2), \ldots, index(v_k)\}$ is contained in $F(v)$.

In ZDD, we apply the following two reduction rules to obtain the minimal graph: (a) Zero-suppress rule: A nonterminal node whose 1-child is the 0-terminal node is removed. (b) Sharing rule: Two or more nonterminal nodes having the same attribute triple are merged together. If we apply the two reduction rules as much as possible, then we obtain a canonical form for a given set family.

We can further reduce the size of ZDDs by using a type of *attributed edges* [11] named *0-element edges*. A 0-element edge is a 1-edge with an *empty set flag*. This edge represents the union of the empty set $\emptyset$ and the set family represented by the node pointed to by the edge. To distinguish 0-element edges and normal 1-edges, each nonterminal node v has an $\emptyset$-*flag empflag(v)*. If $empflag(v) = 1$, the 1-edge from v is an 0-element edge. In the figure, 0-element edges have circles at their starting points.

Table 1 summarizes basic operations of ZDDs. The operations $index(v)$, $zero(v)$, and $one(v)$ do not create new nodes. Therefore they can be done on a static ZDD.

2.2   Succinct Data Structures

For a bit vector $B[0..n-1] \in \{0,1\}^n$, we define the following operations: $rank_c(B, i)$ where $c \in \{0, 1\}$ and $0 \le i < n$ is the number of c's in $B[0..i]$, and $select_c(B, j)$ where $c \in \{0, 1\}$ and $1 \le j \le rank_c(B, n-1)$ is the position of j-th c from the left in B. These operations are done in constant time using a data structure called *FID* of $n + o(n)$ bits on the word-RAM [14]. We also define the successor $succ_c(B, i) := select_c(B, rank_c(B, i) + 1)$ (the position of the leftmost c in $B[i..n-1]$) and the predecessor $pred_c(B, i) := select_c(B, rank_c(B, i))$ (the position of the rightmost c in $B[0..i]$). In practice, these operations are directly implemented without using *rank* and *select* operations to improve performance.

An ordered tree T with n nodes can be represented in $2n + o(n)$ bits such that many tree operations are done in constant time [13]. The tree T is represented by a balanced parentheses sequence (BP) $U[0..2n-1]$. Each node in T is represented by a pair of parentheses in U. The node is identified with the position of the open parenthesis. We use the following operations:

- $depth(U, v)$: depth of node v
- $parent(U, v)$: position in U of the parent node of v
- $level\_ancestor(U, v, d)$: position in U of the ancestor node of v with depth d
- $degree(U, v)$: degree (number of children) of node v
- $child(U, v, i)$: position in U of the i-th child of node v.

They can be done in constant time for static trees and $\mathcal{O}(\log n / \log \log n)$ time for dynamic trees [13]. There are $\mathcal{O}(\log n)$ time implementations of static and dynamic trees [1,5].

3 DenseZDDs

A DenseZDD $DZ = \langle U, M, I \rangle$ is a space-efficient data structure representing a ZDD G [3], and composed of three data structures: a zero-edge tree U, a dummy node vector M, and a one-child array I.

The spanning tree of ZDD G formed by the 0-edges is called the *zero-edge tree* of G. In a zero-edge tree, all 0-edges are reversed and the 0-terminal node becomes the root of the tree. The preorder rank of each node is used to identify the node.

We call nodes in the original ZDD as *real nodes*. We insert dummy nodes on each 0-edge to guarantee that the depth of every real node v in the zero-edge tree equals $index(v)$. We define the depth of the 0-terminal node, the root of this tree, to be 0. We use the balanced parentheses sequence U to encode the zero-edge tree with dummy nodes. An example of a zero-edge tree and its BP are shown in Fig. 1. Black circles are dummy nodes and the number next to each node is its preorder rank.

A bit vector M of the same length as U is used to distinguish dummy nodes and real nodes. The i-th bit is 1 if and only if the i-th parenthesis of U is '(' and its corresponding node is a real node. We construct the FID for M.

An integer array to represent the 1-child of each node is called the *one-child array* and denoted by I. The i-th element of the array is the preorder rank of the 1-child of the nonterminal node whose preorder rank is i. We negate the value if the $\emptyset$-flag of the node is set.

Among the basic operation of ZDDs, $index(v)$, $zero(v)$, and $one(v)$ are easily implemented by using succinct data structures, whereas $getnode(i, v_0, v_1)$ is not trivial. To support it efficiently, children of a node in the zero-edge tree are sorted in the order of preorder ranks of their 1-children. Then we can perform a binary search to find the answer. However, because of this rule, it is difficult to compute the zero-edge tree in space-efficient manner.

Because the above data structure is static, we cannot use it to manipulate dynamic set families. Denzumi et al. [3] proposed a hybrid scheme to combine ordinary ZDDs and DenseZDDs to support dynamic operations. If a ZDD has changed, new nodes are stored as ordinary ZDDs. Children of ordinary ZDD nodes are in either ordinary ZDDs or DenseZDDs, while children of DenseZDD nodes are always in DenseZDDs. If the number of ordinary nodes increases, we *compress* the data structure. That is, we construct the DenseZDD including all the nodes and delete ordinary ZDDs. Note that though this scheme is suggested in their paper, the details are not given. The working memory of their algorithm is larger than the space used if all the nodes are represented by ordinary ZDDs. In this paper, we give a space-efficient algorithm for the compress operation.

4   New Algorithms for Hybrid DenseZDDs

In this section, we present a space-efficient algorithm for converting a multi-rooted Hybrid DenseZDD into a DenseZDD. Our algorithm named *re-compress* takes a list of pointers to nodes in a Hybrid DenseZDD as arguments. The purposes of the arguments are twofold: (1) The pointers show which nodes are currently used, and other nodes are collected as garbage; (2) Since the pointers will change after compression, user programs must know these changes of pointers. This second purpose is not discussed in their original paper [3].

For the sake of simplicity, we define some terms. We refer to the nodes in the DenseZDD-section and the ZDD-section as *DZ-nodes* and *Z-nodes*, respectively. We denote 0-edges from Z-node v to DZ-node u as *junction edges*, v as *Z-junction node* and u as *DZ-junction node*. We represent a non-terminal node v as a 0^r-*child* of $zero(v)$. The numbers of the DZ-nodes and the Z-nodes are represented as m and n, respectively. The identification number within $[FirstZID, \ldots, FirstZID + n - 1]$ is assigned to each Z-node, where $FirstZID = select_1(M, m)$. The identification number of the node v is denoted by $\bar{v}$. The identification number $\bar{v}$ can be obtained easily from v, and *vice versa*. We refer to as T_0 and T_1 a zero-edge tree of DenseZDD before and after the compression, respectively. The preorder rank of the node v on zero-edge tree T is denoted by $prerank(T, v)$. The preorder rank of DZ-node v is calculated as $prerank(T_0, v) = rank_1(M, \bar{v})$.

The algorithm *re-compress* is composed of three parts; (1) We decide which nodes should remain after the compression. (2) We define preorder ranks of the nodes on a new zero-edge tree. (3) We construct a new DenseZDD. Part (3) is simple and omitted due to lack of space.

4.1   Selection of Nodes to Remain After Compression

First, we determine which nodes should remain after the compression. From root nodes, which are the input of *re-compress*, we traverse the Hybrid DenseZDD in depth-first order and set flags on visited nodes. Two FIDs, *dzrefflags* and *zrefflags* are used to store flags. When we visit a DZ-node v, we set a flag on the

($prerank(T_0, v)$)-th bit of *dzrefflags*. When we visit a Z-node u, we set a flag on the ($\bar{u} - FirstZID$)-th bit of *zrefflags*. In the same way, the information about junction nodes is also stored in two FIDs, *dzjunctionflags* and *zjunctionflags*. After traversal, we index these four FIDs.

The direction of the 0-edges in the DenseZDD-section is the opposite to those in the ZDD-section. Therefore, we must reverse the 0-edges in the ZDD-section to compress a Hybrid DenseZDD. On traversal of the Hybrid DenseZDD, we reverse the 0-edges in the ZDD-section. All Z-nodes have a field named *ZERO* to store their 0-children, and a field to represent a chaining hash table for *getnode*. We reuse these fields to reverse the 0-edges for the sake of saving memory. When the traversal finishes, the ZERO field of a Z-node v stores the first 0^r-child c of v, and the field of c in the hash table stores the sibling node of c. Since DZ-nodes have no ZERO fields, we use an array *dzarray* to store the first 0^r-child among the Z-nodes of each DZ-junction node v. In more detail, we store the number $rank_1(zjunctionflags, \bar{c} - FirstZID)$ as the ($rank_1(dzjunctionflags, \bar{v})$)-th element of *dzarray*, where c is the first 0^r-child among the Z-nodes of the DZ-junction node v.

Let m' and n' be the number of the DZ-nodes and the Z-nodes that remain after the compression, respectively. First, we analyze the space complexity for this operation. Two FIDs *dzrefflags* and *dzjunctionflags* use $n + o(n)$ bits, respectively, and *zrefflags* and *zjunctionflags* use $m + o(m)$ bits, respectively. The array *dzarray* uses $n' \log n'$ bits at most, since the number of DZ-junction nodes is n' at most. In total, we need $2(m + n) + o(m + n) + n' \log n'$ bits. Next, let us consider the time complexity. In the traversal of the Hybrid DenseZDD, we visit ($m' + n'$) nodes and it takes only constant time at each node. Therefore the time complexity is $\mathcal{O}(m + n)$.

4.2 Decision of Preorder Ranks

By the traversal of the Hybrid DenseZDD, we obtained a zero-edge tree T' of $m' + n'$ nodes. The new zero-edge tree T_1 can be obtained by ordering the nodes of T' under the constraint of DenseZDDs. For simplicity, we refer to as T_{DZ} a zero-edge tree obtained by removing Z-nodes from T'. It is self-evident that T_{DZ} is the zero-edge tree generated by pruning T_0 of unused nodes. The preorder rank of the DZ-node v in T_{DZ} is calculated as $prerank(T_{DZ}, v) = rank_1(dzrefflags, prerank(T_0, v))$ in constant time. For preparation of the algorithm that assigns new preorder ranks, we prove Theorem 2.

Theorem 2. *Two nodes v_1, v_2 are DZ-nodes that remain after compression. It holds the following equation:*

$$prerank(T_0, v_1) < prerank(T_0, v_2) \Rightarrow prerank(T_1, v_1) < prerank(T_1, v_2).$$

Proof. If v_1 is the ancestor of v_2, this theorem obviously holds. We consider the other cases. We here focus on the nodes p_1 and p_2, which are the child nodes of the lowest common ancestor of v_1 and v_2 and the subtrees of which include v_1

and v_2, respectively. We note that it holds $prerank(T_1, v_1) < prerank(T_1, v_2)$ if and only if $prerank(T_1, p_1) < prerank(T_1, p_2)$ is satisfied.

The assumption of the theorem indicates $index(p_1) \geq index(p_2)$, and we divide it into the following cases.

1. If $index(p_1) > index(p_2)$, it obviously holds that $prerank(T_1, p_1) < prerank(T_1, p_2)$.

2. If $index(p_1) = index(p_2)$, we focus on $prerank(T_0, one(p_1))$ and $prerank(T_0, one(p_2))$. We note that it holds $prerank(T_0, one(p_1)) \geq prerank(T_0, one(p_2))$ from the assumption.

   (a) If $prerank(T_0, one(p_1)) = prerank(T_0, one(p_2))$, it holds that $one(p_1) = one(p_2)$ and $empflag(one(p_1)) > empflag(one(p_2))$. Therefore, the theorem is obviously satisfied.

   (b) If $prerank(T_0, one(p_1)) > prerank(T_0, one(p_2))$. In this case, what we want to prove is the equation $prerank(T_1, one(p_1)) > prerank(T_1, one(p_2))$. We regard $one(p_1)$ and $one(p_2)$ as v_2 and v_1, respectively, and repeat the discussion above. After repeating at most k times, where k indicates the number of items used in the Hybrid DenseZDD, we can prove the equation $prerank(T_1, one(p_1)) > prerank(T_1, one(p_2))$, since $index(v_1) > index(one(p_1))$ and $index(v_2) > index(one(p_2))$ hold, and the number of nodes whose item index is 1 is at most 1.   □

Theorem 2 implies that we need not order nodes in T_{DZ} to obtain T_1.

Calculation of Subtree Sizes. For the efficient assignment of preorder ranks, it is necessary to obtain the size of each subtree in T' efficiently. For this purpose, we prepare two arrays named *zarray1* and *starray*. The array *zarray1* stores the size of the subtree rooted at each Z-node, and the array *starray* is the index used to obtain the size of the subtree rooted at each DZ-node. Algorithms 1 and 2 display how to make *zarray1* and *starray*, and Algorithm 3 exhibits how to calculate the size of the subtree of T' rooted at v. We can execute Algorithm 3 in constant time.

Let us consider the space complexity for *zarray1* and *starray*. The length of *zarray1* is n' and the length of *starray* is at most n', since the number of DZ-junction nodes is at most n'. The elements of these arrays are within $[1, \ldots, n']$. Therefore, *zarray1* and *starray* use $n' \log n'$ and at most $n' \log n'$ bits, respectively.

Algorithm 1. *calcStsize(v)*: store the size of the subtree of T' rooted at v in *zarray1* and return it.

1: $stsize \leftarrow 1$
2: **for** c: 0^r-children of v **do**
3:     $stsize \leftarrow stsize + calcStsize(c)$
4: **end for**
5: $idx \leftarrow rank_1(zrefflags, \bar{v} - FirstZID) - 1$
6: $zarray1[idx] \leftarrow stsize$
7: **return** $stsize$

Algorithm 2. *makeStsizeIndexes*: how to make *zarray1* and *starray*.

1: $sum \leftarrow 0$
2: **for** $i = 1$ **to** number of DZ-junction nodes **do**
3:    $\bar{v} \leftarrow select_1(zjunctionflags, i)$ /* v is the i-th DZ-junction node. */
4:    **for all** 0^r-children c of v that is a Z-node **do**
5:       $sum \leftarrow sum + calcStsize(c)$
6:    **end for**
7:    $starray[i - 1] \leftarrow sum$
8: **end for**

Algorithm 3. $stsize(v)$: gain the size of the subtree of T' rooted at v.

1: **if** v is a Z-node **then**
2:    **return** $zarray1[rank_1(zrefflags, \bar{v})]$
3: **else**
4:    /* v is a DZ-node. */
5:    $v\_rank \leftarrow prerank(T_0, v)$
6:    $open\_p \leftarrow \bar{v}$ /* v's position in U */
7:    $close\_p \leftarrow find\_close(U, open\_p)$
8:    $last\_rank \leftarrow rank_1(M, close\_p)$ /* the preorder rank of the last DZ-node in the subtree rooted at v */
9:    $from \leftarrow rank_1(dzjunctionflags, v\_rank - 1)$ /* the number of DZ-junction nodes before v */
10:   $to \leftarrow rank_1(dzjunctionflags, last\_rank - 1)$ /* the number of DZ-junction nodes before $last$ */
11:   **return** $rank_1(dzrefflags, last\_rank) - rank_1(dzrefflags, v\_rank) + 1 + starray[to] - starray[from]$
12: **end if**

The Algorithm 2 takes $\mathcal{O}(n')$ time because we visit all the DZ-junction nodes and Z-nodes that remain after the compression and it takes constant time at each node.

Storing Preorder Ranks. We assign numbers from 0 to $m' + n' - 1$ to each node as the preorder ranks on T_1. If we simply use an array to store them, $(m' + n') \log(m' + n')$ bits are needed. Here, we propose a more efficient way. We note that m' is much greater than n' after compressing the Hybrid DenseZDD many times.

First, we reuse *zarray1*, which is the index for the calculation of subtree sizes, to store the preorder ranks of Z-nodes. The subtree size rooted at Z-node v is never referred after the assignment of the preorder rank to v. Therefore, we overwrite the element corresponding to Z-node v with the preorder rank of v. To do this, $n' \log(m' + n')$ bits must be assigned for *zarray1*.

Next, we show how to store the preorder ranks of DZ-nodes, which is based on Theorem 2. To store the preorder ranks of DZ-nodes, we use a FID named *order* of length $m' + n'$. When a preorder rank r is assigned to a Z-node, we set a flag on a r-th bit of *order*. When the assignment of all preorder ranks is

Algorithm 4. $prank(v)$: compute the preorder rank of v on T_1.

1: **if** v is a Z-node **then**
2:     **return** $zarray1[rank_1(zrefflags, \bar{v} - FirstZID)]$
3: **else**
4:     /* v is a DZ-node. */
5:     **return** $select_0(order, prank(T_{DZ}, v))$
6: **end if**

Algorithm 5. $lastDZ(v)$: obtain the lastDZ-node of v.

Require: v is a Z-node
1: $r \leftarrow zarray2[rank_1(zrefflags, \bar{v} - FirstZID)]$
2: **if** v is a Z-junction node **then**
3:     $\bar{v}_\ell \leftarrow select_1(dzrefflags, r)$
4:     **return** v_ℓ
5: **else**
6:     $\bar{v}_a \leftarrow select_1(zjunctionflags, r) + FirstZID$ /* v_a is the ancestral Z-junction node of v. */
7:     **return** $lastDZ(v_a)$
8: **end if**

Algorithm 6. $compareSameIndexes(v_1, v_2)$: return whether it holds $prerank(v_1) < prerank(v_2)$ or not, where $index(v_1) = index(v_2)$ is satisfied.

1: **if** Both v_1 and v_2 are DZ-nodes **then**
2:     **return** $\overline{v_1} < \overline{v_2}$
3: **else**
4:     $one1 \leftarrow one(v_1)$
5:     $one2 \leftarrow one(v_2)$
6:     /* The calculation of the preorder ranks of $one1$ and $one2$ have been already done. */
7:     **if** Both $one1$ and $one2$ are DZ-nodes **then**
8:         **return** $\langle \overline{one1}, empflag(one1) \rangle > \langle \overline{one2}, empflag(one2) \rangle$
9:     **else if** Both $one1$ and $one2$ are Z-nodes **then**
10:        **return** $\langle prank(one1), empflag(one1) \rangle > \langle prank(one2), empflag(one2) \rangle$
11:     **else if** $one1$ is Z-node **then**
12:        **return** $\overline{lastDZ(one1)} \geq \overline{one2}$
13:     **else**
14:        /* $one2$ is a Z-node. */
15:        **return** $\overline{one1} > \overline{lastDZ(one2)}$
16:     **end if**
17: **end if**

finished, the number of 1 in $order$ is n'. The preorder rank of the DZ-node v can be calculated as $prerank(T_1, v) = select_0(order, prerank(T_{DZ}, v))$.

Algorithm 4 displays how to obtain the preorder rank of the node v on T_1. Here, we note that we cannot acquire the preorder ranks of DZ-nodes unless the construction of $order$ is finished.

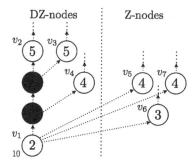

Fig. 2. Assignment of the new prerank orders.

Comparing Preorder Ranks. To obtain T_1, which is the zero-edge tree of the DenseZDD after compression, we need to sort the 0^r-children of each node in the descending dictionary order of their triplets $\langle index(), prank(one()), empflag(one())\rangle$. When there are plural child nodes having the same item index for one parent node, we need to compare $prank(one())$ for the 0^r-children; however, we cannot gain the preorder rank of the DZ-node v on T_1, since the construction of $order$ has not been finished yet.

Now, we discuss how to compare preorder ranks $prank(v_1)$ and $prank(v_2)$ for two nodes v_1 and v_2. When both v_1 and v_2 are DZ-nodes, we just need to compare $prank(T_0, v_1)$ and $prank(T_0, v_2)$ because of Theorem 2. When both v_1 and v_2 are Z-nodes, we can directly calculate $prank(v_1)$ and $prank(v_2)$ and compare them. When v_1 is a DZ-node and v_2 is a Z-node, we compare the two preorder ranks as follows: First, we find a DZ-node visited just before v_2 in the preorder of T_1 and name it v_{last}; next, we compare $prank(T_0, v_1)$ and $prank(T_0, v_{\text{last}})$. We refer to v_{last} as the *lastDZ-node* of v_2.

We prepare an array $zarray2$ as the index for efficiently calculating the lastDZ-node of each Z-node. The index $zarray2$ is the array of size n', and its elements corresponds to each Z-node one to one. If a Z-node v is a Z-junction node, we register the number $rank_1(dzrefflags, \bar{v}_\ell)$ as the element corresponding to v, where v_ℓ is the lastDZ-node of v; otherwise, we store the number $rank_1(zjunctionflags, \bar{v}_a - FirstZID)$, where v_a is the ancestral Z-junction node of v. The space complexity for $zarray2$ is $\mathcal{O}(n' \log(m' + n'))$, since the elements of $zarray2$ are within $[1, \ldots, \max(m', n')]$. We exhibit the algorithm for obtaining the lastDZ-node of a Z-node v using $zarray2$ in Algorithm 5. This processing can be executed in constant time.

When we initialize $zarray2$, we set the number $rank_1(zjunctionflags, \bar{v}_a)$ as the element corresponding to a Z-node v, where v_a is the ancestral Z-junction node of v. This initialization takes $\mathcal{O}(n')$ time. After this, we store the lastDZ-node of each Z-junction node while we allocate preorder ranks to all nodes by traversing T'. We can determine the lastDZ-node of each Z-junction node v when the allocation of the preorder rank to v is finished.

The methodology of comparing $\langle prank(one()), empflag(one())\rangle$ for the two nodes v_1 and v_2 having the same item index are displayed in Algorithm 6. This comparator can be executed in constant time.

Assignment of New Preorder Ranks. We explain the algorithm for allocating preorder ranks on T_1 by taking an example. We display the process of the traversal in Fig. 2. This figure shows a part of T' just after allocating preorder rank 10 to the node v_1.

First, we partition 0^r-children nodes of v_1 in descending order from the node having the largest item index. We sort only the Z-nodes v_5, v_6 and v_7 in the order of larger item indexes, and partition the child nodes while merging DZ-nodes and Z-nodes. By doing this, we gain the partition as $[v_2, v_3], [v_4, v_5, v_7], [v_6]$. We note that, since Z-nodes v_5, v_6 and v_7 are stored in a list, we need to change the connection of the list at this time. Owing to the statement of Theorem 2, we allocate 11 as its preorder rank to v_2. After setting the preorder ranks on the subtree rooted at v_2, the number $(10 + stsize(v_2) + 1)$ will be designated as the preorder rank of v_3.

The node to be traversed next is either one of v_4, v_5, and v_7; however, we cannot decide it because preorder ranks might not have been assigned to three nodes $one(v_4)$, $one(v_5)$ and $one(v_7)$ yet. Therefore, we cannot compare their $prank(one())$ at this time. Thus, we stop the process to these nodes and save the current state. As the memory for saving the state, we use lists $L_1, \ldots, L_k$ initialized empty, where the number k is the item number of the current Hybrid DenseZDD. Since the item indexes v_4, v_5, v_7 are 4, we add $\langle DZleft, DZright, Znodes,\ rank,\ prev,\ isParent \rangle = \langle v_4, v_4, [v_5, v_7], 10 + stsize([v_2, v_3]) + 1, v_1, true \rangle$ to L_4, where $DZleft$ and $DZright$ denote the start and the last of the DZ-nodes the preorder rank allocation of which is deferred, $Znodes$ the array of Z-nodes the allocation of which is deferred, and $rank$ the minimum value of preorder rank to be allocated to the nodes. The area $prev$ and $isParent$ are the storage area for the reconnection of the list of the Z-nodes. In more detail, $prev$ indicates the parent node or the last Z-node in the previous partition and $isParent$ is the flag indicating whether $prev$ is the parent node or not. We pick out the information stored in L_i $(i = 1, \ldots, k)$ from L_1 to L_k and resume the allocation of preorder ranks. Here, it should be noted that we have already finished the allocation of preorder ranks to the nodes the item indexes of which are less than i when we pick up the information from L_i; therefore, we can sort the nodes in all the elements of L_i by comparing their pairs $\langle prank(one()), empflag(one()) \rangle$.

As the elements in $zarray2$ corresponding to the nodes v_5, v_7, we store the last node in the subtree of T_{DZ} rooted at v_3, which is the current provisional lastDZ-node of v_5 and v_7. After that, we allocate preorder rank $(10 + stsize([v_2, v_3, v_4, v_5, v_7]) + 1)$ to v_6, and store the last node in the subtree of T_{DZ} rooted at v_4 as the lastDZ-node of v_6 in $zarray2$.

In the worst case, the number of elements in $L_1, \ldots, L_k$ is n', which is the number of the Z-nodes that remain after compression. Therefore, $L_1, \ldots, L_k$ use $\mathcal{O}(n'(\log m' + \log n'))$.

We discuss the time complexity of the algorithm. First, it takes $\mathcal{O}(n')$ time to initialize $zarray2$ as mentioned above. Next, while traversing T', we visit $m' + n'$ nodes, and it takes only constant time at each node except for sorting the nodes.

Now, let us consider the total time complexity for the sorting. In the algorithm, we sort the nodes in two steps; (1) We sort only Z-nodes. (2) We merge Z-nodes and DZ-nodes. It takes $\mathcal{O}(n' \log n')$ time in the worst case at the first step, while only $\mathcal{O}(m' + n')$ time at the second step. Therefore, we need $\mathcal{O}(m' + n' \log n')$ time for the sorting. Finally, it takes $\mathcal{O}(m' + n')$ time to index *order*.

5   Experimental Results

We give some experimental results to show the effectiveness of our Hybrid DenseZDDs. We compare ours with an efficient implementation of ZDDs [9]. We ran three types of experiments; one is to construct ZDDs representing solutions of the N-queen problem, one is to incorporate our Hybrid DenseZDDs into the LCM-over-ZDD algorithm [12], which is a fast and space-efficient algorithm for the frequent itemset mining problem, and the last one is the weak-division algorithm [6] for converting two-level logic functions to multilevel ones. We used a Linux machine with 64 GB memory and Intel Xeon CPU E5-2650 (2.60 GHz). Our algorithm uses single core. We used g++ compiler 4.7.2.

Our algorithm has a parameter θ. Let m and n be the numbers of nodes in DenseZDD and ordinary ZDD, respectively. We compress the Hybrid DenseZDD into a static DenseZDD every time $n/m > \theta$. The smaller θ is, the more frequent the compress operation is used. Therefore the working space is reduced if θ is small. Our algorithm uses smaller working space than ordinary ones, whereas it takes more running time. We therefore define a measure *improvement ratio* to evaluate effectiveness of our algorithm. Let T_A and S_A denote the running time and working space of an algorithm A, respectively. We define the improvement ratio of our algorithm (A) to an existing one (B) as $\frac{S_B}{S_A} \cdot \frac{T_B}{T_A}$. The greater the value is, the better our algorithm is.

5.1   N-queens

N-queen is a well-known problem of counting the number of solutions for placing N queens on an $N \times N$ board so that no two queens share the same row, column, or diagonal. We can solve the problem by using set operations on ZDDs.

Table 2 shows running time and working space of algorithms using ordinary ZDDs and our Hybrid DenseZDDs. Our algorithm using Hybrid DenseZDDs use less than 6 % of memory of that using ZDDs, though the running time is about five to seven times longer. If $\theta = 0.3$, the improvement ratio is $3.06, 3.66, 4.98$ for $N = 13, 14, 15$, respectively.

5.2   LCM over Hybrid DenseZDD

Frequent itemset mining is the problem to find subsets of a set in database records whose numbers of occurrences are more than a given threshold (minimum support). The LCM-over-ZDD algorithm uses a ZDD to store frequent subsets.

Table 2. Finding solutions of N-queens. $|G|$ and $|\mathcal{F}|$ denote the size of the ZDD and the number of solutions, respectively.

N	13		14		15	
$|G|$, $|\mathcal{F}|$	204781	73712	911420	365596	4796502	2279184
	Time (s)	Space (KB)	Time (s)	Space (KB)	Time (s)	Space (KB)
ZDD	21.19	1107552	132.40	4644772	1059.05	34566152
$\theta = 10$	64.87	297028	432.89	2181084	3136.46	9045172
5	81.19	288768	516.41	1205464	3670.22	8486984
2	91.17	171440	622.45	594828	4220.90	4370688
1.5	96.50	170668	642.42	569704	4486.29	2444740
1	104.14	99908	636.15	327512	4558.42	2310304
0.8	106.04	95036	641.97	329176	4677.31	2308152
0.5	118.79	91376	752.72	340064	5171.41	2317700
0.3	131.23	58424	776.81	215832	5591.37	1312704
0.1	140.43	54204	764.51	213792	6387.34	1315572
0.05	129.95	54604	780.87	215196	5953.22	1279840
0.01	129.59	54524	766.95	215200	6070.43	1279264

Table 3. LCM over ZDD/Hybrid DenseZDD. $|G|$ and $|\mathcal{F}|$ denote the size of the ZDD and the number of frequent subsets, respectively. The suffix ":t" of the database name shows the minimum support t.

Database	T10I4D100K:2		T40I10D100K:100		T40I10D100K:50	
$|G|$, $|\mathcal{F}|$	3270977	19561715	1132008	70470947	6533076	203738242
	Time (s)	Space (KB)	Time (s)	Space (KB)	Time (s)	Space (KB)
ZDD	8.99	165276	104.53	119116	428.41	330004
$\theta = 10$	18.24	170428	132.89	89964	527.87	312132
5	20.82	182692	128.99	121804	546.75	325460
2	24.73	111328	136.94	90080	532.36	231996
1.5	25.71	185292	129.25	85484	602.92	217220
1	25.25	115404	134.29	73140	540.12	218676
0.8	25.46	125688	129.99	71120	535.57	218624
0.5	27.74	134392	131.36	73860	542.18	232304
0.3	34.93	89672	133.14	74780	583.54	166156
0.1	59.61	72356	139.33	60436	593.77	132544
0.05	95.65	64028	146.91	55912	608.54	110372
0.01	387.15	65504	208.45	59996	929.20	110380

Table 4. Logic function minimization.

Data set	16-adder_col		al2_split		apex5	
	Time (s)	Space (KB)	Time (s)	Space (KB)	Time (s)	Space (KB)
ZDD	12.01	1076564	0.24	38216	0.04	3704
$\theta = 10$	46.05	282884	1.20	2508	0.16	2416
5	48.52	162984	0.82	2132	0.07	2072
2	55.33	90476	2.14	2092	0.12	1904
1.5	62.09	96612	14.51	2164	0.15	1864
1	71.09	95424	17.40	2076	0.18	1856
0.8	76.24	52748	23.20	2088	0.19	1840
0.5	92.26	50828	27.27	2092	0.26	1840
0.3	129.23	50800	37.84	2132	0.36	1860
0.1	253.48	51736	59.48	2256	0.64	1872
0.05	406.65	54004	52.22	3560	0.82	1864
0.01	1529.81	75220	154.86	6392	2.76	1920

We replace the ZDD in the algorithm with our Hybrid DenseZDD. The datasets are obtained from Frequent Itemset Mining Dataset Repository[1].

Table 3 shows the results. For the dataset T10I4D100K with threshold 2, our Hybrid DenseZDD ($\theta = 0.3$), the working space is reduced about a half (165276 KB to 89672 KB), at the cost of increase in running time by a factor of 3.88. Therefore the improvement ratio is 0.47. For a larger dataset T40I10D100K, if the threshold is 50, our Hybrid DenseZDD ($\theta = 0.05$) uses 33 % of memory of the ZDD, and the running time increases only 40 %. Therefore the improvement ratio is 2.1.

5.3 Logic Function Minimization

We did experiments of the weak-division algorithm using ZDDs [6] for logic function minimization. The problem is, given a DNF (disjunctive normal form), or sum-of-products, of a logic function, to create a multilevel logics which consists of small number of cubes (products of literals). Because the problem is NP-hard, the algorithm is heuristic. For example, the function $f = abd + ab\bar{e} + ab\bar{g} + cd + c\bar{e} + ch$ can be rewritten as $f = p\bar{d} + pe + ab\bar{g} + ch$ where $p = ab + c$ and the number of cubes is reduced.

We used the data sets in "Collection of Digital Design Benchmarks"[2]. Table 4 shows the results. For the data 16-adder_col, which is the circuit for adding two 16-bit integers, if $\theta = 2.0$, our Hybrid DenseZDD uses about 10 % memory of ZDD, and running time is about 4.6 times longer. Therefore the improvement

---

[1] http://fimi.ua.ac.be/data/.
[2] http://ddd.fit.cvut.cz/prj/Benchmarks/.

ratio is 2.58. For other inputs, the improvement ratio is low because the problem size is small.

6   Concluding Remarks

We have given the first implementation of the Hybrid DenseZDDs. The source codes are available at https://github.com/otita/hdzbdd.git. Our new space-efficient compression algorithm enables us to reduce the working memory greatly. Future work is to give an algorithm for compression whose running time depends only the number of ordinary ZDD nodes.

References

1. Arroyuelo, D., Cánovas, R., Navarro, G., Sadakane, K.: Succinct trees in practice. In: Proceedings of the 11th Workshop on Algorithm Engineering and Experiments (ALENEX), pp. 84–97. SIAM Press (2010)
2. Bryant, R.E.: Graph-based algorithms for Boolean function manipulation. IEEE Trans. Comput. $C-35$(8), 677–691 (1986)
3. Denzumi, S., Kawahara, J., Tsuda, K., Arimura, H., Minato, S.-I., Sadakane, K.: DenseZDD: a compact and fast index for families of sets. In: Gudmundsson, J., Katajainen, J. (eds.) SEA 2014. LNCS, vol. 8504, pp. 187–198. Springer, Heidelberg (2014)
4. Hansen, E.R., Rao, S.S., Tiedemann, P.: Compressing binary decision diagrams. In: Proceedings of the 18th European Conference on Artificial Intelligence (ECAI 2008), pp. 799–800. ACM (2008)
5. Joannou, S., Raman, R.: Dynamizing succinct tree representations. In: Klasing, R. (ed.) SEA 2012. LNCS, vol. 7276, pp. 224–235. Springer, Heidelberg (2012)
6. Minato, S.: Fast factorization method for implicit cube set representation. IEEE Trans. Comput. Aided Des. Integr. Circ. Syst. 15(4), 377–384 (1996)
7. Minato, S.-I.: Zero-suppressed BDDs for set manipulation in combinatorial problems. In: Proceeding of Design Automation Conference (DAC 1993), pp. 272–277. IEEE (1993)
8. Minato, S.-I.: Zero-suppressed BDDs and their applications. J. Softw. Tools Technol. Transf. 3(2), 156–170 (2001)
9. Minato, S.-I.: SAPPORO BDD package. Hokkaido University (2011). unreleased
10. Minato, S.-I., Arimura, H.: Frequent pattern mining and knowledge indexing based on zero-suppressed BDDs. In: Džeroski, S., Struyf, J. (eds.) KDID 2006. LNCS, vol. 4747, pp. 152–169. Springer, Heidelberg (2007)
11. Minato, S.-I., Ishiura, N., Yajima, S.: Shared binary decision diagram with attributed edges for efficient boolean function manipulation. In: Proceedings of the 27th Design Automation Conference (DAC 1990), pp. 52–57. IEEE (1990)
12. Minato, S.-I., Uno, T., Arimura, H.: LCM over ZBDDs: fast generation of very large-scale frequent itemsets using a compact graph-based representation. In: Washio, T., Suzuki, E., Ting, K.M., Inokuchi, A. (eds.) PAKDD 2008. LNCS (LNAI), vol. 5012, pp. 234–246. Springer, Heidelberg (2008)
13. Navarro, G., Sadakane, K.: Fully-functional static and dynamic succinct trees. ACM Trans. Algorithms, 10(3) (2014). Article No. 16. doi:10.1145/2601073
14. Raman, R., Raman, V., Rao, S.S.: Succinct indexable dictionaries with applications to encoding k-ary trees, prefix sums and multisets. ACM Trans. Algorithms 3(4), 43 (2007)

Steiner Tree Heuristic in the Euclidean d-Space Using Bottleneck Distances

Stephan S. Lorenzen and Pawel Winter[✉]

Department of Computer Science, University of Copenhagen,
Universitetsparken 5, 2100 Copenhagen O, Denmark
stephan.lorenzen@gmail.com, pawel@di.ku.dk

Abstract. Some of the most efficient heuristics for the Euclidean Steiner minimal tree problem in the d-dimensional space, $d \geq 2$, use Delaunay tessellations and minimum spanning trees to determine small subsets of geometrically close terminals. Their low-cost Steiner trees are determined and concatenated in a greedy fashion to obtain a low cost tree spanning all terminals. The weakness of this approach is that obtained solutions are topologically related to minimum spanning trees. To avoid this and to obtain even better solutions, bottleneck distances are utilized to determine good subsets of terminals without being constrained by the topologies of minimum spanning trees. Computational experiments show a significant solution quality improvement.

Keywords: Steiner minimal tree · d-dimensional Euclidean space · Heuristic · Bottleneck distances

1 Introduction

Given a set of points $N = \{t_1, t_2, ..., t_n\}$ in the Euclidean d-dimensional space $\mathcal{R}^d$, $d \geq 2$, the *Euclidean Steiner minimal tree* (ESMT) *problem* asks for a shortest connected network $T = (V, E)$, where $N \subseteq V$. The points of N are called *terminals* while the points of $S = V \backslash N$ are called *Steiner points*. The length $|uv|$ of an edge $(u, v) \in E$ is the Euclidean distance between u and v. The length $|T|$ of T is the sum of the lengths of the edges in T. Clearly, T must be a tree. It is called the *Euclidean Steiner minimal tree* and it is denoted by SMT(N). The ESMT problem has originally been suggested by Fermat in the 17-th century. Since then many variants with important applications in the design of transportation and communication networks and in the VLSI design have been investigated. While the ESMT problem is one of the oldest optimization problems, it remains an active research area due to its difficulty, many open questions and challenging applications. The reader is referred to [4] for the fascinating history of the ESMT problem.

The ESMT problem is NP-hard [5]. It has been studied extensively in $\mathcal{R}^2$ and a good exact method for solving problem instances with up to 50.000 terminals is available [13,23]. However, no analytical method can exist for $d \geq 3$ [1]. Furthermore, no numerical approximation seems to be able to solve instances with more

© Springer International Publishing Switzerland 2016
A.V. Goldberg and A.S. Kulikov (Eds.): SEA 2016, LNCS 9685, pp. 217–230, 2016.
DOI: 10.1007/978-3-319-38851-9_15

than 15–20 terminals [7–9,14,21]. It is therefore essential to develop good quality heuristics for the ESMT problem in $\mathcal{R}^d$, $d \geq 3$. Several such heuristics have been proposed in the literature [10,15,17,22]. In particular, [17] builds on the R^2-heuristics [2,20]. They use **D**elaunay tessellations and **M**inimum spanning trees and are therefore referred to as DM-heuristics. The randomized heuristic suggested in [15] also uses Delaunay tessellations. It randomly selects a predefined portion of simplices in the Delaunay tessellation. It adds a centroid for each selected simplex as a Steiner point. It then computes the minimum spanning tree for the terminals and added Steiner points. Local improvements of various kinds are then applied to improve the quality of the solution. It obtains good solutions in R^2. Only instances with $n \leq 100$ are tested and CPU times are around 40 sec. for $n = 100$. The randomized heuristic is also tested for very small problem instances ($n = 10, d = 3, 4, 5$) and for specially structured "sausage" instances ($n < 100, d = 3$). It can be expected that the CPU times increase significantly as d grows since the number of simplices in Delaunay tessellations then grows exponentially.

The goal of this paper is to improve the DM-heuristic in a deterministic manner so that the minimum spanning tree bondage is avoided and good quality solutions for large problem instances can be obtained. Some basic definitions and a resume of the DM-heuristic is given in the remainder of this section. Section 2 discusses how *bottleneck distances* can be utilized to improve the solutions produced by the DM-heuristic. The new heuristic is referred to as the DB-heuristic as it uses both **D**elaunay tessellations and **B**ottleneck distances. Section 3 describes data structures used for the determination of bottleneck distances while Sect. 4 gives computational results, including comparisons with the DM-heuristic.

1.1   Definitions

SMT(N) is a tree with $n - 2$ Steiner points, each incident with 3 edges [12]. Steiner points can overlap with adjacent Steiner points or terminals. Terminals are then incident with exactly 1 edge (possibly of zero-length). Non-zero-length edges meet at Steiner points at angles that are at least 120°. If a pair of Steiner points s_i and s_j is connected by a zero-length edge, then s_i or s_j are also be connected via a zero-length edge to a terminal and the three non-zero-length edges incident with s_i and s_j make 120° with each other. Any geometric network ST(N) satisfying the above degree conditions is called a *Steiner tree*. The underlying undirected graph $\mathcal{ST}(N)$ (where the coordinates of Steiner points are immaterial) is called a *Steiner topology*. The shortest network with a given Steiner topology is called a *relatively minimal Steiner tree*. If ST(N) has no zero-length edges, then it is called a *full Steiner tree* (FST). Every Steiner tree ST(N) can be decomposed into one or more full Steiner subtrees whose degree 1 points are either terminals or Steiner points overlapping with terminals. A reasonable approach to find a good suboptimal solution to the ESMT problem is therefore to identify few subsets $N_1, N_2, \ldots$ and their low cost Steiner trees ST(N_1), ST(N_2), $\ldots$ such that a union of some of them, denoted by ST(N), will be a good approximation of $\mathrm{SMT}(N)$. The selection of the subsets $N_1, N_2, \ldots$ should in particular ensure that $|\mathrm{ST}(N)| \leq |\mathrm{MST}(N)|$ where $\mathrm{MST}(N)$ is the minimum spanning tree of N.

A Delaunay tessellation of N in $\mathcal{R}^d$, $d \geq 2$, is denoted by $\mathrm{DT}(N)$. $\mathrm{DT}(N)$ is a simplicial complex. All its k-simplices, $0 \leq k \leq d+1$, are also called k-*faces* or *faces* if k is not essential. In the following, we will only consider k-faces with $1 \leq k \leq d+1$, as 0-faces are terminals in N. It is well-known that the minimum spanning tree $\mathrm{MST}(N)$ of N is a subgraph of $\mathrm{DT}(N)$ [3]. A face σ of $\mathrm{DT}(N)$ is *covered* if the subgraph of $\mathrm{MST}(N)$ induced by the corners N_σ of σ is a tree. Corners of N_σ are then also said to be *covered*.

The *Steiner ratio* of a Steiner tree $\mathrm{ST}(N)$ is defined by

$$\rho(\mathrm{ST}(N)) = \frac{|\mathrm{ST}(N)|}{|\mathrm{MST}(N)|}$$

The Steiner ratio of N is defined by

$$\rho(N) = \frac{|\mathrm{SMT}(N)|}{|\mathrm{MST}(N)|}$$

It has been observed [23] that for uniformly distributed terminals in a unit square in $\mathcal{R}^2$, $\rho(N)$ typically is between 0.96 and 0.97 corresponding to $3\,\%$–$4\,\%$ length reduction of $\mathrm{SMT}(N)$ over $\mathrm{MST}(N)$. The reduction seems to increase as d grows. The smallest Steiner ratio over all sets N in $\mathcal{R}^d$ is defined by

$$\rho_d = \inf_N \{\rho(N)\}$$

It has been conjectured [11] that $\rho_2 = \sqrt{3}/2 = 0.866025....$ There are problem instances achieving this Steiner ratio; for example three corners of an equilateral triangle. Furthermore, ρ_d seems to decrease as $d \to \infty$. It has also been conjectured that ρ_d, $d \geq 3$ is achieved for infinite sets of terminals. In particular, a *regular 3-sausage* in R^3 is a sequence of regular d-simplices where consecutive ones share a regular 2-simplex (equilateral triangle). It has been conjectured that regular 3-sausages have Steiner ratios decreasing toward 0.7841903733771... as $n \to \infty$ [21].

Let $N_\sigma \subseteq N$ denote the corners of a face σ of $\mathrm{DT}(N)$. Let $\mathrm{ST}(N_\sigma)$ denote a Steiner tree spanning N_σ. Let F be a forest whose vertices are a superset of N. Suppose that terminals of N_σ are in different subtrees of F. The *concatenation* of F *with* $\mathrm{ST}(N_\sigma)$, denoted by $F \oplus \mathrm{ST}(N_\sigma)$, is a forest obtained by adding to F all Steiner points and all edges of $\mathrm{ST}(N_\sigma)$.

Let G be a complete weighted graph spanning N. The *contraction* of G by N_σ, denoted by $G \ominus N_\sigma$, is obtained by replacing the vertices in N_σ by a single vertex n_σ. Loops in $G \ominus N_\sigma$ are deleted. Among any parallel edges of $G \ominus N_\sigma$ incident with n_σ, all but the shortest ones are deleted.

Finally, let $T = \mathrm{MST}(N)$. The *bottleneck contraction* of T by N_σ, denoted by $T \ominus N_\sigma$, is obtained by replacing the vertices in N_σ by a single vertex n_σ. Any cycles in $T \ominus N_\sigma$ are destroyed by removing their longest edges. Hence, $T \ominus N_\sigma$ is a minimum spanning tree of $(N \backslash N_\sigma) \cup \{n_\sigma\}$. Instead of replacing N_σ by n_σ, the vertices of N_σ could be connected by a tree with zero-length edges spanning N_σ. Any cycles in the resulting tree are destroyed by removing their longest edges. We use the same notation, $T \ominus N_\sigma$, to denote the resulting MST on N.

1.2   DM-Heuristic in R^d

The DM-heuristic constructs $DT(N)$ and $MST(N)$ in the preprocessing phase. For corners N_σ of every covered face σ of $DT(N)$ in R^d (and for corners of some covered d-sausages), a low cost Steiner tree $ST(N_\sigma)$ is determined using a heuristic [17] or a numerical approximation of $SMT(N_\sigma)$ [21]. If full, $ST(N_\sigma)$ is stored in a priority queue Q ordered by non-decreasing Steiner ratios. Greedy concatenation, starting with a forest F of isolated terminals in N, is then used to form a tree spanning N.

In the postprocessing phase of the DM-heuristic, a *fine-tuning* is performed. The topology of F is extended to the full Steiner topology $\mathcal{ST}(N)$ by adding Steiner points overlapping with terminals where needed. The numerical approximation of [21] is applied to $\mathcal{ST}(N)$ in order to approximate the relatively minimal Steiner tree $ST(N)$ with the Steiner topology $\mathcal{ST}(N)$.

1.3   Improvement Motivation

The DM-heuristic returns better Steiner trees than its $\mathcal{R}^2$ predecessor [20]. It also performs well for $d \geq 3$. However, both the DM-heuristic and its predecessor rely on covered faces of $DT(N)$ determined by the $MST(N)$. The Steiner topology $\mathcal{ST}(N)$ of $ST(N)$ is therefore dictated by the topology of the $MST(N)$. This is a good strategy in many cases but there are also cases where this will exclude good solutions with Steiner topologies not related to the topology of the $MST(N)$. Consider for example Steiner trees in Fig. 1. In T_{DM} (Fig. 1a) only covered faces of $DT(N)$ are considered. By considering some uncovered faces (shaded), a better Steiner tree T_{DB} can be obtained (Fig. 1b).

We wish to detect useful uncovered faces and include them into the greedy concatenation. Consider for example the uncovered triangle σ of $DT(N)$ in R^2 shown in Fig. 2a. If uncovered faces are excluded, the solution returned will be the $MST(N)$ (red edges in Fig. 2a). The simplex σ is uncovered but it has a very good Steiner ratio. As a consequence, if permitted, $ST(N_\sigma) = SMT(N_\sigma)$ should be in the solution yielding as much better $ST(n)$ shown in Fig. 2b.

Some uncovered faces of $DT(N)$ can however be harmful in the greedy concatenation even though they seem to be useful in a local sense. For example, use of the uncovered 2-simplex σ of $DT(N)$ in $\mathcal{R}^2$ (Fig. 3a) will lead to a Steiner tree longer than $MST(N)$ (Fig. 3b) while the ratio $\rho(SMT(N_\sigma))$ is lowest among all faces of $DT(N)$. Hence, we cannot include all uncovered faces of $DT(N)$.

Another issue arising in connection with using only uncovered faces is that the fraction of covered faces rapidly decreases as d grows. As a consequence, the number of excluded good Steiner trees increases as d grows.

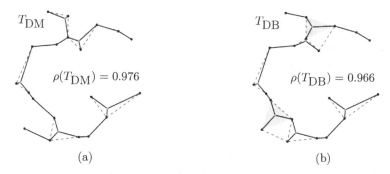

Fig. 1. Uncovered faces of $DT(N)$ can improve solutions. Edges of $MST(N)$ not in Steiner trees are dashed and red. (Color figure online)

Fig. 2. $\rho(SMT(N_\sigma))$ is very low and $SMT(N_\sigma)$ should be included in $ST(N)$. (Color figure online)

Fig. 3. $\rho(SMT(N_\sigma))$ is very low but the inclusion of $SMT(N_\sigma)$ into $ST(N)$ increases the length of $ST(N)$ beyond $|MST(N)|$. (Color figure online)

2   DB-Heuristic in $\mathcal{R}^d$

Let $T = MST(N)$. The *bottleneck distance* $|t_i t_j|_T$ between two terminals $t_i, t_j \in N$ is the length of the longest edge on the path from t_i to t_j in T. Note that $|t_i t_j|_T = |t_i t_j|$ if $(t_i, t_j) \in T$.

The *bottleneck minimum spanning tree* $B_T(N_\sigma)$ of a set of points $N_\sigma \subseteq N$ is defined as the minimum spanning tree of the complete graph with N_σ as its vertices and with the cost of an edge $(t_i, t_j), t_i, t_j \in N_\sigma$, given by $|t_i t_j|_T$. If N_σ

is covered by T, then $|B_T(N_\sigma)| = |\text{MST}(N_\sigma)|$. Easy proof by induction on the size of N_σ is omitted. Note that N is covered. Hence, $|B_T(N)| = |T|$.

Consider a Steiner tree $\text{ST}(N_\sigma)$ spanning $N_\sigma \subseteq N$. The *bottleneck Steiner ratio* $\beta_T(\text{ST}(N_\sigma))$ is given by:

$$\beta_T(\text{ST}(N_\sigma)) = \frac{|\text{ST}(N_\sigma)|}{|B_T(N_\sigma)|}$$

If N_σ is covered by T, then $\beta_T(\text{ST}(N_\sigma)) = \rho(\text{ST}(N_\sigma))$. Note that $\rho_T(\text{ST}(N_\sigma))$ for the 2-simplex σ in Fig. 3 is very high (even if $\text{ST}(N_\sigma) = \text{SMT}(N_\sigma)$) because $|B_T(N_\sigma)|$ is the sum of the lengths of the two red dashed edges shown in Fig. 3b. Hence, $\text{ST}(N_\sigma)$ will be buried deep in the priority queue Q_B. In fact, it will never be extracted from Q_B as $\rho_T(\text{ST}(N_\sigma)) > 1$.

The DB-heuristic consists of three phases: preprocessing, main loop and postprocessing, see Fig. 5. In the preprocessing phase, the DB-heuristic constructs $\text{DT}(N)$ and $T = \text{MST}(N)$. For corners N_σ of each k-face σ of $\text{DT}(N)$, $2 \le k \le d + 1$, a low cost Steiner tree $\text{ST}(N_\sigma)$ is determined using a heuristic [17] or a numerical approximation of $\text{SMT}(N_\sigma)$ [21]. Each full $\text{ST}(N_\sigma)$ is stored in a priority queue Q_B ordered by non-decreasing bottleneck Steiner ratios. If σ is a 1-face, then $\text{ST}(N_\sigma) = \text{SMT}(N_\sigma)$ is the edge connecting the two corners of σ. Such $\text{ST}(N_\sigma)$ is added to Q_B only if its edge is in T. Note that bottleneck Steiner ratios of these 1-faces are 1.

Let F be the forest of isolated terminals from N. Furthermore, let $N_0 = N$. In the main loop of the DB-heuristic, a greedy concatenation is applied repeatedly until F becomes a tree. Consider the i-th loop of the DB-heuristic, $i = 1, 2, \ldots$ Let $\text{ST}(N_{\sigma_i})$ be a Steiner tree with *currently* smallest bottleneck Steiner ratio in Q_B. If a pair of terminals in N_{σ_i} is connected in F, $\text{ST}(N_{\sigma_i})$ is thrown away. Otherwise, let $F = F \oplus \text{ST}(N_{\sigma_i})$ and $T = T \ominus N_{\sigma_i}$, see Fig. 4(a) where the $|B_T(\text{ST}(N_\sigma))| = |e_1| + |e_2|$. Such a bottleneck contraction of T (see Fig. 4(b)) may reduce bottleneck distances between up to $O(n^2)$ pairs of terminals. Hence, bottleneck Steiner ratios of some Steiner trees still in Q_B need to be updated

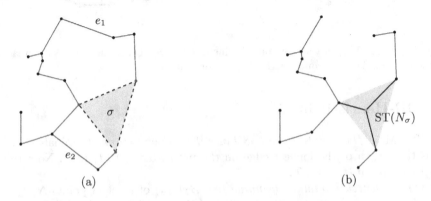

Fig. 4. The insertion of $ST(N_\sigma)$

```
// Preprocessing
    Construct DT(N) and T = MST(N);
    Push Steiner trees of all faces of DT(N) onto Q_B (
        except 1-faces not in T)
    Let F be the forest on N with no edges.
// Main loop
    while (F is not a tree on N) {
        ST(N_σ) = Steiner tree in Q_B with smallest
            bottleneck Steiner ratio w.r.t. T;
        if (no pair of terminals in N_σ is connected in F) {
            F = F ⊕ ST(N_σ);
            T = T ⊖ N_σ;
        }
    }
// Postprocessing
    Fine-tune F;
    return F
```

Fig. 5. DB-heuristic

either immediately or in a lazy fashion. Note that bottleneck Steiner ratios cannot decrease. If they increase beyond 1, the corresponding Steiner trees do not need to be placed back in Q_B. This is due to the fact that all 1-faces (edges) of the MST(N) are in Q_B and have bottleneck Steiner ratios equal to 1. We will return to the updating of bottleneck Steiner ratios in Sect. 3. Fine-tuning (as in the DM-heuristic) is applied in the postprocessing phase.

Unlike the DM-heuristic, d-sausages are not used in the DB-heuristic. In the DB-heuristic all faces of DT(N) are considered. As a consequence, fine-tuning in the postprocessing will in most cases indirectly generate Steiner trees spanning terminals in d-sausages if they are good candidates for subtrees of ST(N).

3   Contractions and Bottleneck Distances

As face-spanning Steiner trees are added to F in the main loop of the DB-heurstic, corners of these faces are bottleneck contracted in the current minimum spanning tree T. Bottleneck contractions will reduce bottleneck distances between some pairs of terminals. As a consequence, bottleneck Steiner ratios of face-spanning Steiner trees still in Q_B will increase. A face-spanning Steiner tree subsequently extracted from Q_B will not necessarily have the smallest bottleneck Steiner ratio (unless Q_B has been rearranged). Hence, appropriate lazy updating has to be carried out. To summarize, a data structure supporting the following operations is needed:

- bc($N_σ$): corners of a face $N_σ \in$ DT(N) are bottleneck contracted in T,
- bd(p,q): bottleneck distance between p and q in current minimum spanning tree T is returned,
- Q_B is maintained as a priority queue ordered by non-decreasing bottleneck Steiner ratios.

Maintaining Q_B could be done by recomputing bottleneck Steiner ratios and rearranging the entire queue after each contraction. Since there may be as many as $O(n^{\lceil d/2 \rceil})$ faces in DT(N) [18], this will be too slow.

To obtain better CPU times, a slightly modified version of dynamic rooted trees [19] maintaining a changing forest of disjoint rooted trees is used. For our purposes, bc-operations and bd-queries require maintaining a changing minimum spanning tree T rather than a forest. Dynamic rooted trees support (among others) the following operations:

- evert(n_j): makes n_j the root of the tree containing n_j.
- link(n_j, n_i, x): links the tree rooted at n_j to a vertex n_i in another tree. The new edge (n_j, n_i) is assigned the cost x.
- cut(n_j): removes the edge from n_j to its parent. n_j cannot be a root.
- mincost(n_j): returns the vertex n_i on the path to the root and closest to its root such that the edge from n_i to its parent has minimum cost. n_j cannot be a root.
- cost(n_j): returns the cost of the edge from n_j to its parent. n_j cannot be a root.
- update(n_j, x): adds x to the weight of every edge on the path from n_j to the root.

For our purposes, a maxcost operation replaces mincost. Furthermore, the update operation is not needed. A root of the minimum spanning tree can be chosen arbitrarily.

Rooted trees are decomposed into paths (see Fig. 6) represented by balanced binary search trees or biased binary trees. The path decomposition can be changed by splitting or joining these binary trees.

By appropriate rearrangement of the paths, all above operations can be implemented using binary search tree operations [19]. Since the update operation is not needed, the values cost and maxcost can be stored directly with n_j. Depending on whether balanced binary search trees or biased binary trees are used for the paths, the operations require respectively $O(\log^2 n)$ and $O(\log n)$ amortized time.

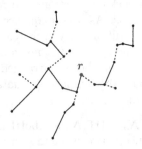

Fig. 6. A rooted tree decomposed into paths.

Using dynamic rooted trees to store the minimum spanning tree, bd-queries and bottleneck contractions can be implemented as shown in Fig. 7. The bd-query makes n_i the new root. Then it finds the vertex n_k closest to n_i such that the edge from n_k to its parent has maximum cost on the path from n_j to n_i. The cost of this edge is returned. The bc-operation starts by running through all pairs of vertices of N_σ. For each pair n_i, n_j, n_i is made the root of the tree (evert(n_i)) and then the edge with the maximum cost on the path from n_j to n_i is found. If n_i and n_j are connected, the edge is cut away. Having cut away all connecting edges with maximum cost, the vertices of N_σ are reconnected by zero-length edges.

```
// t_i and t_j are vertices of the minimum spanning tree T
bd(t_i,t_j) {
   evert(t_i)
   return cost(maxcost(t_j))
}

// N_σ = {t_1, t_2, ..., t_k }, k ≥ 2 is a set of corners of a face
      of DT(N)
bc(t_σ) {
   for(i = 1 .. k) {
      evert(t_i)
      for(j = i+1 .. k)
         if(t_i and t_j are connected) cut(maxcost(t_j));
   }
   evert(t_1)
   for(i = 2 .. k)   link(t_{i-1}, t_i, 0)
}
```

Fig. 7. bd-query and bc-operation

When using balanced binary trees, one bd-query takes $O((\log n)^2)$ amortized time. Since only faces of DT(N) are considered, the bc-operation performs $O(d)$ everts and links, $O(d^2)$ maxcosts and cuts. Hence, it takes $O((d \log n)^2)$ time.

In the main loop of the algorithm, Steiner trees of faces of DT(N) are extracted one by one. A face σ is rejected if some of its corners are already connected in F. Since the quality of the final solution depends on the quality of Steiner trees of faces, these trees should have smallest possible bottleneck Steiner ratios. When a Steiner tree $ST(N_\sigma)$ is extracted from Q_B, it is first checked if $ST(N_\sigma)$ spans terminals already connected in F. If so, $ST(N_\sigma)$ is thrown away. Otherwise, its bottleneck Steiner ratio may have changed since the last time it was pushed onto Q_B. Hence, bottleneck Steiner ratio of $ST(N_\sigma)$ is recomputed. If it increased since last but is still below 1, $ST(N_\sigma)$ is pushed back onto Q_B (with the new bottleneck Steiner ratio). If the bottleneck Steiner ratio did not change, $ST(N_\sigma)$ is used to update F and bottleneck contract T.

4   Computational Results

The DB-heuristic was tested against the DM-heuristic. Both Steiner ratios and CPU times were compared. To get reliable Steiner ratio and computational time comparisons, they were averaged over several runs whenever possible. Furthermore, the results in $\mathcal{R}^2$ were compared to the results achieved by the exact GeoSteiner algorithm [13].

To test and compare the DM- and the DB-heuristic, they were implemented in C++. The code and instructions on how to run the DM- and DB-heuristics can be found in the GitHub repository [16]. All tests have been run on a Lenovo ThinkPad S540 with a 2 GHz Intel Core i7-4510U processor and 8 GB RAM.

The heuristics were tested on randomly generated problem instances of different sizes in $\mathcal{R}^d$, $d = 2, 3, ..., 6$, as well as on library problem instances. Randomly generated instances were points uniformly distributed in $\mathcal{R}^d$-hypercubes.

The library problem instances consisted of the benchmark instances from the 11-th DIMACS Challenge [6]. More information about these problem instances can be found on the DIMACS website [6]. For comparing the heuristics with the GeoSteiner algorithm, we used ESTEIN instances in $\mathcal{R}^2$.

Dynamic rooted trees were implemented using AVL trees. The *restricted numerical optimisation* heuristic [17] for determining Steiner trees of $DT(N)$ faces was used in the experiments.

In order to get a better idea of the improvement achieved when using bottleneck distances, the DM-heuristic does not consider covered d-sausages as proposed in [17]. Test runs of the DM-heuristic indicate that the saving when using d-sausages together with fine-tuning is only around 0.1 % for $d = 2$, 0.05 % for $d = 3$ and less than 0.01 % when $d > 3$. As will be seen below, the savings achieved by using bottleneck distances are more significant.

In terms of quality, the DB-heuristic outperforms the DM-heuristic. The Steiner ratios of obtained Steiner trees reduces by 0.2−0.3 % for $d = 2$, 0.4−0.5 % for $d = 3$, 0.6−0.7 % for $d = 4$, 0.7−0.8 % for $d = 5$ and 0.8−0.9 % for $d = 6$. This is a significant improvement for the ESMT problem as will be seen below, when comparing $\mathcal{R}^2$ results to the optimal solutions obtained by the exact GeoSteiner algorithm [13].

CPU times for both heuristics for $d = 2, 3, ..., 6$, are shown in Fig. 8. It can be seen that the improved quality comes at a cost for $d \geq 4$. This is due to the fact that the DB-heuristic constructs low cost Steiner trees for all $O(n^{\lceil d/2 \rceil})$ faces of $DT(N)$ while the DM-heuristic does it for covered faces only. Later in this section it will be explored how the Steiner ratios and CPU times are affected if the DB-heuristic drops some of the faces.

Figure 9 shows how the heuristics and GeoSteiner (GS) performed on ESTEIN instances in $\mathcal{R}^2$. Steiner ratios and CPU times averaged over all 15 ESTEIN instances of the given size, except for $n = 10000$ which has only one instance. For the numerical comparisons, see Table 1 in the GitHub repository [16]. It can be seen that the DB-heuristic produces better solutions than the DM-heuristic without any significant increase of the computational time.

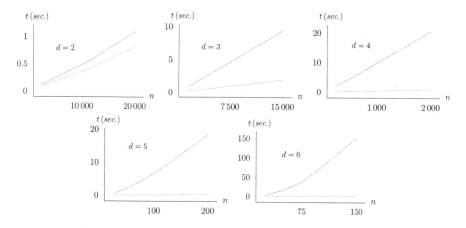

Fig. 8. Comparison of the CPU times for the DB-heuristic (blue) and the DM-heuristic (red) for $d = 2, 3, ..., 6$. (Color figure online)

It is also worth noticing that the DB-heuristic gets much closer to the optimal solutions. This may indicate that the DB-heuristic also produces high quality solutions when $d > 2$, where optimal solutions are only known for instances with at most 20 terminals. For the performance of the DB-heuristic on individual $\mathcal{R}^2$ instances, see Tables 3–7 in the GitHub repository [16].

The results for ESTEIN instances in $\mathcal{R}^3$ are presented in Fig. 10. The green plot for $n = 10$ is the average ratio and computational time achieved by numerical approximation [21]. Once again, the DB-heuristic outperforms the DM-heuristic when comparing Steiner ratios. However, the running times are

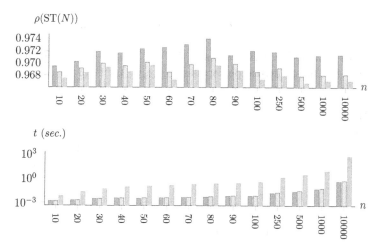

Fig. 9. Averaged ratios and CPU times for ESTEIN instances in $\mathcal{R}^2$. DM-heuristic (red), DB-heuristic (blue), GeoSteiner (green). (Color figure online)

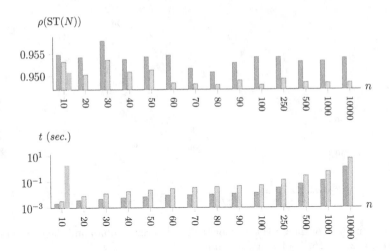

Fig. 10. Averaged ratio and CPU times for ESTEIN instances in $\mathcal{R}^3$. DM-heuristic (red), DB-heuristic (blue), numerical approximation (green). (Color figure online)

now up to four times worse. For the numerical comparisons, see Table 2 in the GitHub repository [16]. For the performance of the DB-heuristic on individual $\mathcal{R}^3$ instances, see Tables 8–12 in the GitHub repository [16].

The DB-heuristic starts to struggle when $d \geq 4$. This is caused by the number of faces of $DT(N)$ for which low cost Steiner trees must be determined. The DB-heuristic was therefore modified to consider only faces with less than k terminals, for $k = 3, 4, ..., d + 1$. Figure 11 shows the performance of this modified DB_k-heuristic with $k = 3, 4, ..., 7$, on a set with 100 terminals in $\mathcal{R}^6$. Note that $DB_7 = DB$.

As expected, the DB_k-heuristic runs much faster when larger faces of $DT(N)$ are disregarded. Already the DB_4-heuristic seems to be a reasonable alternative since solutions obtained by DB_k-heuristic, $5 \leq k \leq 7$, are not significantly better. Surprisingly, the DB_6-heuristic performs slightly better than the DB_7-heuristic.

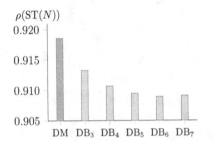

Method	t
DM	0.4714
DB_3	0.6000
DB_4	6.0525
DB_5	26.2374
DB_6	51.3653
$DB_7 = DB$	62.8098

Fig. 11. Results achieved when considering faces of $DT(N)$ with at most $k = 3, 4, ..., 7$ terminals in the concatenation for $d = 6$ and $n = 100$.

This is probably due to the fact that low cost Steiner trees of smaller faces have fewer Steiner points. This in turn causes the fine-tuning step of the DB_6-heuristic to perform better than is the case for DB_7.

5 Summary and Conclusions

The DM-heuristic in $\mathcal{R}^d$ [17] was extended to the DB-heuristic that uses bottleneck distances to determine good candidates for low cost Steiner trees. Computational results show a significant improvement in the quality of the Steiner trees produced by the DB-heuristic.

The CPU times of the DB-heuristic are comparable to the CPU times of the DM-heuristic in $\mathcal{R}^d$, $d = 2, 3$. It runs slower for $d \geq 4$. However, its CPU times can be significantly improved by skipping larger faces of $DT(N)$. This results in only small decrease of the quality of solutions obtained.

References

1. Bajaj, C.: The algebraic degree of geometric optimization problems. Discrete Comput. Geom. **3**, 177–191 (1988)
2. Beasley, J.E., Goffinet, F.: A Delaunay triangulation-based heuristic for the Euclidean Steiner problem. Networks **24**(4), 215–224 (1994)
3. de Berg, M., Cheong, O., van Krevald, M., Overmars, M.: Computational Geometry - Algorithms and Applications, 3rd edn. Springer, Heidelberg (2008)
4. Brazil, M., Graham, R.L., Thomas, D.A., Zachariasen, M.: On the history of the Euclidean Steiner tree problem. Arch. Hist. Exact Sci. **68**, 327–354 (2014)
5. Brazil, M., Zachariasen, M.: Optimal Interconnection Trees in the Plane. Springer, Cham (2015)
6. DIMACS, ICERM: 11th DIMACS Implementation Challenge: Steiner Tree Problems (2014). http://dimacs11.cs.princeton.edu/
7. Fampa, M., Anstreicher, K.M.: An improved algorithm for computing Steiner minimal trees in Euclidean d-space. Discrete Optim. **5**, 530–540 (2008)
8. Fampa, M., Lee, J., Maculan, N.: An overview of exact algorithms for the Euclidean Steiner tree problem in n-space, Int. Trans. OR (2015)
9. Fonseca, R., Brazil, M., Winter, P., Zachariasen, M.: Faster exact algorithms for computing Steiner trees in higher dimensional Euclidean spaces. In: Proceedings of the 11th DIMACS Implementation Challenge, Providence, Rhode Island, USA (2014). http://dimacs11.cs.princeton.edu/workshop.html
10. do Forte, V.L., Montenegro, F.M.T., de Moura Brito, J.A., Maculan, N.: Iterated local search algorithms for the Euclidean Steiner tree problem in n dimensions. Int. Trans. OR (2015)
11. Gilbert, E.N., Pollak, H.O.: Steiner minimal trees. SIAM J. Appl. Math. **16**(1), 1–29 (1968)
12. Hwang, F.K., Richards, D.S., Winter, P.: The Steiner Tree Problem. North-Holland, Amsterdam (1992)
13. Juhl, D., Warme, D.M., Winter, P., Zachariasen, M.: The GeoSteiner software package for computing Steiner trees in the plane: an updated computational study. In: Proceedings of the 11th DIMACS Implementation Challenge, Providence, Rhode Island, USA (2014). http://dimacs11.cs.princeton.edu/workshop.html

14. Laarhoven, J.W.V., Anstreicher, K.M.: Geometric conditions for Euclidean Steiner trees in R^d. Comput. Geom. Theor. Appl. **46**(5), 520–531 (2013)

15. Laarhoven, J.W.V., Ohlmann, J.W.: A randomized Delaunay triangulation heuristic for the Euclidean Steiner tree problem in R^d. J. Heuristics **17**(4), 353–372 (2011)

16. Lorenzen, S.S., Winter, P.: Code and Data Repository at Github (2016). https://github.com/StephanLorenzen/ESMT-heuristic-using-bottleneck-distances/blob/master/README.md

17. Olsen, A., Lorenzen, S. Fonseca, R., Winter, P.: Steiner tree heuristics in Euclidean d-space. In: Proceedings of the 11th DIMACS Implementation Challenge, Providence, Rhode Island, USA (2014). http://dimacs11.cs.princeton.edu/workshop.html

18. Seidel, R.: The upper bound theorem for polytopes: an easy proof of its asymptotic version. Comp. Geom.-Theor. Appl. **5**, 115–116 (1995)

19. Sleator, D.D., Tarjan, R.E.: A data structure for dynamic trees. J. Comput. Syst. Sci. **26**(3), 362–391 (1983)

20. Smith, J.M., Lee, D.T., Liebman, J.S.: An O($n \log n$) heuristic for Steiner minimal tree problems on the Euclidean metric. Networks **11**(1), 23–39 (1981)

21. Smith, W.D.: How to find Steiner minimal trees in Euclidean d-space. Algorithmica **7**, 137–177 (1992)

22. Toppur, B., Smith, J.M.: A sausage heuristic for Steiner minimal trees in three-dimensional Euclidean space. J. Math. Model. Algorithms **4**, 199–217 (2005)

23. Warme, D.M., Winter, P., Zachariasen, M.: Exact algorithms for plane Steiner tree problems: a computational study. In: Du, D.-Z., Smith, J., Rubinstein, J. (eds.) Advances in Steiner Trees, pp. 81–116. Springer, Dordrecht (2000)

Tractable Pathfinding for the Stochastic On-Time Arrival Problem

Mehrdad Niknami[1(⊠)] and Samitha Samaranayake[2]

[1] Electrical Engineering and Computer Science, UC Berkeley, Berkeley, USA
mniknami@berkeley.edu
[2] School of Civil and Environmental Engineering, Cornell University, Ithaca, USA

Abstract. We present a new and more efficient technique for computing the route that maximizes the probability of on-time arrival in stochastic networks, also known as the path-based stochastic on-time arrival (SOTA) problem. Our primary contribution is a pathfinding algorithm that uses the solution to the *policy*-based SOTA problem—which is of pseudo-polynomial-time complexity in the time budget of the journey— as a search heuristic for the optimal path. In particular, we show that this heuristic can be exceptionally efficient in practice, effectively making it possible to solve the path-based SOTA problem as quickly as the policy-based SOTA problem. Our secondary contribution is the extension of policy-based preprocessing to path-based preprocessing for the SOTA problem. In the process, we also introduce Arc-Potentials, a more efficient generalization of Stochastic Arc-Flags that can be used for *both* policy- and path-based SOTA. After developing the pathfinding and preprocessing algorithms, we evaluate their performance on two different real-world networks. To the best of our knowledge, these techniques provide the most efficient computation strategy for the path-based SOTA problem for general probability distributions, both with and without preprocessing.

1 Introduction

Modern advances in graph theory and empirical computational power have essentially rendered deterministic point-to-point routing a solved problem. While the ubiquity of routing and navigation tools in our everyday lives is a testament to the success and usefulness of deterministic routing technology, inaccurate predictions remain a fact of life, resulting in missed flights, late arrivals to meetings, and failure to meet delivery deadlines. Recent research in transportation engineering, therefore, has focused on the collection of traffic data and the incorporation of *uncertainty* into traffic models, allowing for the optimization of relevant *reliability* metrics desirable for the user.

The point-to-point *stochastic on-time arrival* problem [1], or SOTA for short, concerns itself with this reliability aspect of routing. In the SOTA problem, the network is assumed to have uncertainty in the travel time across each link, represented by a strictly positive random variable. The objective is then to

© Springer International Publishing Switzerland 2016
A.V. Goldberg and A.S. Kulikov (Eds.): SEA 2016, LNCS 9685, pp. 231–245, 2016.
DOI: 10.1007/978-3-319-38851-9_16

maximize the probability of on-time arrival when traveling between a given origin-destination pair with a fixed time budget.[1] It has been shown that the SOTA solution can appreciably increase the probability of on-time arrival compared to the classical *least expected travel time* (LET) path [3], motivating the search for efficient solutions to this problem.

1.1 Variants

There exist two primary variants of the SOTA problem. The path-based SOTA problem, which is also referred to as the *shortest-path problem with on-time arrival reliability* (SPOTAR) [4], consists of finding the a-priori most reliable path to the destination. The policy-based SOTA problem, on the other hand, consists of computing a *routing policy*—rather than a fixed path—such that, at every intersection, the choice of the next direction depends on the current state (i.e., the remaining time budget).[2] While a policy-based approach provides better reliability when online navigation is an option, in some situations it can be necessary to determine the entire path prior to departure.

The policy-based SOTA problem, which is generally solved in discrete-time, can be solved via a successive-approximation algorithm, as shown by Fan and Nie [5]. This approach was subsequently improved by Samaranayake et al. [3] to a pseudo-polynomial-time label-setting algorithm based on dynamic-programming with a sequence of speedup techniques and the use of *zero-delay convolution* [6, 7]. It was then demonstrated in Sabran et al. [8] that graph preprocessing techniques such as Reach [9] and Arc-Flags [10] can be used to further reduce query times for this problem.

In contrast with the policy-based problem, however, no polynomial-time solution is known for the general path-based SOTA problem [4]. In the special case of normally-distributed travel times, Nikolova et al. [11] present an $O(n^{O(\log n)})$-algorithm for computing exact solutions, while Lim et al. [12] present a polylogarithmic-time algorithm for approximation solutions. To allow for more general probability distributions, Nie and Wu [4] develop a label-correcting algorithm that solves the problem by utilizing the first-order stochastic dominance property of paths. While providing a solution method for general distributions, the performance of this algorithm is still insufficient to be of practical use in many real-world scenarios; for example, while the stochastic dominance approach provides a reasonable computation time (on the order of half a minute per instance) for networks of a few hundred to a thousand vertices, it fails to perform well on metropolitan road networks, which easily exceed tens of thousands of vertices. In contrast, our algorithm easily handles networks of tens of thousands of edges in approximately the same amount of time *without* any kind of preprocessing.[3]

---

[1] The target objective can in fact be generalized to utility functions other than the probability of on-time arrival [2] with little effect on our algorithms, but for our purposes, we limit our discussion to this scenario.

[2] In this article, we only consider time-invariant travel-time distributions. The problem can be extended to incorporate time-varying distributions as discussed in [3].

[3] Parmentier and Meunier [13] have concurrently also developed a similar approach concerning stochastic shortest paths with risk measures.

With preprocessing, our techniques further reduce the running time to less than half a second, making the problem tractable for larger networks.[4]

1.2 Contributions

Our primary contribution in this article is a practically efficient technique for solving the path-based SOTA problem, based on the observation that the solution to the policy-based SOTA problem is in practice itself an extremely efficient heuristic for solving the path-based problem.

Our secondary contribution is to demonstrate how graph preprocessing can be used to speed up the computation of the policy heuristic, and thus the optimal path, while maintaining correctness.[5] Toward this goal, we present Arc-Potentials, a more efficient generalization of the existing preprocessing technique known as Stochastic Arc-Flags that can be used for both policy- and path-based preprocessing.

After presenting these techniques, we evaluate the performance of our algorithms on two real-world networks while comparing the trade-off between their scalability (in terms of memory and computation time) and the speedups achieved. Our techniques, to the best of our knowledge, provide the most efficient computation strategy for the path-based SOTA problem with general probability distributions, both with and without preprocessing.

2 Preliminaries

We are given a stochastic network in the form of a directed graph $G = (V, E)$ where each edge $(i, j) \in E$ has an associated probability distribution $p_{ij}(\cdot)$ representing the travel time across that edge.[6] The source is denoted by $s \in V$, the destination by $d \in V$, and the travel time budget by $T \in \mathbb{R}^+$.

For notational simplicity, we present the SOTA problem in continuous-time throughout this article, with the understanding that the algorithms are applied after discretization with a time interval of Δt.

Definition 1 (SOTA Policy). *Let $u_{ij}(t)$ be the probability of arriving at the destination d with time budget t when first traversing edge $(i, j) \in E$ and subsequently following the optimal policy. Let $\delta_{ij} > 0$ be the minimum travel time along edge (i, j), i.e. $\min\{\tau : p_{ij}(\tau) > 0\}$. Then, the on-time arrival probability $u_i(t)$ and the policy (optimal subsequent node) $w_i(t)$ at node i, can be defined via the dynamic programming equations below [1]. Note that Δt must satisfy $\Delta t \leq \delta_{ij} \ \forall (i, j) \in E$.*

---

[4] It should be noted that the largest network we consider only has approximately 71,000 edges and is still much smaller than networks used to benchmark deterministic shortest path queries, which can have millions of edges [14].

[5] As explained later, there is a potential pitfall that must be avoided when the preprocessed policy is to be used as a heuristic for the path.

[6] We assume that at most one edge exists between any pair of nodes in each direction.

$$u_{ij}(t) = \int_{\delta_{ij}}^{t} u_j(t - \tau)p_{ij}(\tau)\, d\tau \qquad\qquad u_i(t) = \max_{j:\, (i,j)\in E} u_{ij}(t)$$

$$u_d(\cdot) = 1 \qquad\qquad\qquad\qquad\qquad w_i(t) = \arg\max_{j:\, (i,j)\in E} u_{ij}(t)$$

The solution to the policy-based SOTA problem can be obtained by solving this system of equations using dynamic programming as detailed in [3]. This requires evaluating a set of convolution integrals. The computation of the policy $u_s(\cdot)$ for a source s and time budget T using direct convolution takes $O(|E|T^2)$ time, as computing $u_s(T)$ could in the worst case require convolutions of length $O(T)$ for each edge in the graph. However, by using an online convolution technique known as *zero-delay convolution* (ZDC) [6,15], the time complexity can be reduced to $O(|E|T \log^2 T)$. Justifications for the results and time complexity, including details on how to apply ZDC to the SOTA problem, can be found in [3,7].

Assumptions. Our work, as with other approaches to both the policy-based and path-based SOTA problems, makes a number of assumptions about the nature of the travel time distributions. The three major assumptions are that the travel time distributions are (1) *time invariant*, (2) *exogenous* (not impacted by individual routing choices), and (3) *independent*. The time-invariance assumption—which prevents accounting for traffic variations throughout the day—can be relaxed under certain conditions as described in [3]. Furthermore, the exogeneity assumption is made even in the case of deterministic shortest path problems. This leaves the independence assumption as a major concern for this problem.

It might, in fact, be possible to partially relax this assumption [3] to allow for conditional distributions at the cost of increasing the computation time by a factor linear in the number of states to be conditioned on. (If we assume the Markov property for road networks, the number of conditioning states becomes the in-degree of each vertex, a small enough constant that may make generalizations in this direction practical.) Nevertheless, we will only focus on the independent setting and make no claim to have solved the path-based SOTA problem in full generality, as the problem already lacks efficient solution methods even in this simplified setting. Our techniques should, however, provide a foundation that allows for relaxing these assumptions in the future.

3   Path-Based SOTA

In the deterministic setting, efficient solution strategies (from Dijkstra's algorithm to state-of-the-art solutions) generally exploit the *sub-path optimality* property: namely, the fact that any optimal route to a destination node d that includes some intermediate node i necessarily includes the optimal path from i to d. Unfortunately, this does not hold in the stochastic setting.

Algorithm 1. Algorithm for computing the optimal SOTA path

Notation: $*$ is the convolution operator and $\parallel$ is the concatenation operator

for all $i \in V, 0 \le t \le T$ **do**

    Compute the optimal policy's reliability[8] $u_i(t)$

$Q \leftarrow$ PriorityQueue()

Q.Push $(u_s(T), ([1.0], [s]))$ ▷ Push (reliability, (cost dist., initial path))

while Q is not empty **do**

    $(r, (q, P)) \leftarrow Q$.PopMax()   ▷ Extract most reliable path so far

    $i \leftarrow P[|P| - 1]$ ▷ Get the last node in the path

    if $i = d$ **then**

        return P

    for all $j \in E$.Neighbors(i) **do**

        Q.Push $((q * u_j)[T], (q * E.\text{Cost}(i, j), P \parallel [j]))$ ▷ Append new edge.

return nil   ▷ No path found

Furthermore, blind enumeration of all possible paths in the graph is absolutely intractable for all but the most trivial networks, as the number of simple paths grows exponentially with the number of nodes in the graph. Naturally, this leads us to seek a heuristic to guide us toward the optimal path efficiently, while not compromising its optimality.

3.1 Algorithm

Consider a fixed path P from the source s to node i. Let $q_{si}^P(t)$ be the travel time distribution along P from node s to node i, i.e., the convolution of the travel time distributions of every edge in P. Upon arriving at node i at time t, let the user follow the optimal policy toward d, therefore reaching d from s with probability density $q_{si}^P(t)u_i(T - t)$. The reliability of following path P to node i and subsequently following the optimal policy toward d is[7]:

$$r_{si}^P(T) = \int_0^T q_{si}^P(t)u_i(T - t)\, dt$$

Note that the route from $s \to i$ is a fixed path while that from $i \to d$ is a policy.

    The optimal path is found via the procedure in Algorithm 1. Briefly, starting at the source s, we add the hybrid (path + policy) solution $r_{si}^P(T)$ for each neighbor i of s to a priority queue. Each of these solutions gives an upper bound on the solution (success probability). We then dequeue the solution with the highest upper bound, repeating this process until a path to the destination is found.

    Essentially, Algorithm 1 performs an A^* search for the destination, using the policy as a heuristic. While it is obvious that the algorithm would find the

---

[7] The bounds of this integral can be slightly tightened through inclusion of the minimum travel times, but this has been omitted for simplicity.

[8] Can be limited to those i and t reachable from s in time T, and can be further sped up through existing policy preprocessing techniques such as Arc-Flags.

optimal path *eventually* if the search were allowed to continue indefinitely, it is less obvious that the *first* path found will be optimal. We show this by showing that the policy is an admissible heuristic for the path, and consequently, by the optimality of A^* [16], the first returned path must be optimal.

Proposition 1 (Admissibility). *The solution to policy-based SOTA problem is an admissible heuristic for the optimal solution to the path-based SOTA problem using Algorithm 1.*

Proof. When finding a minimum cost path, an admissible heuristic is a heuristic that never overestimates the actual cost [17]. In our context, since the goal is to maximize the reliability (success probability), this corresponds to a heuristic that never underestimates the reliability of a routing strategy. The reliability of an optimal SOTA policy clearly provides an upper bound on the reliability of any fixed path with the same source, destination, and travel budget. (Otherwise, a better policy would be to simply follow the fixed path irrespective of the time remaining, contradicting the assumption that the policy is optimal.) Therefore, the SOTA policy is an admissible heuristic for the optimal SOTA path.

3.2   Analysis

The single dominant factor in this algorithm's (in)efficiency is the length of the priority queue (i.e., the number of paths considered by the algorithm), which in turn depends on the travel time distribution along each road. As long as the number of paths considered is approximately linear in length of the optimal path, the path computation time is easily dominated by the policy computation time, and the algorithm finds the optimal path very quickly. In the worst-case scenario for the algorithm, the optimal path at a node corresponds to the direction for the worst policy at that node. Such a scenario, or even one in which the optimal policy frequently chooses a suboptimal path, could result in a large (even exponential) running time as well as space usage. However, it is difficult to imagine this happening in practice. As shown later, experimentally, we came across very few cases in which the path computation time dominated the policy computation time, and even in those cases, they were still quite reasonable and extremely far from such a worst-case scenario. We conjecture that such situations are extremely unlikely to occur in real-world road networks.

An interesting open problem is to define characteristics (network structure, shape of distributions, etc.) that guarantee pseudo-polynomial running times in stochastic networks, similar in nature to the *Highway Dimension* property [18] in deterministic networks, which guarantees logarithmic query times when networks have a low *Highway Dimension*.

4   Preprocessing

In deterministic pathfinding, preprocessing techniques such as *Arc-Flags* [10], *reach-based routing* [9,19], *contraction hierarchies* [20], and *transit node routing*

[21] have been very successfully used to decrease query times by many orders of magnitude by exploiting structural properties of road networks. Some of these approaches allow for pruning the search space based solely on the destination node, while others also take the source node into account, allowing for better pruning at the cost of additional preprocessing. The structure of the SOTA problem, however, makes it more challenging to apply such techniques to it. Previously, Arc-Flags and Reach have been successfully adapted to the policy-based problem in [8], resulting in Stochastic Arc-Flags and Directed Reach. While at first glance one may be tempted to directly apply these algorithms to the computation of the policy heuristic for the path-based problem, a naive application of source-dependent pruning (such as Directed Reach or source-based Arc-Flags) can result in an incorrect solution, as the policy needs to be recomputed for source nodes that correspond to different source regions. This effectively limits any preprocessing of the policy heuristic to destination-based (i.e., source-independent) techniques such as Stochastic Arc-Flags, precluding the use of source-based approaches such as Directed Reach for the policy computation.

With sufficient preprocessing resources (as explained in Sect. 5.2), however, one can improve on this through the direct use of *path-based* preprocessing—that is, pruning the graph to include only those edges which may be part of the most reliable path. This method allows us to simultaneously account for both source and destination regions, and generally results in a substantial reduction of the search space on which the policy needs to be computed. However, as path-based approaches require computing paths between all $\approx |V|^2$ pairs of vertices in the graph, this approach may become computationally prohibitive for medium- to large-scale networks. In such cases, we would then need to either find alternate approaches (e.g. approximation techniques), or otherwise fall back to the less aggressive policy-based pruning techniques, which only require computing $|V|$ separate policies (one per destination).

4.1 Efficient Path-Based Preprocessing

Path-based preprocessing requires finding the optimal paths for each time budget up to the desired time budget T for all source-destination pairs. Naively, this can be done by placing Algorithm 1 in a loop, executing it for all time budgets from 1 to T. This requires T times the work of finding the path for a single time budget, which is clearly prohibitive for any reasonable value of T. However, we can do far better by observing that many of the computations in the algorithm are independent of the time budget and can be factored out when the path does not change with T.

To improve the efficiency of the naive approach in this manner, we make two observations. First, we observe that, in Algorithm 1, *only* the computation of the path's reliability (priority) in the priority queue ever requires knowledge of the time budget. Crucially, the convolution $q * E.\text{Cost}(i, j)$ only depends on the *maximum* time budget T for truncation purposes, which is a fixed value. This means that the travel time distribution of any path under consideration can be computed once for the maximum time budget, and re-used for all lower time budgets thereafter. Second, we observe that when a new edge is appended, the priority of the new path is the inner product of the vector q and (the reverse of) the vector u_j, shifted by T. As noted in the algorithm itself, this quantity in fact the convolution of the two aforementioned vectors evaluated at T. Thus, when a new edge is appended, instead of recomputing the inner product, we can simply convolve the two vectors once, and thereafter look up the results instantly for other time budgets.

Together, these two observations allow us to compute the optimal paths for all budgets far faster than would seem naively possible, making path-based preprocessing a practical option.

4.2   Arc-Potentials

As noted earlier, Arc-Flags, a popular method for graph preprocessing, has been adapted to the SOTA problem as Stochastic Arc-Flags [8]. Instead of applying it directly, however, we present *Arc-Potentials*, a more natural generalization of Arc-Flags to SOTA that can still be directly applied to the policy- and path-based SOTA problems alike, while allowing for more efficient preprocessing.

Consider partitioning the graph G into R regions (we choose $R = O(\log |E|)$, described below), where R is tuned to trade off storage space for pruning accuracy. In the deterministic setting, Arc-Flags allow us to preprocess and prune the search space as follows. For every arc (edge) $(i, j) \in E$, Arc-Flags defines a bit-vector of length R that denotes whether or not this arc belongs to an optimal path ending at some node in region R. We then pre-compute these Arc-Flags, and store them for pruning the graph at query time. (This approach has been extended to the dynamic setting [22] in which the flags are updated with low recomputation cost after the road network is changed.)

Sabran et al. [8] apply Arc-Flags to the policy-based SOTA problem as follows: each bit vector is defined to represent whether or not its associated arc is *realizable*, meaning that it *belongs to an optimal policy* to some destination in the target region associated with each bit. The problem with this approach, however, is that it requires computing arc-flags for *all* target budgets (or more, practically, some ranges of budgets), each of which takes a considerable amount of space. Instead, we propose a more efficient alternative Definition 2, which we call *Arc-Potentials*.

Definition 2 (Arc-Potentials). *For a given destination region D, we define the arc activation potential ϕ_{ij} of the edge from node i to node j to be the minimum time budget at which the arc becomes part of an optimal policy to some destination $d \in D$.*

The Arc-Potentials pruning algorithm only stores the *"activation"* potential of every edge. As expected, this implies that for large time budgets, every edge is potentially active. We could have further generalized the algorithm to allow for asymptotically *exact* pruning at relatively low cost by simply storing the actual potential *intervals* during which the arc is active, rather than merely their first activation potential. However, in our experiments this was deemed unnecessary as Arc-Potentials were already sufficient for significant pruning in the time budgets of interest in our networks.

The computation of the set of realizable edges (and nodes) under a given policy is essentially equivalent to the computation of the policy itself, except that updates are performed in the reverse order (from the source to the destination). The activation potentials ϕ can then be obtained from this set. As with Arc-Flags, we limit the space complexity to $O(|E|R) = O(|E| \log |E|)$ by choosing R to be proportional $\log |E|$, tuning it as desired to increase the pruning accuracy. In our experiments, we simply used a rectangular grid of size $\sqrt{R} \times \sqrt{R}$. Note, however, that the preprocessing time does not depend on R, as the paths between all $\approx |V|^2$ pairs of nodes must be eventually computed.

5 Experimental Results

We evaluated the performance of our algorithms on two real-world test networks: a small San Francisco network with 2643 nodes and 6588 edges for which real-world travel-time data was available as a Gaussian mixture model [23], and a second (relatively larger) Luxembourg network with 30647 nodes and 71655 edges for which travel-time distributions were synthesized from road speed limits, as real-world data was unavailable. The algorithms were implemented in C++ (2003) and executed on a cluster of 1.9 GHz AMD Opteron^{TM} 6168 CPUs. The SOTA queries were executed on a single CPU and the preprocessing was performed in parallel as explained below.

The SOTA policies were computed as explained in [3,7] using zero-delay convolution with a discretization interval of $\Delta t = 1\,\mathrm{s}$.[9] To generate random problem instances, we independently picked a source and a destination node uniformly at random from the graph and computed the least expected travel-time (LET) path between them. We then evaluated our pathfinding algorithm for budgets chosen uniformly at random from the 5th to 95th percentile of LET path travel times (those of practical interest) on $10,000$ San Francisco and 1000 Luxembourg problems instances.

First, we discuss the speed of our pathfinding algorithm, and afterward, we evaluate the effectiveness and scalability of our preprocessing strategies.

5.1 Evaluation

We first evaluate the performance of our path-based SOTA algorithm without any graph preprocessing. Experimental results, as can be seen in Fig. 1, show that the run time of our solution is dominated by the time taken to obtain the solution to the policy-based SOTA problem, which functions as a search heuristic for the optimal path.

The stochastic-dominance (SD) approach [4], which to our knowledge is the fastest published solution for the path-based SOTA problem with general probability distributions, takes, on average, between 7 and 18 s (depending on the

---

[9] Recall that we must have $\Delta t \leq \min_{(i,j) \in E} \delta_{ij}$, which is $\approx 1\,\mathrm{s}$ for our networks.

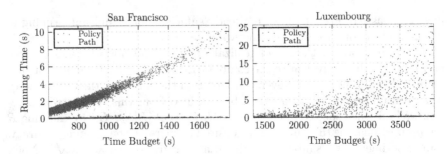

Fig. 1. Running time of the pathfinding algorithm as a function of the travel time budget for random unpruned (i.e., non-preprocessed) instantiations of each network. We can see that the path computation time is dominated by the policy computation time, effectively reducing the path-based SOTA problem to the policy-based SOTA problem in terms of computation time.

variance of the link travel time distributions) to compute the optimal path for 100 time-step budgets. For comparison, our algorithm solves for paths on the San Francisco network with budgets of up to 1400 s (= 1400 time-steps) in ≈ 7 s, even achieving query times below 1 s for budgets less than 550 s without any pre-processing at all. Furthermore, it also handles most queries on the 71655-edge Luxembourg network in ≈ 10 s (almost all of them in 20 s), where the network and time budgets are more than an order of magnitude larger than the 2950-edge network handled by the SD approach in the same amount of time.

Of course, this speedup—which increases more dramatically with the problem size—is hardly surprising or coincidental; indeed, it is quite fundamental to the nature of the algorithm: by drawing on the optimal policy as an upper bound (and quite often an accurate one) for the reliability of the final path, it has a very clear and fundamental informational advantage over any search algorithm that lacks any knowledge of the final solution. This allows the algorithm to direct itself toward the final path in an intelligent manner.

It is, however, less clear and more difficult to see how one might compare the performance of our generic discrete-time approach with Gaussian-restricted, continuous-time approaches [12,24]. Such approaches operate under drastically different assumptions and, in the case of [12], use approximation techniques, which we have yet to employ for additional performance improvements. When the true travel times cannot be assumed to follow Gaussian distributions, however, our method, to the best of our knowledge, presents the most efficient means for solving the path-based SOTA problem.

As we show next, combining our algorithm with preprocessing techniques allows us to achieve even further reductions in query time, making it more tractable for industrial applications on appropriately sized networks.

Preprocessing. Figure 2 demonstrates policy-based and path-based preprocessing using Arc-Potentials for two random San Francisco and Luxembourg problem instances. As can be seen in the figure, path-based preprocessing is in general much more effective than policy-based preprocessing.

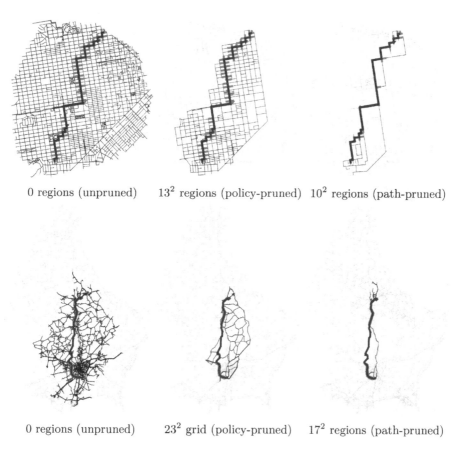

0 regions (unpruned) 13^2 regions (policy-pruned) 10^2 regions (path-pruned)

0 regions (unpruned) 23^2 grid (policy-pruned) 17^2 regions (path-pruned)

Fig. 2. Policy- vs. path-based pruning for random instances of San Francisco ($T = 837$ s, source at top) and Luxembourg ($T = 3165$ s, source at bottom). Light-gray edges are pruned from the graph and blue edges belong to the optimal path, whereas red edges belong to (sub-optimal) paths that were on the queue at the termination of the algorithm. (Color figure online)

Figure 3, summarized in Table 1, shows how the computation times scale with the preprocessing parameters. As expected, path-based preprocessing performs much better than purely policy-based preprocessing, and both become faster as we use more fine-grained regions. Nevertheless, we see that the majority of the speedup is achieved via a small number of regions, implying that preprocessing can be very effective even with low amounts of storage. (For example, for a 17×17 grid in Luxembourg, this amounts to $71655 \times 17^2 \approx 21$ M floats.)

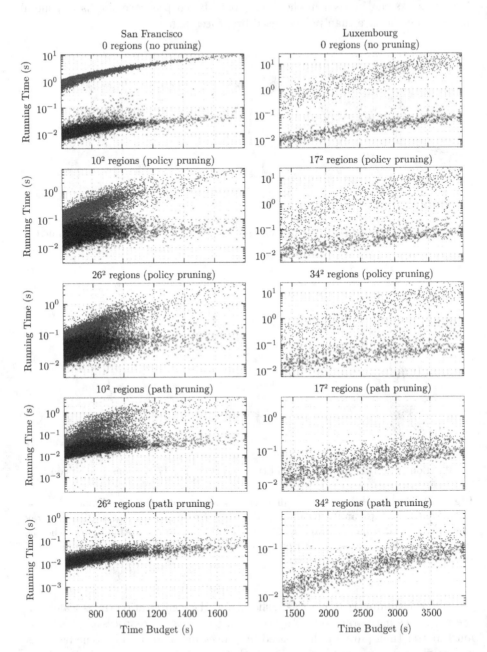

Fig. 3. Running time of pathfinding algorithm as a function of the time budget for each network. Red dots represent the computation time of the policy, and blue crosses represent the computation of the path using that policy. (Color figure online)

Table 1. The average query time with both policy-based and path-based pruning at various grid sizes and time budgets on the San Francisco network (left) and the Luxembourg network (right). We can see that in both cases, most of the speedup occurs at low granularity (and thus low space requirements).

Grid/	Time budget (seconds)					Grid/	Time budget (seconds)				
Pruning	800	1000	1200	1400	1600	Pruning	1500	2000	2500	3000	3500
Unpruned	1.81	3.00	4.10	5.11	5.23	Unpruned	0.54	1.29	3.53	6.27	9.95
10×10, policy	0.30	0.69	1.11	1.66	1.72	17×17, policy	0.25	0.83	2.31	4.57	7.97
26×26, policy	0.17	0.40	0.63	0.93	0.97	34×34, policy	0.21	0.71	2.09	3.79	7.07
10×10, path	0.11	0.38	0.63	0.87	0.90	17×17, path	0.03	0.06	0.09	0.13	0.18
26×26, path	0.02	0.04	0.06	0.07	0.08	34×34, path	0.02	0.04	0.06	0.08	0.12

5.2 Scalability

Path-based preprocessing requires routing between all $\approx |V|^2$ pairs of vertices, which is quadratic in the size of the network and intractable for moderate size networks. In practice, this meant that we had to preprocess every region lazily (i.e. on-demand), which on our CPUs took 9000 CPU-hours. It is therefore obvious that this becomes intractable for large networks, leaving policy-based preprocessing as the only option. One possible approach for large-scale path-based preprocessing might be to consider the boundary of each region rather than its interior [8]. While currently uninvestigated, such techniques may prove to be extremely useful in practice, and are potentially fruitful topics for future exploration.

6   Conclusion and Future Work

We have presented an algorithm for solving the path-based SOTA problem by first solving the easier policy-based SOTA problem and then using its solution as a search heuristic. We have also presented two approaches for preprocessing the underlying network to speed up computation of the search heuristic and path query, including a generalization of the Arc-Flags preprocessing algorithm that we call Arc-Potentials. We have furthermore applied and implemented these algorithms on moderate-sized transportation networks and demonstrated their potential for high efficiency in real-world networks.

While unobserved in practice, there remains the possibility that our algorithm may perform poorly on stochastic networks in which the optimal policy is a poor heuristic for the path reliability. Proofs in this direction have remained elusive, and determining whether such scenarios can occur in realistic networks remains an important step for future research. In the absence of theoretical advances, however, our algorithm provides a more tractable alternative to the state-of-the-art techniques for solving the path-based SOTA problem.

While our approach is tractable for larger networks than were possible with previous solutions, it does not scale well enough to be used with regional or continental sized networks that modern deterministic shortest path algorithms can handle with ease. In the future, we hope to investigate how our policy-based approach might be combined with other techniques such as the first-order stochastic dominance [4] and approximation methods such approximate Arc-Flags [8] for further speedup, and to also look into algorithms that allow for at least a partial relaxation of the independence assumption. We hope that our techniques will provide a strong basis for even better algorithms to tackle this problem for large-scale networks in the foreseeable future.

References

1. Fan, Y., Robert Kalaba, J.E., Moore, I.I.: Arriving on time. J. Optim. Theor. Appl. **127**(3), 497–513 (2005)
2. Flajolet, A., Blandin, S., Jaillet, P.: Robust adaptive routing under uncertainty (2014). arXiv:1408.3374
3. Samaranayake, S., Blandin, S., Bayen, A.: A tractable class of algorithms for reliable routing in stochastic networks. Transp. Res. Part C **20**(1), 199–217 (2012)
4. Nie, Y.M., Wu, X.: Shortest path problem considering on-time arrival probability. Trans. Res. Part B Methodol. **43**(6), 597–613 (2009)
5. Fan, Y., Nie, Y.: Optimal routing for maximizing travel time reliability. Netw. Spat. Econ. **6**(3–4), 333–344 (2006)
6. Dean, B.C.: Speeding up stochastic dynamic programming with zero-delay convolution. Algorithmic Oper. Res. **5**(2), 96 (2010)
7. Samaranayake, S., Blandin, S., Bayen, A.: Speedup techniques for the stochastic on-time arrival problem. In: ATMOS, pp. 83–96 (2012)
8. Sabran, G., Samaranayake, S., Bayen, A.: Precomputation techniques for the stochastic on-time arrival problem. In: SIAM, ALENEX, pp. 138–146 (2014)
9. Gutman, R.: Reach-based routing: a new approach to shortest path algorithms optimized for road networks. In: ALENEX/ANALC, pp. 100–111 (2004)
10. Hilger, M., Köhler, E., Möhring, R., Schilling, H.: Fast point-to-point shortest path computations with Arc-Flags. Ninth DIMACS Implementation Challenge **74**, 41–72 (2009)
11. Nikolova, E., Kelner, J.A., Brand, M., Mitzenmacher, M.: Stochastic shortest paths via quasi-convex maximization. In: Azar, Y., Erlebach, T. (eds.) ESA 2006. LNCS, vol. 4168, pp. 552–563. Springer, Heidelberg (2006)
12. Lim, S., Sommer, C., Nikolova, E., Rus, D.: Practicalroute planning under delay uncertainty: stochastic shortest path queries. Robot. Sci. Syst. **8**(32), 249–256 (2013)
13. Parmentier, A., Meunier, F.: Stochastic shortest paths and risk measures (2014). arXiv:1408.0272
14. Delling, D., Sanders, P., Schultes, D., Wagner, D.: Engineering route planning algorithms. In: Lerner, J., Wagner, D., Zweig, K.A. (eds.) Algorithmics of Large and Complex Networks. LNCS, vol. 5515, pp. 117–139. Springer, Heidelberg (2009)
15. Gardner, W.G.: Efficient convolution without input/output delay. In: Audio engineering society convention 97. Audio Engineering Society (1994)
16. Dechter, R., Pearl, J.: Generalized best-first search strategies and the optimality of A^*. J. ACM (JACM) **32**(3), 505–536 (1985)
17. Russell, S.J., Norvig, P.: Artificial Intelligence: A Modern Approach. Prentice-Hall Inc., London (1995). ISBN 0-13-103805-2

18. Abraham, I., Fiat, A., Goldberg, A., Werneck, R.: Highway dimension, shortest paths, and provably efficient algorithms. In: Proceedings of the Twenty-First Annual ACM-SIAM Symposium on Discrete Algorithms, pp. 782–793. Society for Industrial and Applied Mathematics (2010)
19. Goldberg, A., Kaplan, H., Werneck, R.: Reach for A^*: efficient point-to-point shortest path algorithms. In: ALENEX, vol. 6, pp. 129–143. SIAM (2006)
20. Geisberger, R., Sanders, P., Schultes, D., Delling, D.: Contraction hierarchies: faster and simpler hierarchical routing in road networks. In: McGeoch, C.C. (ed.) WEA 2008. LNCS, vol. 5038, pp. 319–333. Springer, Heidelberg (2008)
21. Bast, H., Funke, S., Matijevic, D.: Transit: ultrafast shortest-path queries with linear-time preprocessing. In: 9th DIMACS Implementation Challenge [1] (2006)
22. D'Angelo, G., Frigioni, D., Vitale, C.: Dynamic arc-flags in road networks. In: Pardalos, P.M., Rebennack, S. (eds.) SEA 2011. LNCS, vol. 6630, pp. 88–99. Springer, Heidelberg (2011)
23. Hunter, T., Abbeel, P., Bayen, A.M.: The path inference filter: model-based low-latency map matching of probe vehicle data. In: Frazzoli, E., Lozano-Perez, T., Roy, N., Rus, D. (eds.) Algorithmic Foundations of Robotics X. STAR, vol. 86, pp. 591–607. Springer, Heidelberg (2013)
24. Lim, S., Balakrishnan, H., Gifford, D., Madden, S., Rus, D.: Stochastic motion planning and applications to traffic. Int. J. Robot. Res. 3–13 (2010)

An Experimental Evaluation of Fast Approximation Algorithms for the Maximum Satisfiability Problem

Matthias Poloczek[(⊠)] and David P. Williamson

School of Operations Research and Information Engineering,
Cornell University, Ithaca, NY 14853, USA
{poloczek,davidpwilliamson}@cornell.edu

Abstract. We evaluate the performance of fast approximation algorithms for MAX SAT on the comprehensive benchmark sets from the SAT and MAX SAT contests. Our examination of a broad range of algorithmic techniques reveals that greedy algorithms offer particularly striking performance, delivering very good solutions at low computational cost. Interestingly, their relative ranking does not follow their worst case behavior. Johnson's deterministic algorithm is constantly better than the randomized greedy algorithm of Poloczek et al. [31], but in turn is outperformed by the derandomization of the latter: this 2-pass algorithm satisfies more than 99 % of the clauses for instances stemming from industrial applications. In general it performs considerably better than non-oblivious local search, Tabu Search, WalkSat, and several state-of-the-art complete and incomplete solvers, while being much faster. But the 2-pass algorithm does not achieve the excellent performance of Spears' computationally intense simulated annealing. Therefore, we propose a new hybrid algorithm that combines the strengths of greedy algorithms and stochastic local search to provide outstanding solutions at high speed: in our experiments its performance is as good as simulated annealing, achieving an average loss with respect to the best known value of less that 0.5 %, while its speed is comparable to the greedy algorithms.

1 Introduction

In the maximum satisfiability problem (MAX SAT) we are given a set of clauses over Boolean variables and want to find an assignment that satisfies a maximum number of clauses. Besides its prominent role in theoretical computer science, in particular in the hardness of approximation, this problem has been studied intensely due to its large variety of applications that spread across computer science, mathematical logic, and artificial intelligence. In general, encoding a problem as a possibly unsatisfiable CNF formula allows us to handle inconsistencies stemming from incomplete and/or noisy data, or to deal with contradictory objectives, where no solution can satisfy all constraints: an example for the former scenario arises in the study of protein interaction [37]; the latter task occurs for instance in assigning resources (e.g., [24], also see [25] for a comprehensive list

© Springer International Publishing Switzerland 2016
A.V. Goldberg and A.S. Kulikov (Eds.): SEA 2016, LNCS 9685, pp. 246–261, 2016.
DOI: 10.1007/978-3-319-38851-9_17

of MAX SAT formulations). The recent increase in applications of MAX SAT has been fueled by the progress on scalable algorithms that allow solving larger instances than before [22].

We conduct an extensive experimental evaluation of fast approximation algorithms, many of which provide provable guarantees on how far their solution may deviate from the optimum. The first class of algorithms under consideration are greedy algorithms that have a particularly simple design: they perform one or two passes over the variables and decide each variable by looking only at the clauses that it appears in. Johnson's algorithm [7,8,13] satisfies at least $\frac{2}{3}$Opt clauses, where Opt is the number of clauses satisfied by an optimal assignment. The randomized greedy algorithm [30,32] and the deterministic 2-pass algorithm [31] both achieve $\frac{3}{4}$Opt (the former in expectation); all these guarantees hold for any input.

Our primary goal is to explore how these algorithms perform on clause sets that arise in real-world applications, given that their development was driven by rather theoretical considerations [30,36]. Therefore we examine them with respect to the quality of their solutions and their running time on the instances of the SAT (SC) and MAX SAT (MS) contests in 2015 and 2014 [3–5,12]. Additional categories of these benchmark sets are random instances and crafted instances. Our experiments reveal a superb performance of the greedy algorithms on the industrial benchmarks. All algorithms satisfy a much higher fraction of the clauses than their respective worst case guarantees: for the category SC-APP of application instances Johnson's algorithms satisfies 98.7% and the 2-pass algorithm even 99.5% of the clauses on average.

In the light of these results the average fraction of 97.4% achieved by the randomized greedy algorithm is somewhat unexpected, given its superior worst case performance compared to Johnson's algorithm – in this case designing an algorithm with a considerably improved guarantee did not lead to better results in practice.

The local search paradigm is known to yield practical algorithms for MAX SAT, many of which build on algorithmic ideas originally developed to solve large SAT instances (cp. Sect. 7 in the survey of Gu, Purdom, Franco, and Wah [11]). We studied a variety of these methods in order to contrast our results for greedy algorithms. In the course of our examination we both confirmed and extended observations made by Pankratov and Borodin [28] in their comprehensive study. They reported a good performance of non-oblivious local search [18] for random instances, which we confirm. Interestingly, we noticed that the 2-pass algorithm produces equally good solutions in much shorter time for these instances. Furthermore, Pankratov and Borodin proposed a combination of non-oblivious local search with Tabu Search, which provides a deterministic strategy to escape local optima, and showed that it obtains very good results for instances from the MAX SAT 2007 competition. At first glance, our experiments did not confirm this: on the industrial benchmark SC-APP their algorithm satisfied merely 90.5% of the clauses on average. A closer examination revealed an explanation for this mediocre performance (see Sect. 3.1).

Of particular interest is Spears' simulated annealing [34] that was found in [28] and also in our study to produce excellent solutions in all benchmark categories. However, the downside of this method is its extremely high computational cost, which becomes especially evident in contrast to greedy algorithms: Johnson's algorithm for example solved even the largest instance of the benchmark sets, that contains about 14,000,000 variables in 56,000,000 clauses, in less than ten seconds, whereas simulated annealing exceeded the time limit of $120\,\mathrm{s}$ already for instances with 26,000 variables in less than 100,000 clauses.

Therefore, we propose a new hybrid algorithm that combines the strengths of both paradigms: our algorithm achieves the outstanding quality of Spears' algorithm while preserving the speed of a greedy algorithm: for example, it satisfies 99.75 % of the clauses in $4.7\,\mathrm{s}$ averaged over the instances of the SC-APP category, whereas simulated annealing obtains a 99.77 %-average in $104.9\,\mathrm{s}$. Moreover, our algorithm is easy to implement.

Finally, we also measure the performance of our algorithm with respect to the "best known value" (see Sect. 6). The hybrid algorithm achieves more than 99.5 % on average of the best known value and presents itself as an "all-rounder" that delivers excellent approximative solutions for all types of MAX SAT instances. Moreover, it proves to be very robust in the sense that its worst relative performance seen for any industrial instance is a "loss" of 1.88 % with respect to the best known value; the largest loss over all 2032 benchmark instances is only 3.18 % and attained for a crafted formula.

The Structure of the Article. Our selection of algorithms and the related work are presented in Sect. 2. Then we evaluate the performance of greedy algorithms and local search methods on benchmark instances in Sect. 3. In Sect. 4 we propose two algorithms that obtain excellent solutions at low computational cost. Section 5 describes our experiences with two modifications of greedy algorithms. Here we witness a curious phenomenon that illustrates the pitfalls when exploiting theoretical observations in practice. Section 6 examines the performance of the algorithms relative to the "best known value". Our conclusions are given in Sect. 7.

2   The Selected Algorithms and Related Work

We describe our selection of approximation algorithms in closer detail and motivate our choices. First we present the greedy algorithms and their derivatives; they can all be implemented in linear time (see [32] for details). Then we describe the methods that utilize (stochastic) local search, before we present our selection of complete solvers for MAX SAT.

Our Performance Metric. In our experiments we measure the ratio of clauses that an algorithm has satisfied divided by the total number of clauses in the instance. This measure is known as *totality ratio* or *performance ratio*. For randomized algorithms we take the mean over ten iterations, unless stated otherwise. For comparison, note that the performance indicator of an approximation algorithm typically used in theoretical computer science is the *approximation ratio*, where we divide by the optimum instead of the total number of clauses. However, the approximation ratio is impractical, since for many instances no optimal assignment is known. We point out that the totality ratio lower-bounds the approximation ratio, hence if the totality ratio is close to one, then the approximation ratio is as well.

A Simple Variant of Johnson's Algorithm. Johnson's greedy algorithm [13] (in the following abbreviated as JA) from 1974 was the first approximation algorithm for MAX SAT. We use the following slightly simpler version proposed in [29]; both variants guarantee a $\frac{2}{3}$-approximation [7,29]. This deterministic algorithm considers the variables in a fixed order $x_1, \ldots, x_n$. In order to decide x_i, the algorithm iterates over all clauses that contain x_i and are not satisfied by the setting of $x_1, \ldots, x_{i-1}$. These clauses are partitioned into disjoint sets P_1, P_2, N_1, and N_2 as follows: if x_i has a positive sign in clause c, then c is contained in P_1 if x_i is the last unfixed variable in c and in P_2 otherwise. If x_i occurs with negative sign in c, the clause is analogously either added to N_1 or N_2. Then the algorithm sets $x_i = 1$ if $2 \cdot |P_1| + |P_2| \geq 2 \cdot |N_1| + |N_2|$, and $x_i = 0$ otherwise.

The Randomized Greedy Algorithm. Poloczek and Schnitger [30] gave a greedy algorithm (RG) that also considers the variables in a fixed order, but makes carefully biased random decisions to achieve a worst case guarantee of $\frac{3}{4}$. We use the following elegant assignment probabilities suggested by van Zuylen [38]: the algorithm determines P_1, P_2, N_1, and N_2 as above, and then sets $x_i = 1$ with probability $\frac{|P_1| + |P_2| - |N_1|}{|P_2| + |N_2|}$ and to zero otherwise (see [32] for more details).

The 2-Pass Algorithm. The 2-pass algorithm (2Pass) is a derandomization of RG and proposed in [31]. It performs two passes over the input. During the first pass it computes an assignment probability σ_i for each variable x_i, but does not fix any variable yet: when considering x_i, it pretends that $x_1, \ldots, x_{i-1}$ were fixed independently according to their respective assignment probability, and then chooses σ_i similarly to the probability used by RG. Afterwards, in a second pass over the variables, this virtual random assignment is derandomized via the method of conditional expectations to achieve a $\frac{3}{4}$-approximation deterministically.

Non-oblivious Local Search. Khanna, Motwani, Sudan, and Vazirani [18] proposed non-oblivious local search for MAX SAT (NOLS). Starting from an arbitrary assignment b it searches the 1-flip environment of b for a better solution, where the 1-flip environment of b consists of all assignments where exactly one variable is flipped with respect to b. The value of an assignment is assessed by the following objective function: if k is the maximum clause length, then let S_i for $0 \leq i \leq k$ be the number of clauses for which exactly i literals evaluate to one under some assignment b. Then the objective value of b is $\sum_{i=0}^{k} c_i \cdot S_i$, where the coefficient c_i is defined recursively as $c_0 = 0$ and $c_i = \Delta_i + c_{i-1}$ for $i \geq 1$ with $\Delta_i = \frac{1}{(k-i+1) \cdot \binom{k}{i-1}} \cdot \sum_{j=0}^{k-i} \binom{k}{j}$. Non-oblivious local search flips in each iteration the variable of largest gain according to this function and stops if no variable offers a positive gain. Thus, NOLS does not look greedily in direction of the assignment satisfying a maximum number of clauses. Khanna et al. showed that it satisfies a $(1 - 2^{-k})$-fraction of all clauses if all clauses have length exactly k, and hence performs better than the local search that simply tries to maximize the number of satisfied clauses. In [28] NOLS has an excellent performance if the total number of flips is limited.

Tabu Search. In each iteration Tabu Search first invokes non-oblivious local search to find a locally optimal assignment (according to the underlying objective function). If there are unsatisfied clauses, it tries to improve this solution in a second stage. To this end, it maintains the so called taboo list, that contains the last t variables that were flipped, and repeatedly applies the following rules (in the given order): If there is a variable x such that flipping x increases the number of satisfied clauses, then it flips the variable with maximum gain. Otherwise it determines the (possibly negative) gain for each variable that appears in an unsatisfied clause. Then it flips the best such variable that is not contained in the taboo list. If all these variables are listed, the algorithm flips the variable that was least recently flipped. All ties are broken arbitrarily. It is important to note that the second rule might flip variables with negative gain, i.e. it allows for a decrease in satisfied clauses. If an assignment is obtained that is better than the one found by the first stage, then the algorithm starts a new iteration beginning with non-oblivious local search on this assignment. If the second stage fails to construct a better assignment within t steps, then Tabu Search terminates. Following [23,28], we set t to the number of variables in the instance.

Pankratov and Borodin [28] found a good performance of NOLS+TS for their benchmark instances, where it performed better than WalkSat and was comparable to Spears' simulated annealing.

Remark. For the sake of completeness we mention two natural extensions of these local search algorithms. The effects of restarting the local search algorithms multiple times were studied in [28] with different initial assignments drawn uniformly at random. While this increases the already large running time even further, the authors report that it only improved the returned solution if the number of variables was small, the reason apparently being that *"good initializations are not very dense."* They also studied the effects of lengthening the taboo list in order to allow a more exhaustive exploration of the search space. However, they found that this variation did *not* yield a larger number of satisfied clauses despite increasing the running time further. Therefore, we do not include these variants in our examination.

Simulated Annealing. We examine the following stochastic local search method due to Spears [34] called SA in the sequel. Starting with an arbitrary assignment, in each iteration it considers all variables according to a fixed ordering. For every variable x_i it computes the change δ_i in satisfied clauses if the variable were flipped. Then the variable is flipped with probability p that depends on δ_i and a noise parameter called the *temperature* T: Spears set $p = 1/(1 + e^{-\delta_i/T})$. The intention is that initially the temperature T is high and the algorithm is likely to flip a variable even if this worsens the assignment.

After each iteration the temperature is cooled down, and hence the algorithm explores the solution space in a more goal-oriented way. We use the parameters proposed by Spears that were also used in [28]: the initial temperature of 0.3 is multiplied by $e^{-1/n}$ in each iteration, where n is the number of variables. The search terminates when the temperature drops below 0.01. Note that the running time is $O(n^2)$, where the constant factor is comparable to the greedy algorithms.

WalkSat and Dist. Selman, Kautz, and Cohen [33] studied the following random walk. Given an assignment to the variables, their algorithm chooses an unsatisfied clause and flips one of its literals, thereby satisfying the respective clause. If there still is an unsatisfied clause the algorithm iterates. Both the selection of the clause and of the literal give rise to various heuristics developed over the last decades. In our evaluation we study the latest release of WalkSat [17] with default settings. Pankratov and Borodin [28] found that WalkSat performed well for random 3CNF formulae, but was outperformed for their benchmark instances by Tabu Search and by Spears' simulated annealing.

We also evaluate a local search called Dist that was developed by Cai, Luo, Thornton, and Su [6]. For the type of MAX SAT instances we consider, where all clauses are unweighted, Dist performs a random walk similar to WalkSat and employs a sophisticated heuristic to select the next variable to flip. Dist was found to be among the best incomplete solvers for the MS-CRAFTED and MS-RANDOM categories at the MAX SAT 2014 contest. Our evaluation confirms a great performance on these two categories, but on the corresponding sets of the SAT competitions and on all industrial instances it is considerably worse. That is why we omit Dist in this evaluation. Our findings are reported in the full version of this paper.

Open-WBO, Clasp, EvaSolver, and AHMAXSAT-LS. For the algorithms discussed so far one can be sure to have found an optimal assignment only if it satisfies all clauses. The goal of complete solvers is to certify the optimality of their assignments even if the clause set is unsatisfiable.

We tested three solvers that were among the best in the MS-APP category at the MAX SAT competitions, since industrial instances are our primary interest. These solvers utilize unsatisfiable core strategies, that seem particularly successful for instances stemming from industrial applications. The first solver, Open-WBO [20–22], was developed by Martins, Manquinho, and Lynce and ranked among the best in the MS-APP category both in 2015 and 2014. It has a modular design that allows the user to substitute the underlying SAT-solver: we chose Glucose 3.0, following a recommendation of Martins [19]. The second solver is Clasp [16], an award-winning answer set solver based on work by Andres, Gebser, Kaminski, Kaufmann, Romero, and Schaub [2,9,10]. It features a large variety of optimization strategies that are useful for MAX SAT instances (see their articles and the source code for details). We experimented with various parameter settings and algorithms, following suggestions of Kaufmann [15], and report our results for two complementary strategies that performed best. The third candidate, EvaSolver [26], was developed by Narodytska and Bacchus and is a refinement of [27]. This solver employs an alternative approach to express the subproblems arising for a core-guided strategy. EvaSolverwas awarded "the best non-portfolio solver" at the MAX SAT 2014 competition.

These solvers are unmatched in the discipline of providing certified optimal solutions for industrial instances. However, this comes at the expense of outputting no solution at all more often than not in our evaluation. In particular, we found the instances from the SAT competitions out of reach. Therefore, following suggestions of Kaufmann [15] and Martins [19], we also tested simpler strategies of their respective solvers, for example based on branch-and-bound or linear search, that return some solution in most cases, albeit a sub-optimal one.

Complementing the above selection, we also evaluated AHMAXSAT-LS 1.55 (AHMSLS) by Abramé and Habet [1]. This sophisticated branch-and-bound based algorithm uses local search pre-processing and was the best complete solver on crafted and random instances at the MAX SAT competitions in 2015 and 2014.

Complete solvers aim for a certificate of optimality, and therefore have a much larger conceptual complexity than the algorithms above. We observed that the core-guided solvers do not output any solution for a large fraction of the instances, while those relying on simpler and faster methods had a mediocre performance in terms of "average fraction of satisfied clauses". That is why we omit the complex solvers in this evaluation and report their experimental results in the full version of this paper.

3   The Experimental Results

The benchmark instances were taken from the SAT Competitions and the MAX SAT evaluations in 2015 and 2014 [3–5,12]. Our focus is on the collections stemming from industrial applications, called SC-APP and MS-APP. They contain 300 and 55 instances respectively that encode problems arising for example in formal verification of processors, attacks on cryptographic algorithms, or transportation planning. Furthermore, the sets SC-CRAFTED and MS-CRAFTED contain 300 and 402 clause sets, originating for example from instances of maximum cut, graph isomorphism, or clique coloring. The last category of benchmark instances are random formulae, the crucial parameter being the ratio of clauses to variables: a larger ratio implies that the optimal assignment satisfies a smaller fraction of clauses. SC-RANDOM contains 225 randomly generated CNFs formulae with a low ratio, MS-RANDOM contains 750 instances with a very high ratio. The main distinguishing trait of MS-CRAFTED and MS-RANDOM is that most of these instances are highly unsatisfiable, i.e. no algorithm can obtain a large fraction of the clauses. Summing up, our evaluation comprises over 2000 instances in six categories; we will omit 156 uncategorized clause sets from [12] in the sequel.

The sizes of the instances vary a lot: the smallest has 27 variables in 48 clauses, whereas the largest contains 14 million variables in over 53 million clauses. The mean values over the two industrial sets are roughly half a million variables and two million clauses. Moreover, the SC sets contain much larger instances than their MS counterparts. The benchmark instances are all unweighted, although repeating clauses allows for implicit weighting.

For our evaluation the algorithms were implemented in C++ using GCC 4.7.2. All experiments ran under Debian wheezy on a Dell Precision 490 workstation (Intel Xeon 5140 2.33 GHz with 8 GB RAM). Since the local search methods and the complete solvers can take a very long time to terminate, we set a time limit of 120 s, after which the best assignment found so far (or its value resp.) is returned.

3.1 The Greedy Algorithms

The greedy algorithms produced solutions of high quality at very low computational cost in our experimental evaluation. The loss compared to the respective best algorithm is never larger than a few percent for any of the benchmark categories. The randomized greedy algorithm (RG) never left more than five percent of the clauses unsatisfied on average, except for the highly unsatisfiable instances of MS-CRAFTED and MS-RANDOM. Together with Johnson's algorithm (JA), RG is the fastest algorithm in our selection. The average fraction of clauses satisfied by JA is larger than RG's for all sets; looking more closely, we see that JA obtained a better solution than RG for almost all instances individually, which is somewhat surprising given the superior worst case guarantee of RG.

Therefore, it is even more interesting that the deterministic 2-pass algorithm (2Pass), that is a derandomization of RG and in particular relies on the same algorithmic techniques, outperforms JA (and RG) on all categories. On the instances from the SAT Competition it even satisfies more than 99 % of the clauses on average. Its running time is about three times larger than for JA, and 2Pass computes a better solution than JA in three out of every four instances. The results are summarized in Table 1.

3.2 The Local Search Methods

The random walks of WalkSat typically converged quickly, making it the fastest local search variant in our line-up. We confirm the observation of [28] that WalkSat performs well on random instances: for SC-RANDOM it found very

Table 1. The performance of greedy algorithms

	RG		JA		2Pass	
	%sat	∅time	%sat	∅time	%sat	∅time
SC-APP	97.42	0.38 s	98.71	0.38 s	99.48	1.11 s
MS-APP	95.69	0.25 s	97.97	0.23 s	98.08	0.63 s
SC-CRAFTED	97.40	0.17 s	98.37	0.17 s	99.04	0.46 s
MS-CRAFTED	80.33	0.00 s	82.69	0.00 s	82.97	0.00 s
SC-RANDOM	97.58	1.39 s	98.72	1.38 s	99.19	5.38 s
MS-RANDOM	84.61	0.00 s	87.30	0.00 s	88.09	0.00 s

good assignments (98.7 %) and its running times were particularly short. For the application benchmarks WalkSat's performance exhibited a large discrepancy: its average fraction of satisfied clauses for SC-APP was only slightly worse than RG, while the average running time of about two seconds was rather high. But for the MS-APP instances it merely averaged 89.9 % of satisfied clauses, which is significantly worse than any of the greedy algorithms.

The second candidate is the combination of non-oblivious local search with Tabu Search (NOLS+TS), as suggested by Pankratov and Borodin [28]; we started it from a random assignment. Its forte were the random instances, where it was comparable to JA, and it also performed well on MS-CRAFTED. But on the application-based benchmarks it showed a poor performance: it only averaged 90.5 % for SC-APP and 83.6 % for MS-APP, which surprises because of its good performance reported in [28].

A closer examination reveals that NOLS+TS satisfied 98.9 % of the clauses on average for the SC-APP benchmark, *if* it finished before the timeout. For SC-CRAFTED we observed a similar effect (98.4 %); on the MS-APP the time bound was always violated. NOLS+TS returns the best assignment it has found in any iteration, if its running time exceeds the time limit; hence we interpret our findings that the escape strategy of Tabu Search (with the parameters suggested in Sect. 2) does not find a good solution quickly for two thirds of the SC-APP instances.

Therefore, we looked at non-oblivious local search, where the initial assignment was either random or obtained by 2Pass. The latter combination gave better results, therefore we focus on it in this exposition: 2Pass+NOLS achieved a higher mean fraction of satisfied clauses than WalkSat and NOLS+TS. However, comparing it to 2Pass, it turns out that the improvement obtained by the additional local search stage itself is marginal and comes at a high computational cost. For the instances of SC-APP the average running time was increased by a factor of 40, for MS-APP even by a factor of 130 compared to 2Pass.

Spears' simulated annealing (SA) finds excellent assignments, achieving the peak fraction of satisfied clauses over all benchmarks: for example, its average fraction of satisfied clauses is 99.8 % for SC-APP and 99.4 % for MS-APP. However, these results come at a extremely high computational cost; in our experiments SA almost constantly violated the time bound.

In this context another series of experiments is noteworthy: when we set the time bound of SA for each instance individually to the time that 2Pass needed to obtain its solution for that instance[1], then the average fraction of satisfied clauses of SA was decreased significantly: for example, for the SC-APP category its mean fraction of satisfied clauses dropped to 99.28 %, whereas 2Pass achieved 99.48 %. Our empirical data for local search based algorithms is summarized in Table 2.

---

[1] This analysis technique was proposed by Pankratov and Borodin [28] and is called "normalization".

Table 2. The performance of local search methods

	NOLS+TS		2Pass+NOLS		SA		WalkSat	
	%sat	Øtime	%sat	Øtime	%sat	Øtime	%sat	Øtime
SC-APP	90.53	93.59 s	99.54	45.14 s	99.77	104.88 s	96.50	2.16 s
MS-APP	83.60	120.14 s	98.24	82.68 s	99.39	120.36 s	89.90	0.48 s
SC-CRAFTED	92.56	61.07 s	99.07	22.65 s	99.72	70.07 s	98.37	0.66s
MS-CRAFTED	84.18	0.65 s	83.47	0.01 s	85.12	0.47 s	82.56	0.06 s
SC-RANDOM	97.68	41.51 s	99.25	40.68 s	99.81	52.14 s	98.77	0.94 s
MS-RANDOM	88.24	0.49 s	88.18	0.00 s	88.96	0.02 s	87.35	0.06 s

4 A Hybrid Algorithm that Achieves Excellent Performance at Low Cost

Among the algorithms considered so far, Spears' simulated annealing produced the best solutions. But given that the greedy algorithms were not far off in terms of satisfied clauses and only needed a fraction of the running time, the question is if it is possible to improve their solutions while preserving their speed.

Therefore, we combine the deterministic 2-pass algorithm with ten rounds of simulated annealing (ShortSA); in particular, we utilize the last ten rounds of Spears' algorithm, during which the temperature is low and hence the random walk is very goal-oriented. Here it is advantageous that below the hood both algorithms are very similar, in particular they consider the variables one-by-one and iterate for each variable over its set of clauses. Thus, the implementation of our hybrid variant requires very little additional effort. To the best of our knowledge, the combination of a greedy algorithm with only a few steps of simulated annealing is novel; in particular, the rationale and characteristics differ from using a greedy algorithm to produce a starting solution for local search, as it is common for example for TSP [14]. Moreover, our experiments demonstrate that using the 2-pass algorithm to provide an initial solution in standard local search for MAX SAT does not achieve both goals simultaneously (cp. Sect. 3.2).

The empirical running time of our linear-time algorithm scales even better than expected, averaging at 4.7 s for SC-APP and 3.9 s for MS-APP. Therefore its speed is comparable to the greedy algorithms and much faster than NOLS or SA; the latter took 104.88 s and 120.38 s respectively on average for these sets.

In terms of satisfied clauses our hybrid algorithm achieves the excellent performance of SA: for the SC-APP category 2Pass+ShortSA satisfies 97.75 % of the clauses, and hence the difference to SA is only marginal (0.02 %). Also for the other categories the additional local search stage essentially closes the gap, the maximum difference being 0.4 % for MS-CRAFTED. Like SA, it dominates strictly the other algorithms on the overwhelming majority of the instances.

In order to study the effect of the initial assignment provided by 2Pass, we contrasted the performance of our hybrid algorithm by starting ShortSA from the all-zero assignment. It turns out that the 2Pass assignment bridges about

Table 3. Comparison of the hybrid algorithm with 2Pass, SA, and ShortSA

	2Pass+ShortSA		SA		ShortSA		2Pass	
	%sat	∅time	%sat	∅time	%sat	∅time	%sat	∅time
SC-APP	99.75	4.72 s	99.77	104.88 s	99.63	3.60 s	99.48	1.11 s
MS-APP	99.20	3.29 s	99.39	120.36 s	99.68	2.45 s	98.08	0.63 s
SC-CRAFTED	99.56	2.20 s	99.72	70.07 s	99.55	1.74 s	99.04	0.46 s
MS-CRAFTED	84.69	0.00 s	85.12	0.47 s	84.63	0.00 s	82.97	0.00 s
SC-RANDOM	99.71	19.60 s	99.81	52.14 s	99.57	14.18 s	99.19	5.38 s
MS-RANDOM	88.84	0.00 s	88.96	0.02 s	88.67	0.00 s	88.09	0.00 s

half of the gap between ShortSA and SA, which reveals ShortSA to be another practical algorithm with excellent performance; typically, it is slightly worse than 2Pass+ShortSAand SA, with the notable exception that it beats all others on the 55 instances of MS-APP. We refer to Table 3 for a summary. In Sect. 6 we compare these algorithms relative to the best known value and also examine the reliability of our algorithm.

5   Two Variants of Greedy Algorithms

Since Johnson's algorithm performs very well for the benchmark instances of SC-APP, we wonder if we can improve the quality of its solutions further without sacrificing its main advantages: its speed and conceptual simplicity. In this section we discuss two ideas and make a curious observation about the structure of some of the benchmark instances.

5.1   Random Variable Orders

Replacing the lexicographical variable order by a random permutation seems to be a promising direction. This modification is also motivated by a result of Costello, Shapira, and Tetali [8]: they proved that this modification improves the expected approximation ratio of Johnson's algorithm by a small constant. Thus, for every instance there is a substantial number of orderings for which the algorithm performs better than in the worst case.

Therefore, we examined Johnson's algorithm and the randomized greedy algorithm for variable orders that were drawn uniformly at random among all permutations; for instance, such a random order can be computed efficiently via the Fisher-Yates-Shuffle. In the sequel these variants are denoted by JARO and RGRO respectively. Since RG and JA are typically about three times faster than the deterministic 2-pass algorithm and roughly ten to twelve times faster than our new variants derived from the 2-pass algorithm, we restarted the algorithms ten times on each instance. Testing a non-negligible fraction of the set of possible permutations would run counter to our primary design goals.

The behavior we witnessed on the industrial benchmark instances came as a surprise: not only did the random variable order fail to improve the average performance; it became even worse! In particular, the average fraction of satisfied clauses for the SC-APP instances dropped from 98.7 % to 97.8 %, when the lexicographical variable order was replaced by a permutation drawn uniformly at random. For the randomized greedy algorithm the decrease was even bigger: 96.0 % instead of 97.4 %. We observed the same effect for the instances of MS-APP: for Johnson's algorithm the average fraction of satisfied clauses changed from 98.0 % to 95.8 %, and from 95.7 % to 92.8 % for the randomized greedy algorithm.

An explanation for this odd phenomenon seems to be the origin of these instances: they stem from industrial applications, and apparently the respective encodings cause a lexicographical variable order to be advantageous. Accordingly, one would not expect such a structure to emerge for clause sets that are created randomly. And indeed, averaged over the instances of SC-RANDOM and MS-RANDOM the mean fractions of satisfied clauses for JA and JARO differed by less than 0.01 %; the same holds for RG and RGRO. Therefore we wonder: can we obtain an improvement by picking the best sampled variable order? Apparently, the random formulae are very symmetric, since best order sampled by JARO was only better by 0.06 % than the mean over all ten random permutations; in case of RGRO it was only 0.12 %. Only for MS-CRAFTED this modification yielded an improvement, as RGRO and JARO both gained about 1 %: thus, the best order of JARO typically was better than 2Pass, but still worse than our hybrid algorithm.

5.2   Multiple Restarts of the Randomized Greedy Algorithm

A single run of the randomized greedy algorithm comes at very cheap computational cost, both in time and memory consumption: among the algorithms under consideration, Johnson's algorithm and the randomized greedy algorithm have the lowest running time, which reflects their extremely efficient design. Therefore, we wonder if one can obtain a better solution by running the randomized greedy algorithm several times on each instance and returning the best solution. Again we set the number of repetitions to ten.

Our findings suggest that even with multiple restarts RG cannot compete with the other algorithms: the relative improvement of the best found assignment over the instance-specific sample mean was only 0.07 % on average over the 300 instances of SC-APP; the largest relative improvement for any of these instances was 0.6 %. Finally, we remark that the randomized greedy algorithm with multiple repetitions still typically performs worse than Johnson's algorithm: for example, of the 300 instances of SC-APP JA found in 273 cases an assignment that was better than the best found in any repetition of RG. The behavior found for the other benchmark sets is similar, in case of the MS-APP category Johnson's algorithm was even better on all inputs.

6   The Performance Relative to the Best Known Value

In our evaluation we report the totality ratio, which equals the average number of satisfied clauses divided by the total number of clauses in the instance. The reason is that the value of an optimal assignment is not known for a large part of the testbed. We point out that restricting the evaluation to instances where the optimum has been determined would incorporate an inherent measurement bias: usually the optimum is known for an unsatisfiable formula only if a complete solver was able to provide a proof of optimality. However, it seems that the latter obtain a certified optimal assignment only for certain types of instances (this aspect is discussed in detail in the full version of this paper). To the best of our knowledge, there is no complete understanding what structural properties make a MAX SAT instance solvable. Thus, the interpretation of such a restricted evaluation would be limited.

In this section we give an "educated guess" of the approximation ratio, where we estimate an unknown optimum of an instance by the *best known value*. This is the best solution returned by any algorithm. Here we include the complete solvers, that are particularly good at industrial instances, and Dist that is excellent on MS-CRAFTED and MS-RANDOM. Further, if a certified optimum is listed on the websites of the SAT or MAX SAT competitions, then we use that value.

We focus on the algorithms with best overall performance in our evaluation: our hybrid algorithm (2Pass+ShortSA), Spears' simulated annealing (SA), the short SA (ShortSA), and the 2-pass algorithm (2Pass). The relative performance ratios are summarized in Table 4. We recall that for the instances of MS-CRAFTED and MS-RANDOM typically no algorithm came close to satisfying all clauses (e.g., Table 3). The performance of the aforementioned algorithms relative to the best known value, however, is close to one: these values are highlighted. Note that the average running times have not changed with respect to Table 3 and are given for completeness.

We note that our new algorithm achieves an average relative performance of more than 99.5 %: that is, the "loss" with respect to the best known value is 0.46 % when averaging over the whole set of instances, and only 0.30 % when restricting ourselves to the application-based instances of SC-APP and MS-APP.

Table 4. The Performances relative to the best known optimum

	2Pass+ShortSA		SA		ShortSA		2Pass	
	%sat	Øtime	%sat	Øtime	%sat	Øtime	%sat	Øtime
SC-APP	99.79	4.72 s	99.81	104.88 s	99.67	3.60 s	99.52	1.11 s
MS-APP	99.22	3.29 s	99.41	120.36 s	99.70	2.45 s	98.10	0.63 s
SC-CRAFTED	99.61	2.20 s	99.77	70.07 s	99.60	1.74 s	99.09	0.46 s
MS-CRAFTED	**98.93**	0.00 s	**99.42**	0.47 s	**98.86**	0.00 s	**96.94**	0.00 s
SC-RANDOM	99.82	19.60 s	99.92	52.14 s	99.68	14.18 s	99.31	5.38 s
MS-RANDOM	**99.65**	0.00 s	**99.78**	0.02 s	**99.46**	0.00 s	**98.83**	0.00 s

Furthermore, our algorithm turns out to be very reliable and does not allow for large outliers: the worst relative performance for any industrial instance is a "loss" of 1.88 % with respect to the best known value; the largest loss over all 2032 benchmark instances is 3.18 % and attained for a crafted formula.

7 Conclusions

We found the performance of simple greedy algorithms and their derivatives on benchmark instances stemming from industrial applications to be impressive. In particular, the achieved performance ratios were much better than the respective worst case guarantees of the algorithms let us expect. Their approximation ratios are known to be tight, i.e. there are inputs for which the performance of the algorithm is not better. But apparently the approximation ratio is not a useful indicator for the worst case performance on the benchmark instances we considered. On the one hand, for none of the instances the experimentally found performance was nearly as bad as in the worst case. On the other hand, Johnson's algorithm that has a relatively bad worst case behavior dominated the randomized greedy algorithm.

Therefore, in order to bridge the gap between theory and practice, it seems interesting to study *and also design* approximation algorithms for MAX SAT from a "non-worst case" perspective. More specifically, we wonder if ideas from the "smoothed analysis" devised by Spielman and Teng can be applied to this end (see their survey [35] for examples).

Moreover, we propose a hybrid algorithm that achieves an excellent performance at low running time, paralleling the accuracy of the much more costly simulated annealing. In particular, its average "loss" relative to the best known assignment is only 0.46 % over all instances and 0.30 % for industrial instances. Therefore, our algorithm reliably produces excellent solutions for all types of MAX SAT instances. Note that the number of rounds for the stochastic local search allows a trade-off between speed and quality, and we leave it to future work to fine-tune this parameter.

Our comparison of Johnson's algorithm with either a lexicographical or a random variable order indicates that instances have some inner semantics that significantly affect the performance of greedy algorithms. Therefore, an interesting direction would be to identify parameters of the input that allow a better prediction of the algorithms' practical performance. Subsequently, this structure could be exploited to design fast algorithms tailored for specific applications.

Acknowledgments. The authors would like to thank Allan Borodin for his valuable comments, and Benjamin Kaufmann and Ruben Martins for their help with optimizing the parameters of their solvers for our setting.

The first author was supported by the Alexander von Humboldt Foundation within the Feodor Lynen program, and by NSF grant CCF-1115256, and AFOSR grants FA9550-15-1-0038 and FA9550-12-1-0200. The second author was supported by NSF grant CCF-1115256.

References

1. Abrame, A., Habet, D.: Ahmaxsat: Description and evaluation of a branch and bound Max-SAT solver. J. Satisfiability, Boolean Model. Comput. **9**, 89–128 (2015). www.lsis.org/habetd/Djamal_Habet/MaxSAT.html. Accessed on 02 Feb 2016
2. Andres, B., Kaufmann, B., Matheis, O., Schaub, T.: Unsatisfiability-based optimization in clasp. In: ICLP 2012, pp. 211–221 (2012)
3. Argelich, J., Li, C.M., Manyà, F., Planes, J.: MAX-SAT 2014: Ninth Max-SAT evaluation. www.maxsat.udl.cat/14/. Accessed on 12 Jan 2016
4. Argelich, J., Li, C.M., Manyà, F., Planes, J.: MAX-SAT 2015: Tenth Max-SAT evaluation. www.maxsat.udl.cat/15/. Accessed on 02 Feb 2016
5. Belov, A., Diepold, D., Heule, M.J., Järvisalo, M.: Proc. of SAT COMPETITION 2014: Solver and Benchmark Descriptions (2014). http://satcompetition. org/edacc/sc14/. Accessed on 28 Jan 2016
6. Cai, S., Luo, C., Thornton, J., Su, K.: Tailoring local search for partial MaxSAT. In: AAAI, pp. 2623–2629 (2014). the code is available at http://lcs.ios.ac.cn/caisw/ MaxSAT.html. Accessed on 25 Jan 2016
7. Chen, J., Friesen, D.K., Zheng, H.: Tight bound on Johnson's algorithm for maximum satisfiability. J. Comput. Syst. Sci. **58**, 622–640 (1999)
8. Costello, K.P., Shapira, A., Tetali, P.: Randomized greedy: new variants of some classic approximation algorithms. In: SODA, pp. 647–655 (2011)
9. Gebser, M., Kaminski, R., Kaufmann, B., Romero, J., Schaub, T.: Progress in clasp Series 3. In: Calimeri, F., Ianni, G., Truszczynski, M. (eds.) LPNMR 2015. LNCS, vol. 9345, pp. 368–383. Springer, Heidelberg (2015)
10. Gebser, M., Kaminski, R., Kaufmann, B., Schaub, T.: Multi-criteria optimization in answer set programming. In: ICLP, pp. 1–10 (2011)
11. Gu, J., Purdom, P.W., Franco, J., Wah, B.W.: Algorithms for the satisfiability (SAT) problem: A survey. In: Satisfiability Problem: Theory and Applications, pp. 19–152 (1996)
12. Heule, M., Weaver, S. (eds.): SAT 2015. LNCS, vol. 9340. Springer, Heidelberg (2015)
13. Johnson, D.S.: Approximation algorithms for combinatorial problems. J. Comput. Syst. Sci. **9**(3), 256–278 (1974)
14. Johnson, D.S., McGeoch, L.A.: The traveling salesman problem: A case study in local optimization. Local Search Comb. Optim. **1**, 215–310 (1997)
15. Kaufmann, B.: Personal communication
16. Kaufmann, B.: Clasp: A conflict-driven nogood learning answer set solver (version 3.1.3). http://www.cs.uni-potsdam.de/clasp/. Accessed on 28 Jan 2016
17. Kautz, H.: Walksat (version 51). www.cs.rochester.edu/u/kautz/walksat/, see the source code for further references. Accessed on 27 Jan 2016
18. Khanna, S., Motwani, R., Sudan, M., Vazirani, U.V.: On syntactic versus computational views of approximability. SIAM J. Comput. **28**(1), 164–191 (1998)
19. Martins, R.: Personal communication
20. Martins, R., Joshi, S., Manquinho, V., Lynce, I.: Incremental cardinality constraints for MaxSAT. In: O'Sullivan, B. (ed.) CP 2014. LNCS, vol. 8656, pp. 531–548. Springer, Heidelberg (2014)
21. Martins, R., Manquinho, V., Lynce, I.: Open-WBO: An open source version of the MaxSAT solver WBO (version 1.3.0). http://sat.inesc-id.pt/open-wbo/index. html. Accessed on 25 Jan 2016

22. Martins, R., Manquinho, V., Lynce, I.: Open-WBO: a modular MaxSAT solver. In: Sinz, C., Egly, U. (eds.) SAT 2014. LNCS, vol. 8561, pp. 438–445. Springer, Heidelberg (2014)

23. Mastrolilli, M., Gambardella, L.M.: MAX-2-SAT: How good is Tabu Search in the worst-case? In: AAAI, pp. 173–178 (2004)

24. Miyazaki, S., Iwama, K., Kambayashi, Y.: Database queries as combinatorial optimization problems. In: CODAS, pp. 477–483 (1996)

25. Morgado, A., Heras, F., Liffiton, M.H., Planes, J., Marques-Silva, J.: Iterative and core-guided MaxSAT solving: A survey and assessment. Constraints 18(4), 478–534 (2013)

26. Narodytska, N., Bacchus, F.: EvaSolver. https://www.cse.unsw.edu.au/ninan/. Accessed on 04 Jan 2016

27. Narodytska, N., Bacchus, F.: Maximum satisfiability using core-guided MaxSAT resolution. In: AAAI 2014, pp. 2717–2723 (2014)

28. Pankratov, D., Borodin, A.: On the relative merits of simple local search methods for the MAX-SAT problem. In: Strichman, O., Szeider, S. (eds.) SAT 2010. LNCS, vol. 6175, pp. 223–236. Springer, Heidelberg (2010)

29. Poloczek, M.: Bounds on greedy algorithms for MAX SAT. In: Demetrescu, C., Halldórsson, M.M. (eds.) ESA 2011. LNCS, vol. 6942, pp. 37–48. Springer, Heidelberg (2011)

30. Poloczek, M., Schnitger, G.: Randomized variants of Johnson's algorithm for MAX SAT. In: SODA, pp. 656–663 (2011)

31. Poloczek, M., Schnitger, G., Williamson, D.P., van Zuylen, A.: Greedy algorithms for the maximum satisfiability problem: Simple algorithms and inapproximability bounds, In Submission

32. Poloczek, M., Williamson, D.P., van Zuylen, A.: On some recent approximation algorithms for MAX SAT. In: Pardo, A., Viola, A. (eds.) LATIN 2014. LNCS, vol. 8392, pp. 598–609. Springer, Heidelberg (2014)

33. Selman, B., Kautz, H.A., Cohen, B.: Local search strategies for satisfiability testing. In: Cliques, Coloring, and Satisfiability: Second DIMACS Implementation Challenge, pp. 521–532 (1993)

34. Spears, W.M.: Simulated annealing for hard satisfiability problems. In: Cliques, Coloring and Satisfiability: Second DIMACS Implementation Challenge, pp. 533–558 (1993)

35. Spielman, D.A., Teng, S.: Smoothed analysis: an attempt to explain the behavior of algorithms in practice. Commun. ACM 52(10), 76–84 (2009)

36. Williamson, D.P.: Lecture notes in approximation algorithms, Fall 1998. IBM Research Report RC 21409, IBM Research (1999)

37. Zhang, Y., Zha, H., Chu, C.H., Ji, X.: Protein interaction interference as a Max-Sat problem. In: Proceedings of the IEEE CVPR 2005 Workshop on Computer Vision Methods for Bioinformatics (2005)

38. van Zuylen, A.: Simpler 3/4-approximation algorithms for MAX SAT. In: Solis-Oba, R., Persiano, G. (eds.) WAOA 2011. LNCS, vol. 7164, pp. 188–197. Springer, Heidelberg (2012)

Experimental Analysis of Algorithms
for Coflow Scheduling

Zhen Qiu, Clifford Stein, and Yuan Zhong[✉]

Department of IEOR, Columbia University, New York, NY 10027, USA
yz2561@columbia.edu

Abstract. Modern data centers face new scheduling challenges in optimizing job-level performance objectives, where a significant challenge
is the scheduling of highly parallel data flows with a common performance goal (e.g., the shuffle operations in MapReduce applications).
Chowdhury and Stoica [6] introduced the coflow abstraction to capture
these parallel communication patterns, and Chowdhury et al. [8] proposed effective heuristics to schedule coflows efficiently. In our previous
paper [18], we considered the strongly NP-hard problem of minimizing the total weighted completion time of coflows with release dates,
and developed the first polynomial-time scheduling algorithms with
$O(1)$-approximation ratios.

In this paper, we carry out a comprehensive experimental analysis on
a Facebook trace and extensive simulated instances to evaluate the practical performance of several algorithms for coflow scheduling, including
our approximation algorithms developed in [18]. Our experiments suggest
that simple algorithms provide effective approximations of the optimal,
and that the performance of the approximation algorithm of [18] is relatively robust, near optimal, and always among the best compared with
the other algorithms, in both the offline and online settings.

1 Introduction

Data-parallel computation frameworks such as MapReduce [9], Hadoop [1,5,19],
Spark [21], Google Dataflow [2], etc., are gaining tremendous popularity as they
become ever more efficient in storing and processing large-scale data sets in
modern data centers. This efficiency is realized largely through massive parallelism. Typically, a datacenter job is broken down into smaller tasks, which
are processed in parallel in a *computation stage*. After being processed, these
tasks produce intermediate data, which may need to be processed further, and
which are transferred between groups of servers across the datacenter network,
in a *communication stage*. As a result, datacenter jobs often alternate between
computation and communication stages, with parallelism enabling the fast completion of these large-scale jobs. While this massive parallelism contributes to
efficient data processing, it presents many new challenges for network scheduling.

Research partially supported by NSF grant CCF-1421161.

A.V. Goldberg and A.S. Kulikov (Eds.): SEA 2016, LNCS 9685, pp. 262–277, 2016.
DOI: 10.1007/978-3-319-38851-9_18

In particular, traditional networking techniques focus on optimizing flow-level performance such as minimizing flow completion times[1], and ignore job-level performance metrics. However, since a computation stage often can only start after all parallel dataflows within a preceding communication stage have finished [7,10], *all* these flows need to finish early to reduce the processing time of the communication stage, and of the entire job.

To faithfully capture application-level communication requirements, Chowdhury and Stoica [6] introduced the *coflow* abstraction, defined to be a collection of parallel flows with a common performance goal. Effective scheduling heuristics were proposed in [8] to optimize coflow completion times. In our previous paper [18], we developed scheduling algorithms with constant approximation ratios for the strongly NP-hard problem of minimizing the total weighted completion time of coflows with release dates, and conducted preliminary experiments to examine the practical performance of our approximation algorithms. These are the first $O(1)$-approximation algorithms for this problem. In this paper, we carry out a systematic experimental study on the practical performance of several coflow scheduling algorithms, including our approximation algorithms developed in [18]. Similar to [18], the performance metric that we focus on in this paper is the total weighted coflow completion time. As argued in [18], it is a reasonable user-oriented performance objective. It is also natural to consider other performance objectives, such as the total weighted *flow time*[2], which we leave as future work. Our experiments are conducted on real-world data gathered from Facebook and extensive simulated data, where we compare our approximation algorithm and its modifications to several other scheduling algorithms in an offline setting, and evaluate their relative performances, and compare them to an LP-based lower bound. The algorithms that we consider in this paper are characterized by several main components, such as the coflow order in which the algorithms follow, the grouping of the coflows, and the backfilling rules. We study the impact of each such component on the algorithm performance, and demonstrate the robust and near-optimal performance of our approximation algorithm [18] and its modifications in the offline setting, under the case of zero release times as well as general release times. We also consider online variants of the offline algorithms, and show that the online version of our approximation algorithm has near-optimal performance on real-world data and simulated instances.

The rest of this section is organized as follows. In Sect. 1.1, we quickly recall the problem formulation of coflow scheduling, the approximation algorithm of [18] as well as its approximation ratio. Section 1.2 gives an overview of the experimental setup and the main findings from our experiments. A brief review of related works is presented in Sect. 1.3.

---

[1] In this paper, the term "flow" refers to data flows in computer networking, and is not to be confused with the notion of "flow time," commonly used in the scheduling literature.

[2] Here "flow time" refers to the length of time from the release time of a coflow to its completion time, as in scheduling theory.

1.1 Coflow Model and Approximation Algorithm

We consider a discrete-time system where n *coflows* need to be scheduled in an $m \times m$ datacenter network with m *inputs* and m *outputs*. For each $k \in \{1, 2, \cdots, n\}$, coflow k is released at time r_k, and can be represented by an $m \times m$ matrix $D^{(k)} = \left(d_{ij}^{(k)} \right)_{i,j=1}^{m}$, where $d_{ij}^{(k)}$ is the number of data units (a.k.a. *flow size*) that need to be transferred from input i to output j. The network has the so-called non-blocking switch architecture [3,4,12,16], so that a data unit that is transferred out of an input is immediately available at the corresponding output. We also assume that all inputs and outputs have unit capacity. Thus, in a time slot, each input/output can process at most one data unit; similar to [18], these restrictions are called *matching constraints*. Let C_k denote the completion time of coflow k, which is the time when all data units from coflow k have finished being transferred. We are interested in developing efficient (offline) scheduling algorithms that minimize $\sum_{k=1}^{n} w_k C_k$, the total weighted completion time of coflows, where w_k is a weight parameter associated with coflow k.

A main result of [18] is the following theorem.

Theorem 1 [18]. *There exists a deterministic polynomial time 67/3-approximation algorithm for the coflow scheduling problem, with the objective of minimizing the total weighted completion time.*

The approximation algorithm of [18] consists of two related stages. First, a *coflow order* is computed by solving a polynomial-sized interval-indexed linear program (LP) relaxation, similar to many other scheduling algorithms (see e.g., [11]). Then, we use this order to derive an actual schedule. To do so, we define a *grouping* rule, under which we partition coflows into a polynomial number of groups, based on the minimum required completion times of the ordered coflows, and schedule the coflows in the same group as a single coflow optimally, according to an integer version of the Birkhoff-von Neumann decomposition theorem. The detailed description of the algorithm can be found in Algorithm 4 of the Appendix in [17]. Also see [18] for more details. From now on, the approximation algorithm of [18] will be referred to as the LP-based algorithm.

1.2 Overview of Experiments

Since our LP-based algorithm consists of an ordering and a scheduling stage, we are interested in algorithmic variations for each stage and the performance impact of these variations. More specifically, we examine the impact of different ordering rules, coflow grouping and backfilling rules, in both the offline and online settings. Compared with the very preliminary experiments we did in [18], in this paper we conduct a substantially more comprehensive study by considering many more ordering and backfilling rules, and examining the performance of algorithms on general instances in addition to real-world data. We also consider the offline setting with general release times, and online extensions of algorithms, which are not discussed in [18].

Workload. Our evaluation uses real-world data, which is a Hive/MapReduce trace collected from a large production cluster at Facebook [7,8,18], as well as extensive simulated instances.

For real-world data, we use the same workload as described in [8,18]. collected on a 3000-machine cluster with 150 racks, so the datacenter in the experiments can be modeled as a 150×150 network switch (and coflows be represented by 150×150 matrices). We select the time unit to be $1/128$ s (see [18] for details) so that each port has the capacity of 1MB per time unit. We filter the coflows based on the number of non-zero flows, which we denote by M', and we consider three collections of coflows, filtered by the conditions $M' \geq 25$, $M' \geq 50$ and $M' \geq 100$, respectively.

We also consider synthetic instances in addition to the real-world data. For problem size with $k = 160$ coflows and $m = 16$ inputs and outputs, we randomly generate 30 instances with different numbers of non-zero flows involved in each coflow. For instances 1–5, each coflow consists of m flows, which represent sparse coflows. For instances 5–10, each coflow consists of m^2 flows, which represent dense coflows. For instances 11–30, each coflow consists of u flows, where u is uniformly distributed on $\{m, \cdots, m^2\}$. Given the number k of flows in each coflow, k pairs of input and output ports are chosen randomly. For each pair of (i, j) that is selected, an integer processing requirement $d_{i,j}$ is randomly selected from the uniform distribution on $\{1, 2, \cdots, 100\}$.

Our main experimental findings are as follows:

- Algorithms with coflow grouping consistently outperform those without grouping. Similarly, algorithms that use backfilling consistently outperform those that do not use backfilling. The benefit of backfilling can be further improved by using a balanced backfilling rule (see Sect. 3.2 for details).
- The performance of the LP-based algorithm and its extensions is relatively robust, and among the best compared with those that use other simpler ordering rules, in the offline setting.
- In the offline setting with general release times, the magnitude of inter-arrival times relative to the processing times can have complicated effects on the performance of various algorithms (see Sect. 4.1 for details).
- The LP-based algorithm can be extended to an online algorithm and has near-optimal performance.

1.3 Related Work

There has been a great deal of success over the past 20 years on combinatorial scheduling to minimize average completion time, see e.g., [11,14,15,20], typically using a linear programming relaxation to obtain an ordering of jobs and then using that ordering in some other polynomial-time algorithm. There has also been success in shop scheduling. We do not survey that work here, but note that traditional shop scheduling is not "concurrent". In the language of our problem, that would mean that two flows in the same coflow could *not* be processed simultaneously. The recently studied concurrent open shop problem removes this restriction and models flows that can be processed in parallel. There is a close connection between concurrent open shop problem and coflow scheduling problem. When all coflow matrices are diagonal, coflow scheduling is equivalent to a concurrent open shop scheduling problem [8,18]. Leung et al. [13] presented heuristics for the total completion time objective and conducted an empirical analysis to compare the performance of different heuristics for concurrent open shop problem. In this paper, we consider a number of heuristics that include natural extensions of heuristics in [13] to coflow scheduling.

2   Preliminary Background

In [18], we formulated the following interval-indexed linear program (LP) relaxation of the coflow scheduling problem, where τ_l's are the end points of a set of geometrically increasing intervals, with $\tau_0 = 0$, and $\tau_l = 2^{l-1}$ for $l \in \{1, 2, \ldots, L\}$. Here L is such that $\tau_L = 2^{L-1}$ is an upper bound on the time that all coflows are finished processing under any optimal algorithm.

$$(LP) \quad \text{Minimize} \sum_{k=1}^{n} w_k \sum_{l=1}^{L} \tau_{l-1} x_l^{(k)} \qquad \text{subject to}$$

$$\sum_{u=1}^{l} \sum_{k=1}^{n} \sum_{j'=1}^{m} d_{ij'}^{(k)} x_u^{(k)} \leq \tau_l, \text{ for } i = 1, \ldots, m, \ l = 1, \ldots, L; \tag{1}$$

$$\sum_{u=1}^{l} \sum_{k=1}^{n} \sum_{i'=1}^{m} d_{i'j}^{(k)} x_u^{(k)} \leq \tau_l, \text{ for } j = 1, \ldots, m, \ l = 1, \ldots, L; \tag{2}$$

$$x_l^{(k)} = 0 \text{ if } r_k + \sum_{j'=1}^{m} d_{ij'}^{(k)} > \tau_l \text{ or } r_k + \sum_{i'=1}^{m} d_{i'j}^{(k)} > \tau_l; \tag{3}$$

$$\sum_{l=1}^{L} x_l^{(k)} = 1, \text{ for } k = 1, \ldots, n;$$

$$x_l^{(k)} \geq 0, \text{ for } k = 1, \ldots, n, \ l = 1, \ldots, L.$$

For each k and l, $x_l^{(k)}$ can be interpreted as the LP-relaxation of the binary decision variable which indicates whether coflow k is scheduled to complete within the interval $(\tau_{l-1}, \tau_l]$. Constraints (1) and (2) are the *load constraints* on the inputs and outputs, respectively, which state that the total amount of work completed on each input/output by time τ_l cannot exceed τ_l, due to matching constraints. Contraint (3) takes into account of the release times.

(LP) provides a lower bound on the optimal total weighted completion time of the coflow scheduling problem. If, instead of being end points of geometrically increasing time intervals, τ_l are end points of the discrete time units, then (LP) becomes exponentially sized (which we refer to as (LP-EXP)), and gives a tighter lower bound, at the cost of longer running time. (LP) computes an approximated completion time $\bar{C}_k = \sum_{l=1}^{L} \tau_{l-1} \bar{x}_l^{(k)}$, for each k, based on which we re-order and index the coflows in a nondecreasing order of $\bar{C}_k$, i.e.,

$$\bar{C}_1 \le \bar{C}_2 \le \ldots \le \bar{C}_n. \tag{4}$$

3 Offline Algorithms with Zero Release Time

In this section, we assume that all the coflows are released at time 0. We compare our LP-based algorithm with others that are based on different ordering, grouping, and backfilling rules.

3.1 Ordering Heuristics

An intelligent ordering of coflows in the ordering stage can substantially reduce coflow completion times. We consider the following five greedy ordering rules, in addition to the LP-based order (4), and study how they affect algorithm performance.

Definition 1. *The First in first (FIFO) heuristic orders the coflows arbitrarily (since all coflows are released at time 0).*

Definition 2. *The Shortest Total Processing Time first (STPT) heuristic orders the coflows based on the total amount of processing requirements over all the ports, i.e., $\sum_{i=1}^{m} \sum_{j=1}^{m} d_{ij}$.*

Definition 3. *The Shortest Maximum Processing Time first (SMPT) heuristic orders the coflows based on the load ρ of the coflows, where $\rho = \max\{ \max_{i=1,\ldots,m} \eta_i,$ $\max_{j=1,\ldots,m} \theta_j\}$, $\eta_i = \{\sum_{j'=1}^{m} d_{ij'}\}$ is the load on input i, and $\theta_j = \{\sum_{i'=1}^{m} d_{i'j}\}$ is the load on output j.*

Definition 4. *To compute a coflow order, the Smallest Maximum Completion Time first (SMCT) heuristic treats all inputs and outputs as $2m$ independent machines. For each input i, it solves a single-machine scheduling problem where n jobs are released at time 0, with processing times $\eta_i^{(k)}$, $k = 1, 2, \cdots, n$, where $\eta_i^{(k)}$ is the ith input load of coflow k. The jobs are sequenced in the order of increasing $\eta_i^{(k)}$, and the completion times $C^{(i)}(k)$ are computed. A similar problem is solved for each output j, where jobs have processing times $\theta_j^{(k)}$, and the completion times $C_{(j)}(k)$ are computed. Finally, the SMCT heuristic computes a coflow order according to non-decreasing values of $C'(k) = \max_{i,j}\{C^{(i)}(k), C_{(j)}(k)\}$.*

Definition 5. *The Earliest Completion Time first (ECT) heuristic generates a sequence of coflow one at a time; each time it selects as the next coflow the one that would be completed the earliest[3].*

---

[3] These completion times depend on the scheduling rule used. Thus, ECT depends on the underlying scheduling algorithm. In Sect. 3.2, the scheduling algorithms are described in more detail.

3.2 Scheduling via Birkhoff-Von Neumann Decomposition, Backfilling and Grouping

The derivation of the actual sequence of schedules in the scheduling stage of our LP-based algorithm relies on two key ideas: scheduling according to an optimal (Birkhoff-von Neumann) decomposition, and a suitable grouping of the coflows. It is reasonable to expect grouping to improve algorithm performance, because it may consolidate skewed coflow matrices to form more balanced ones that can be scheduled more efficiently. Thus, we compare algorithms with grouping and those without grouping to understand its effect. The particular grouping procedure that we consider here is the same as that in [18] (also see Step 2 of Algorithm 4 of the Appendix in [17]), and basically groups coflows into geometrically increasing groups, based on aggregate demand. Coflows of the same group are treated as a single, *aggregated* coflow, and this consolidated coflow is scheduled according to the Birkhoff-von Neumann decomposition (see [18] or Algorithm 5 of the Appendix in [17]).

Backfilling is a common strategy used in scheduling for computer systems to increase resource utilization (see, e.g. [8]). While it is difficult to analytically characterize the performance gain from backfilling in general, we evaluate its performance impact experimentally. We consider two backfilling rules, described as follows. Suppose that we are currently scheduling coflow D. The schedules are computed using the Birkhoff-von Neumann decomposition, which in turn makes use of a related, *augmented* matrix $\tilde{D}$, that is component-wise no smaller than D. The decomposition may introduce unforced idle time, whenever $D \neq \tilde{D}$. When we use a schedule that matches input i to output j to serve the coflow with $D_{ij} < \tilde{D}_{ij}$, and if there is no more service requirement on the pair of input i and output j for the coflow, we backfill in order from the flows on the same pair of ports in the subsequent coflows. When grouping is used, backfilling is applied to the aggregated coflows. The two backfilling rules that we consider – which we call *backfilling* and *balanced backfilling* – are only distinguished by the *augmentation* procedures used, which are, respectively, the augmentation used in [18] (Step 1 of Algorithm 5 in [17]) and the balanced augmentation described in Algorithm 1.

The balanced augmentation (Algorithm 1) results in less skewed matrices than the augmentation step in [18], since it first "spreads out" the unevenness among the components of a coflow. To illustrate, let

$$D = \begin{pmatrix} 10\ 0\ 0 \\ 10\ 0\ 0 \\ 10\ 0\ 0 \end{pmatrix}, B = \begin{pmatrix} 10\ 10\ 10 \\ 10\ 10\ 10 \\ 10\ 10\ 10 \end{pmatrix}, \text{ and } C = \begin{pmatrix} 10\ 20\ 0 \\ 10\ 0\ 20 \\ 10\ 10\ 10 \end{pmatrix}.$$

Under the balanced augmentation, D is augmented to B and under the augmentation of [18], D is augmented to C.

3.3 Scheduling Algorithms and Metrics

We consider 30 different scheduling algorithms, which are specified by the ordering used in the ordering stage, and the actual sequence of schedules used in the

---

Algorithm 1. Balanced Coflow Augmentation

Data: A single coflow $D = (d_{ij})_{i,j=1}^{m}$.

Result: A matrix $\tilde{D} = \left(\tilde{d}_{ij}\right)_{i,j=1}^{m}$ with equal row and column sums, and $D \leq \tilde{D}$.

Let ρ be the load of D.

$p_i \leftarrow \rho - \sum_{j'=1}^{m} d_{ij'}$, for $i = 1, 2, \ldots, m$.

$q_i \leftarrow \rho - \sum_{i'=1}^{m} d_{i'j}$, for $j = 1, 2, \ldots, m$.

$\Delta \leftarrow m\rho - \sum_{i=1}^{m} \sum_{j=1}^{m} d_{ij}$.

$d'_{ij} = \lfloor d_{ij} + p_i q_i / \Delta \rfloor$.

Augment $D' = (d'_{ij})$ to a matrix $\tilde{D}$ with equal row and column sums (see Step 1 of Algorithm 5 of the Appendix in [17]; also see [18]).

---

scheduling stage. We consider 6 different orderings described in Sect. 3.1, and the following 5 cases in the scheduling stage:

- (a) without grouping or backfilling, which we refer to as the base case;
- (b) without grouping but with backfilling;
- (c) without grouping but with balanced backfilling;
- (d) with grouping and with backfilling;
- (e) with grouping and with balanced backfilling.

We will refer to these cases often in the rest of the paper. Our LP-based algorithm corresponds to the combination of LP-based ordering and case (d).

For ordering, six different possibilities are considered. We use H_A to denote the ordering of coflows by heuristic A, where A is in the set {FIFO, STPT, SMPT, SMCT, ECT}, and H_{LP} to denote the LP-based coflow ordering. Note that in [18], we only considered orderings H_{FIFO}, H_{SMPT} and H_{LP}, and cases (a), (b) and (d) for scheduling, and their performance on the Facebook trace.

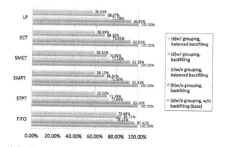
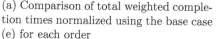

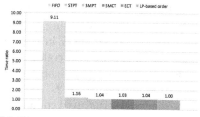

(a) Comparison of total weighted completion times normalized using the base case (e) for each order

(b) Comparison of 6 orderings with zero release times on Facebook data.

Fig. 1. Facebook data are filtered by $M' \geq 50$. Weights are equal.

3.4   Performance of Algorithms on Real-World Data

We compute the total weighted completion times for all 6 orders in the 5 different cases (a)–(e) described in Sect. 3.3, through a set of experiments on filtered coflow data. We present representative comparisons of the algorithms here.

Figure 1a plots the total weighted completion times as percentages of the base case (a), for the case of equal weights. Grouping and backfilling both improve the total weighted completion time with respect to the base case for all 6 orders. In addition to the reduction in the total weighted completion time from backfilling, which is up to 7.69 %, the further reduction from grouping is up to 24.27 %, while the improvement from adopting the balanced backfilling rule is up to 20.31 %. For 5 non-arbitrary orders (excluding FIFO), scheduling with both grouping and balanced backfilling (i.e., case (e)) gives the smallest total weighted completion time.

We then compare the performances of different coflow orderings. Figure 1b shows the comparison of total weighted completion times evaluated on filtered coflow data in case (e) where the scheduling stage uses both grouping and balanced backfilling. Compared with H_{FIFO}, all other ordering heuristics reduce the total weighted completion times of coflows by a ratio between 7.88 and 9.11, with H_{LP} performing consistently better than other heuristics.

3.5   Cost of Matching

The main difference between our coflow scheduling problem and the well-studied concurrent open shop problem we discussed in Sect. 1.3 is the presence of matching constraints on paired resources, i.e. inputs and outputs, which is the most challenging part in the design of approximation algorithms [18]. Since our approximation algorithm handles matching constraints, it is more complicated than scheduling algorithms for concurrent open shop problem. We are interested in how much we lose by imposing these matching constraints.

To do so, we generate two sets of coflow data from the Facebook trace. For each coflow k, let the coflow matrix $D^{(k)}$ be a diagonal matrix, which indicates that coflow k only has processing requirement from input i to output i, for $i = 1, \ldots, m$. The processing requirement $D_{i,i}^{(k)}$ is set to be equal to the sum of all dataflows of coflow k in the Facebook trace that require processing from input i. We then construct coflow matrix $\tilde{D}^{(k)}$ such that $\tilde{D}^{(k)}$ is not diagonal and has the same row sum and column sum as $D^{(k)}$. The details of the generation is described as in Algorithm 2.

The diagonal structured coflow matrices can reduce the total completion time of by a ratio up to 2.09, which indicates the extra processing time introduced by the matching constraints.

3.6   Performance of Algorithms on General Instances

In previous sections, we present the experimental results of several algorithms on the Facebook trace. In order to examine the consistency of the performance of these algorithms, we consider more instances, including examples where certain algorithms behave badly.

Bad Instances for Greedy Heuristics. We consider the following examples which illustrate instances on which the ordering heuristics do not perform well.

Example 1. *Consider a 2×2 network and n coflows with $D = \begin{pmatrix} 10 & 0 \\ 0 & 0 \end{pmatrix}$, n coflows with $D = \begin{pmatrix} 0 & 0 \\ 0 & 10 \end{pmatrix}$, and a \cdot n coflows with $D = \begin{pmatrix} 9 & 0 \\ 0 & 9 \end{pmatrix}$. The optimal schedule in this case is to schedule the orders with the smallest total processing time first, i.e., the schedule is generated according to the STPT rule. The limit of the ratio $\frac{\sum_{k=1}^{n} C_k(ECT\&SMCT\&SMPT)}{\sum_{k=1}^{n} C_k(STPT)}$ is increasing in n and when $n \to \infty$ it becomes $\frac{a^2+4a+2}{a^2+2a+2}$. This ratio reaches its maximum of $\sqrt{2}$ when $a = \sqrt{2}$.*

We can generalize this counterexample to an arbitrary number of inputs and outputs m. To be more specific, in an $m \times m$ network, for $j = 1, 2, \cdots, m$, we have n coflows only including flows to be transferred to output j, i.e., $d_{ij} = 10$. We also have $a \cdot n$ coflows with equal transfer requirement on all pairs of inputs and outputs, i.e., $d_{ij} = 9$ for $i, j = 1, 2, \cdots, m$. The ratio

$$\lim_{n \to \infty} \frac{\sum_{k=1}^{n} C_k(ECT\&SMCT\&SMPT)}{\sum_{k=1}^{n} C_k(STPT)} = \frac{a^2 + 2ma + m}{a^2 + 2a + m}$$

has a maximum value of $\sqrt{m}$ when $a = \sqrt{m}$. Note that in the generalized example, we need to consider the matching constraints when we actually schedule the coflows.

Example 2. *Consider a 2×2 network and n coflows with $D = \begin{pmatrix} 1 & 0 \\ 0 & 10 \end{pmatrix}$, and a \cdot n coflows with $D = \begin{pmatrix} 10 & 0 \\ 0 & 0 \end{pmatrix}$. The optimal schedule in this case is to schedule the orders with the Smallest Maximum Completion Time first, i.e., the schedule is generated according to the SMCT rule. The ratio $\frac{\sum_{k=1}^{n} C_k(STPT)}{\sum_{k=1}^{n} C_k(SMCT)}$ is increasing in n and when $n \to \infty$ it becomes $\frac{a^2+2a}{a^2+1}$ This ratio reaches its maximum of $\frac{\sqrt{5}+1}{2}$ when $a = \frac{\sqrt{5}+1}{2}$.*

This counterexample can be generalized to an arbitrary number of inputs and outputs m. In an $m \times m$ network, for each $i = 2, 3, \cdots, m$, we have n coflows with two nonzero entries, $d_{11} = 1$ and $d_{ii} = 10$. We also have $a \cdot n$ coflows with only one zero entry $d_{11} = 10$. The limit of the ratio

$$\lim_{n \to \infty} \frac{\sum_{k=1}^{n} C_k(STPT)}{\sum_{k=1}^{n} C_k(SMCT)} = \frac{a^2 + 2(m-1)a}{a^2 + m - 1}$$

has a maximum value of $1/2 + \sqrt{m - 3/4}$ when $a = 1/2 + \sqrt{m - 3/4}$.

General Instances. We compare total weighted completion time for 6 orderings and 5 cases on general simulated instances as described in Sect. 1.2 (details in Tables 1 to 5 of [17]), normalized with respect to the LP-based ordering in case (c), which performs best on all of the instances. We have the similar observation from the general instances that both grouping and backfilling reduce the completion time. However, under balanced backfilling, grouping does not improve performance much. Both grouping and balanced backfilling form less skewed matrices that can be scheduled more efficiently, so when balanced backfilling is used, the effect of grouping is less pronounced. It is not clear whether case (c) with balanced backfilling only is in general better than case (e) with both balanced backfilling and grouping, as we have seen Facebook data on which case (e) gives the best result. As for the performance of the orderings, on the one hand, we see very close time ratios among all the non-arbitrary orderings on instances 6–30, and a better performance of H_{ECT} on sparse instances 1–5 over other orderings (Table 3, Appendix [17]); on the other hand, there are also instances where ECT performs poorly (e.g., see Sect. 3.6).

Besides their performance, the running times of the algorithms that we consider are also important. The running time of an algorithm consists of two main parts; computing the ordering and computing the schedule. On a Macbook Pro with 2.53 GHz two processor cores and 6 G memory, the five ordering rules, FIFO, STPT, SMPT, SMCT and ECT, take less than 1 s to compute, whereas the LP-based order can take up to 90 s. Scheduling with backfilling can be computed in around 1 min, whereas balanced backfilling computes the schedules with twice the amount of time, because the balanced augmented matrices have more non-zero entries. Besides improving performance, grouping can also reduce the running time by up to 90 %.

Algorithm 2. Construction of coflow data

Data: A single diagonal coflow $D = (d_{ij})_{i,j=1}^{m}$.

Result: Another coflow $\tilde{D} = \left(\tilde{d}_{ij}\right)_{i,j=1}^{m}$, such that row and column sums of the two matrices are all equal.

Let $\eta(\tilde{D}) = \sum_{i,j=1}^{m} \tilde{d}_{ij}$ be the sum of all entries in $\tilde{D}$. Similarly, $\eta(D) = \sum_{i=1}^{m} d_{ii}$.

$\tilde{D} \leftarrow 0$.

while $(\eta(\tilde{D}) < \eta(D))$ **do**

   $S_i \leftarrow \{i : \sum_{j'=1}^{m} \tilde{D}_{ij'} < d_{ii}\}$; $S_j \leftarrow \{j : \sum_{i'=1}^{m} \tilde{D}_{i'j} < d_{jj}\}$. Randomly pick i^* from set S_i and j^* from set S_j. $\tilde{D} \leftarrow \tilde{D} + pE$, where $p = \min\{d_{i^*i^*} - \sum_{j'=1}^{m} \tilde{D}_{i^*j'}, d_{j^*j^*} - \sum_{i'=1}^{m} \tilde{D}_{i'j^*}\}$, $E_{ij} = 1$ if $i = i^*$ and $j = j^*$, and $E_{ij} = 0$ otherwise.

   $\eta \leftarrow \sum_{i,j=1}^{m} \tilde{d}_{ij}$

end

---

4   Offline Algorithms with General Release Times

In this section, we examine the performances of the same class of algorithms and heuristics as that studied in Sect. 3, when release times can be general. We first extend descriptions of various heuristics to account for release times.

The FIFO heuristic computes a coflow order according to non-decreasing release time r. (Note that when all release times are distinct, FIFO specifies a unique ordering on coflows, instead of any arbitrary order in the case of zero release times.) The STPT heuristic computes a coflow order according to non-decreasing values of $\sum_{i=1}^{m}\sum_{j=1}^{m} d_{ij} + r$, the total amount of processing requirements over all the ports plus the release time. The SMPT heuristic computes a coflow order according to non-decreasing values of $\rho + r$, the sum of the coflow load and release time. Similar to the case of zero release times, the SMCT heuristic first sequences the coflows in non-decreasing order of $\sum_{j'} d_{ij'} + r$ on each input i and $\sum_{i'} d_{i'j} + r$ on each output j, respectively, and then computes the completion times $C^{(i)}$ and $C_{(j)}$, treating each input

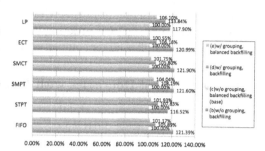

(a) Comparison of total weighted completion times normalized using the base case (c) for each order.

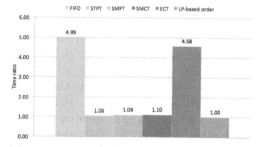

(b) Comparison of 6 orderings with general release times on Facebook data.

Fig. 2. Facebook data are filtered by $M' \geq 50$. Weights are equal.

and output as independent machines. Finally, the coflow order is computed according to non-decreasing values of $C' = \max_{i,j}\{C^{(i)}, C_{(j)}\}$. The ECT heuristic generates a sequence of coflows one at a time; each time it selects as the next coflow the one that has been released and is after the preceding coflow finishes processing and would be completed the earliest.

We compute the total weighted completion time for 6 orderings (namely, the LP-based ordering (4) and the orderings from definitions with release times and cases (b)–(e) (recall the description of these cases at the beginning of Sect. 3.3), normalized with respect to the LP-based ordering in case (c). The results for Facebook data are illustrated in Fig. 2a and b. For general instances, we generate the coflow inter-arrival times from uniform distribution [1, 100]. Performance ratios can be found in Tables 6 to 9 in the Appendix of [17]. As we can see from e.g., Fig. 2a, the effects of backfilling and grouping on algorithm performance are similar to those noted in Sect. 3.3, where release times are all zero. The STPT and LP-based orderings STPT appear to perform the best among all the ordering rules (see Fig. 2b), because the magnitudes of release times have a greater effect on FIFO, SMPT, SMCT and ECT than they do on STPT.

By comparing Figs. 1b and 2b, we see that ECT performs much worse than it does with common release times. This occurs because with general release times, ECT only schedules a coflow after a preceding coflow completes, so it does not back-fill. While we have kept the ECT ordering heuristic simple and reasonable to compute, no backfilling implies larger completion times, hence the worse performance.

4.1   Convergence of Heuristics with Respect to Release Times

In order to have a better understanding of release times, we scale the release times of the coflows and observe the impact of release time distribution on the performance of different heuristics. For general instances, recall that we generated the inter-arrival times with an upper bound of 100. Here we also consider inter-arrival time distributions that are uniform over $[0, 0]$, $[0, 25]$, $[0, 50]$, $[0, 200]$, $[0, 400]$, $[0, 800]$ and $[0, 1600]$, respectively. We compute the total weighted completion time with the adjusted release times in each case for 250 samples and take the average ratio with respect to the LP-based order.

As we can see from Fig. 3a to c, all the heuristics converge to FIFO as the inter-arrival time increases. This is reasonable as the release times dominate the ordering when they are large. The speed of convergence is higher in Fig. 3a where the coflow matrices in the instance are sparse and release times are more influential in all heuristics. On the contrary, when the coflow matrices are dense, release times weigh less in heuristics, which converge slower to FIFO as shown in Fig. 3c. We also note that for heuristics other than FIFO, the relative performance of an ordering heurstic with respect to the LP-based order may deteriorate and then

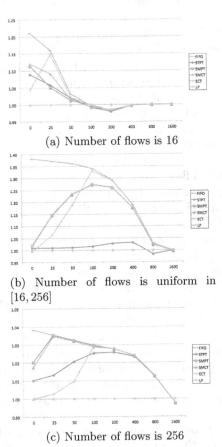

(a) Number of flows is 16

(b) Number of flows is uniform in $[16, 256]$

(c) Number of flows is 256

Fig. 3. Comparison of total weighted completion times with respect to the upper bound of inter-arrival time for each order on general instances. Network size is 16. Number of Coflow is 160.

improve, as we increase the inter-arrival times. This indicates that when inter-arrival times are comparable to the coflow sizes, they can have a significant impact on algorithm performance and the order obtained.

5 Online Algorithms

We have discussed the experimental results of our LP-based algorithm and several heuristics in the offline setting, where the complete information of coflows is revealed at time 0. In reality, information on coflows (i.e., flow sizes) is often only revealed at their release times, i.e., in an online fashion. It is then natural to consider online modifications of the offline algorithms considered in earlier sections. We proceed as follows. For the ordering stage, upon each coflow arrival, we re-order the coflows according to their remaining processing requirements. We consider all six ordering rules described in Sect. 3. For example, the LP-based order is modified upon each coflow arrival, by re-solving the (LP) using the remaining coflow sizes (and the newly arrived coflow) at the time. We describe the online algorithm with LP-based ordering in Algorithm 3. For the scheduling stage, we use case (c) the balanced backfilling rule without grouping, because of its good performance in the offline setting.

Algorithm 3. Online LP-based Approximation

Data: Coflows $\left(d_{ij}^{(k)} \right)_{i,j=1}^{m}$ with different release times, for $k = 1, \ldots, n$.
Result: A scheduling algorithm that uses at most a polynomial number of different matchings.

- Step 1: Given n_a coflows in the system, $n_a \leq n$, solve the linear program (LP). Let an optimal solution be given by $\bar{x}_l^{(k)}$, for $l = 1, 2, \ldots, L$ and $k = 1, 2, \ldots, n_a$. Compute the approximated completion time $\bar{C}_k$ by

$$\bar{C}_k = \sum_{l=1}^{L} \tau_{l-1} \bar{x}_l^{(k)}.$$

Order and index the coflows according to

$$\bar{C}_1 \leq \bar{C}_2 \leq \ldots \leq \bar{C}_{n_a}.$$

- Step 2: Schedule the coflows in order using the Birkhoff-von Neumann decomposition (see [18] or Algorithm 5 of the Appendix in [17])) until an release of a new coflow. Update the job requirement with the remaining job for each coflow in the system and go back to Step 1.

We compare the performance of the online algorithms and we compare the online algorithms to the offline algorithms. We improve the time ratio for all the orderings except FIFO by allowing re-ordering and preemption in the online algorithm compared with the static offline version. Note that we do not preempt with FIFO order. While several ordering heuristics perform as well as LP-based ordering in the online algorithms, a natural question to ask is how close H_A's are to the optimal, where $A \in \{STPT, SMPT, SMCT, ECT, LP\}$. In order to get a tight lower bound of the coflow scheduling problem, we solve (LP-EXP) for sparse instances. Since it is extremely time consuming to solve (LP-EXP) for dense instances, we consider a looser lower bound, which is computed as follows. We first aggregate the job requirement on each input and output and solve a single machine scheduling problem for the total weighted completion time, on each input/output. The lower bound is obtained by taking the maximum of the results (see the last column of Table 11, [17]). The ratio of the lower bound over the weighted completion time under H_{LP} is in the range of 0.91 to 0.97, which indicates that it provides a good approximation of the optimal.

6 Conclusion

We have performed comprehensive experiments to evaluate different scheduling algorithms for the problem of minimizing the total weighted completion time of coflows in a datacenter network. We also generalize our algorithms to an *online* version for them to work in real-time. For additional interesting directions in experimental analysis of coflow scheduling algorithms, we would like to come up with structured approximation algorithms that take into consideration other metrics and the

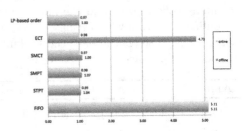

Fig. 4. Comparison of total weighted completion times with respect to the base case for each order under the offline and online algorithms. Data are filtered by $M' \geq 50$. Weights are equal.

addition of other realistic constraints, such as precedence constraints, and distributed algorithms that are more suitable for implementation in a data center. These new algorithms can be used to design other implementable, practical algorithms.

Acknowledgment. Yuan Zhong would like to thank Mosharaf Chowdhury and Ion Stoica for numerous discussions on the coflow scheduling problem, and for sharing the Facebook data.

References

1. Apache hadoop. http://hadoop.apache.org
2. Google dataflow. https://www.google.com/events/io
3. Alizadeh, M., Yang, S., Sharif, M., Katti, S., McKeown, N., Prabhakar, B., Shenker, S.: pfabric: Minimal near-optimal datacenter transport. SIGCOMM Comput. Commun. Rev. **43**(4), 435–446 (2013)

4. Ballani, H., Costa, P., Karagiannis, T., Rowstron, A.: Towards predictable data-center networks. SIGCOMM Comput. Commun. Rev. **41**(4), 242–253 (2011)
5. Borthakur, D.: The hadoop distributed file system: Architecture and design. Hadoop Project Website (2007)
6. Chowdhury, M., Stoica, I.: Coflow: A networking abstraction for cluster applications. In: HotNets-XI, pp. 31–36 (2012)
7. Chowdhury, M., Zaharia, M., Ma, J., Jordan, M.I., Stoica, I.: Managing data transfers in computer clusters with orchestra. SIGCOMM Comput. Commun. Rev. **41**(4), 98–109 (2011)
8. Chowdhury, M., Zhong, Y., Stoica, I.: Efficient coflow scheduling with Varys. In: SIGCOMM (2014)
9. Dean, J., Ghemawat, S.: Mapreduce: Simplified data processing on large clusters. In: OSDI, p. 10 (2004)
10. Dogar, F., Karagiannis, T., Ballani, H., Rowstron, A.: Decentralized task-aware scheduling for data center networks. Technical Report MSR-TR-96 2013
11. Hall, L.A., Schulz, A.S., Shmoys, D.B., Wein, J.: Scheduling to minimize average completion time: Off-line and on-line approximation algorithms. Math. Oper. Res. **22**(3), 513–544 (1997)
12. Kang, N., Liu, Z., Rexford, J., Walker, D.: Optimizing the "one big switch" abstraction in software-defined networks. In: CoNEXT, pp. 13–24 (2013)
13. Leung, J.Y., Li, H., Pinedo, M.: Order scheduling in an environment with dedicated resources in parallel. J. Sched. **8**(5), 355–386 (2005)
14. Phillips, C.A., Stein, C., Wein, J.: Minimizing average completion time in the presence of release dates. Math. Program. **82**(1–2), 199–223 (1998)
15. Pinedo, M.: Scheduling: Theory, Algorithms, and Systems, 3rd edn. Springer, New York (2008)
16. Popa, L., Krishnamurthy, A., Ratnasamy, S., Stoica, I.: Faircloud: Sharing the network in cloud computing. In: HotNets-X, pp. 22:1–22:6 (2011)
17. Qiu, Z., Stein, C., Zhong, Y.: Experimental analysis of algorithms for coflow scheduling. arXiv (2016). http://arxiv.org/abs/1603.07981
18. Qiu, Z., Stein, C., Zhong, Y.: Minimizing the total weighted completion time of coflows in datacenter networks. In: ACM Symposium on Parallelism in Algorithms and Architectures, pp. 294–303 (2015)
19. Shvachko, K., Kuang, H., Radia, S., Chansler, R.: The hadoop distributed file system. In: MSST, pp. 1–10 (2010)
20. Skutella, M.: List scheduling in order of α-points on a single machine. In: Bampis, E., Jansen, K., Kenyon, C. (eds.) Efficient Approximation and Online Algorithms. LNCS, vol. 3484, pp. 250–291. Springer, Heidelberg (2006)
21. Zaharia, M., Chowdhury, M., Das, T., Dave, A., Ma, J., McCauley, M., Franklin, M.J., Shenker, S., Stoica, I.: Resilient distributed datasets: A fault-tolerant abstraction for in-memory cluster computing. In: NSDI, p. 2 (2012)

An Empirical Study of Online Packet Scheduling Algorithms

Nourhan Sakr[✉] and Cliff Stein

Industrial Engineering and Operations Research,
Columbia University, New York, NY 10027, USA
n.sakr@columbia.edu, cliff@ieor.columbia.edu

Abstract. This work studies online scheduling algorithms for buffer management, develops new algorithms, and analyzes their performances. Packets arrive at a release time r, with a non-negative weight w and an integer deadline d. At each time step, at most one packet is scheduled. The modified greedy (MG) algorithm is 1.618-competitive for the objective of maximizing the sum of weights of packets sent, assuming agreeable deadlines. We analyze the empirical behavior of MG in a situation with arbitrary deadlines and demonstrate that it is at a disadvantage when frequently preferring maximum weight packets over early deadline ones. We develop the MLP algorithm, which remedies this problem whilst mimicking the behavior of the offline algorithm. Our comparative analysis shows that, although the competitive ratio of MLP is not as good as that of MG, it performs better in practice. We validate this by simulating the behavior of both algorithms under a spectrum of parameter settings. Finally, we propose the design of three additional algorithms, which may help in improving performance in practice.

1 Introduction

Efficient buffer management at a network router is a critical issue that motivates the online packet scheduling problem. Kesselman et al. [13] introduce a buffer management delay model and give algorithms to minimize end-to-end delay. We adopt a similar model to analyze the empirical behavior of the modified greedy (MG) algorithm introduced in [12], and propose new algorithms that do not have as strong worst-case guarantees, yet perform better in our simulated settings.

Model. For simplicity, we investigate a network router with two nodes. Studying a two node router is a first step towards understanding more complicated and realistic models.In Sect. 7, we briefly discuss possible model modifications. At each integer time step, packets are buffered upon arrival at the source node, then at most one packet is chosen from the buffer to be sent to the target node. A packet (r, d, w) arrives at a release date r, has a non-negative weight w, and needs to be sent by an integer deadline d. A packet not sent by time d expires,

---

N. Sakr—Supported in part by NSF grant CCF-1421161.

C. Stein—Supported in part by NSF grant CCF-1421161.

A.V. Goldberg and A.S. Kulikov (Eds.): SEA 2016, LNCS 9685, pp. 278–293, 2016.
DOI: 10.1007/978-3-319-38851-9_19

and is dropped from the buffer. The objective of a packet-scheduling algorithm A is to maximize its weighted throughput, ζ_A, defined as the total weight of packets sent by A. It is easy to relate our model to an online version of the classical offline unit-job scheduling problem where the input is a set of n unit-length jobs, each specified by a similar triple (r, d, w) and the objective is to maximize weighted throughput, i.e. the total weight of jobs that are processed before their deadlines.

Parameters. We will typically be generating our input according to some type of distribution. Let T denote the number of time steps during which the system can generate arriving packets, and let λ denote an arrival rate. We choose values for T and λ from a predefined set. Then at each integer time step $t = 1, \dots, T$, we generate the number of arriving packets according to a Poisson distribution with rate λ. For each arriving packet, we set $r = t$ and generate w from a uniform (integer) distribution $U(1, w_{\max})$. To find d, we first generate τ, a time to expire, from a uniform (integer) distribution $U(0, d_{\max})$, and set $d = r + \tau$. We call this Model 1. We also consider a bimodal distribution for τ with weights p and $1 - p$, respectively, for two distinct distributions centered on different means and call this Model 2. Although a network may induce correlations between packets, we use i.i.d. distributions as a first step in modeling the behavior of our algorithms.

In order to evaluate the performance of an online scheduling algorithm (A), we use an offline algorithm (OFF) for comparison, which given all future arrivals and packet characteristics, is able to statically find the optimal solution (e.g. using maximum-weight bipartite matching). Its solution gives the highest possible throughput the system can achieve. The online algorithm is k-competitive if ζ_A on any instance is at least $1/k$ of ζ_{OFF} on this instance. The smallest k for which an algorithm is k-competitive is called the competitive ratio [3]. According to [11], k will be at most 2 for any algorithm that uses a static priority policy. In this paper, we will simulate the online algorithm and evaluate the ratio ζ_A/ζ_{OFF}. The average of these ratios across each batch of simulations will be denoted by ρ_A, where A is the corresponding online algorithm.

Related Work. The literature is rich with works that acknowledge the importance of buffer management and present algorithms aiming at better router performance. Motivated by [7,13] gives a randomized algorithm, RMIX, while [4] proves that it remains $\frac{e}{e-1}$-competitive against an adaptive-online adversary. Many researchers attempt to design algorithms with improved competitive ratios. The best lower bound on the competitive ratio of deterministic algorithms is the golden ratio ϕ [5,11]. A simple greedy algorithm that schedules a maximum-weight pending packet for an arbitrary deadline instance is 2-competitive [11,13]. Chrobak et al. [8] introduce the first deterministic algorithm to have a competitive ratio strictly less than 2, namely 1.939. Li et al. [15] use the idea of dummy packets in order to design the DP algorithm with competitive ratio at most 1.854. Independently, [10] gives a 1.828-competitive algorithm. Further research considers natural restrictions on packet deadlines with hopes of improving the competitive ratio. One type of restriction is the agreeable deadline model considered in [12], i.e. deadlines are (weakly) increasing in their release times. Motivated by a more general greedy algorithm, EDF_α [7], that schedules the earliest-deadline pending packet with weight at least $1/\alpha$ of the maximum-weight pending packet, [12] develop the MG algorithm which we describe in Sect. 2.

Our Contribution. We observe that while MG is ϕ-competitive for the case of agreeable deadlines, it may not be the best option to apply in practice. We demonstrate the undesirable performance of MG under certain scenarios, e.g. frequently preferring maximum weight (late deadline) packets over early deadline ones. Our proposed MLP algorithm remedies this drawback, as it outperforms MG on most simulated instances. However, we are able to develop hard instances to prove that MLP does not provide better worst-case guarantees, whereas on those instances MG would produce the same results as an offline solution. Contrasting the advantages of MG and MLP motivates us to explore further algorithmic adjustments which may improve performance, at least in practice, as supported by our preliminary analysis. Finally, we justify that a two-node model with an infinite buffer is a sufficient model for our analysis. Moreover, extending the model to multiple nodes or imposing a threshold on the capacity of the buffer does not significantly alter the performance of the online algorithms.

2   Modified Greedy Algorithm (MG)

MG is a ϕ-competitive deterministic online algorithm for the agreeable deadline model [12]. It focuses on two packets at each time step: the earliest deadline non-dominated packet e (i.e. maximum weight among all earliest-deadline packets) and the maximum weight non-dominated packet h (i.e. earliest deadline among all maximum-weight packets in the buffer). Packet e is chosen if $w_e \geq \frac{w_h}{\phi}$ ($\phi \approx$ 1.618) and packet h is chosen otherwise. While [12] consider an agreeable deadline model, we relax this assumption and explore MG in a more general setting.

MG Analysis. Although MG has the best competitive ratio among deterministic online algorithms, we believe that by better understanding MG, we can improve on it in practice. Intuitively, if MG, at early stages, chooses packets with longer deadlines (due to higher weights) over those with early deadlines, then as time passes, many early packets expire while most of the heavy later-deadline packets will have already been sent. Therefore, the algorithm may resort to choosing packets with even smaller weights, thereby wasting an opportunity to send a higher weight packet that has already expired. We, hence, explore the decisions made by MG by observing its relative frequency of choosing h over e.

In order to consider a diverse set of instances, we set T to 200 and define ranges [0.7,20], [1,20] and [1,40] for λ, w_{max} and d_{max}, respectively. Under the assumptions of Model 1, we run a batch of 200 simulations each for 8000 sampled parameter combinations. Given each parameter combination, we calculate the relative frequency of choosing h over e and average the frequencies over λ to obtain the empirical probability of $P(w_h > 1.618w_e)$, denoted by ψ.

We suspect that when ψ is high, especially if h expires at later deadlines, MG will be at a major disadvantage. Figure 1 plots ψ vs. w_{max}, where each curve corresponds to a fixed level for d_{max}. In general, ψ increases with w_{max} and d_{max}. The decreasing curve slope implies that ψ is more sensitive to lower values of w_{max}. Further analysis shows that at any level of w_{max}, MG will choose h over e at most 66 % of the time. We also observe that regardless the average number of packets in the buffer, if w_{max} is small (≤ 3), the event of interest occurs at most 40 % of the time.

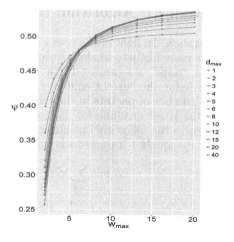

Fig. 1. ψ vs. w_{max}, colored by d_{max} (See [16] for colored figure)

From this probability analysis, we conclude that unless w_{max} or λ is small, MG tends to choose packet h too frequently. To show that this property may "fire back", we construct Scenario 1, forcing MG to favor later deadlines. We reuse the data of the generated packets above and adjust the weight of each packet by multiplying it by its deadline. We let MG run on the new data and plot ρ_{MG} against different parameters. Figure 2 depicts lower ratios for the new dataset. While ρ_{MG} originally increased with λ, it now decreases with λ and d_{max}. w_{max} does not affect the performance much. A gradient boosted tree predictive model for ρ_{MG} (i.e. a sequence of decision tree models where the next model is built upon the residuals of the previous one) shows that d_{max} is the most important factor under Scenario 1, as it explains 30 % of the variability in the model.

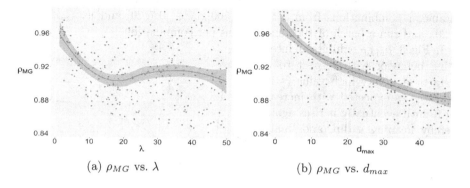

(a) ρ_{MG} vs. λ (b) ρ_{MG} vs. d_{max}

Fig. 2. Performance of MG under Scenario 1

3 Mini LP Algorithm (MLP)

Description. In light of the previous analysis, we develop a new online algorithm that is more likely to send early deadline packets. The mini LP algorithm (MLP) runs a "mini" assignment LP at each time step in order to find the optimal schedule for the current content of the buffer, assuming no more arrivals. If n_t is the number of packets in the buffer at current time t, we search for the packet with the latest deadline ($d_{t,max}$) and set a timeline from $\hat{t} = 0$ to $d_{t,max} - t$. We then solve the following optimization problem, where w_i is the weight of packet i and $x_{i\hat{t}}$ equals 1 if packet i is sent at time $\hat{t}$ and 0 otherwise:

$$\min \sum_{i,\hat{t}} w_i x_{i\hat{t}}$$

$$s.t. \quad \sum_i x_{i\hat{t}} \leq 1 \quad \hat{t} = 0, \dots, d_{t,max} - t$$

$$\sum_i x_{i\hat{t}} \leq 1 \quad i = 1, \dots, n_t$$

$$x_{i\hat{t}} \geq 0 \quad i = 1, \dots, n_t$$

MLP then uses the optimal solution to send the packet that receives the first assignment, i.e. the packet i for which $x_{i0} = 1$, while the rest of the schedule is ignored and recomputed in subsequent time steps.

3.1 Initial Analysis

Similar to the MG analysis, we compute ρ_{MLP} and are interested in its behavior as the load varies. A way to measure load is to define the average number of packets in the buffer as $\bar{n}$, which is a byproduct of λ. We expect a higher ρ_{MLP} at low $\bar{n}$, since the online algorithm would not have many packets to choose from and hence, is more likely to choose the same packet as the offline algorithm at each time step. However, we expect ρ_{MLP} to decrease as $\bar{n}$ increases, since the discrepancy between online and offline solutions increases. To test this, we sample parameter combinations from $T \in [100, 500]$, $\lambda \in [0.7, 20]$ and $w_{max}, d_{max} \in [1, 20]$, and run a batch of 1000 simulations per combination.

In Fig. 3, ρ_{MLP} exhibits an interesting behavior with respect to $\bar{n}$: it starts at a very high value (≈ 1), decreases as expected with increasing $\bar{n}$ until eventually it rises again, thereby forming a dip, and finally converges to 1. Our claim is true at first, when λ is relatively low, as the first range for λ is quite sensitive ($\lambda = 2.3$ vs. 2.8 makes a difference). However, when λ increases, the problem loses its sensitivity. An explanation for such behavior may

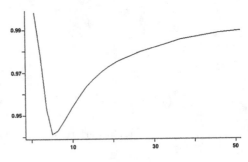

Fig. 3. ρ_{MLP} vs. $\bar{n}$

be that as $\bar{n}$ increases (with increasing λ), we are more likely to have multiple packets achieving maximum weight, in which case both the online and offline algorithms are likely to choose those packets and have less discrepancy between their choices, especially if the weight or deadline ranges are not wide. We conclude that when the system is heavily or lightly loaded, both algorithms perform well. The dip happens in the interesting area. Consequently, we will investigate how the dip moves and the effect that parameter choices will have on it.

3.2   Parameter Effect on MLP Behavior

Changing the parameters to generate different graphs did not change the structure of the dip-shaped graph that we have seen in Fig. 3. Nonetheless, the dip gets narrower/wider and/or shifts to the left/right, as parameters change. In this section, we will only focus on a restricted range for the values of λ, namely 0.7 to 10. However, we believe that the restriction does not mask any interesting results, since MLP converges at higher values of λ, as we have seen before. Therefore, a heavily loaded system is not significant for our analysis.

Arrival Rates. The graph inevitably depends on λ, as it directly affects $\bar{n}$, i.e. the x-axis. However, λ does not have a direct effect on the shape of the graph. By tuning the range for λ, we are able to "zoom in" onto the dip area and monitor the behavior more accurately where the system is neither lightly nor heavily loaded. The range for such sensitive values is on average between 1.3 and 4.2. Figure 4a zooms in on the dip where λ is most sensitive.

Weight Ranges. The range of the weights moves the dip to the right (left), as it gets narrower (wider). Very narrow ranges (i.e. low values for w_{max}) are the most influential. As w_{max} increases, its impact decreases. In fact, this result

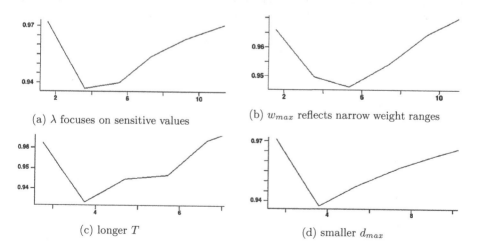

(a) λ focuses on sensitive values

(b) w_{max} reflects narrow weight ranges

(c) longer T

(d) smaller d_{max}

Fig. 4. ρ_{MLP} vs. $\bar{n}$

seems intuitive and one can see an example in Fig. 4b where the weight range is designed to be very narrow (w_{max} is set at 2). Some experimentation led us to the explanation of this phenomenon: When there are few options for weights, both algorithms converge together. For instance, if weights are only 1 and 2, then the higher the $\bar{n}$, the more likely we will have packets of weight 2. In this case, both algorithms find the optimal choice to be the packet with the higher weight (we don't have much choice here so it must be 2). Hence, both behave alike. We note that it is not in particular the range of distinct weights that has this effect but rather the number of distinct weights available, i.e. choosing between weights 1 and 2 vs. 100 and 200, would depict the same behavior.

Time Period and Deadline Range. T and d_{max} have a combined effect. Figure 4c and d give two examples: Allowing a longer timeline T results in a second but higher dip and slows down convergence, such that suddenly higher values of λ become slightly more interesting. Meanwhile, lower d_{max} values (combined with shorter T's) result in a sharp dip as well as much faster convergence to 1.

3.3   Influence of Maximum-Weight Packets

The motivation of MLP was mainly to remedy the drawback we observed for MG when later deadline packets are preferred. Therefore, it is essential to verify that MLP outperforms MG under Scenario 1. In fact, one-sided 99 % confidence intervals (CI) imply that ρ_{MG} is at most 91.98 % while ρ_{MLP} is at most 96.28 %. The difference in performance between both algorithms increases with λ. Figure 5a shows the behavior of ρ against $\bar{n}$ for both algorithms. While MLP is not influenced by $\bar{n}$ under this scenario, the performance of MG gets worse as $\bar{n}$ increases. A 99 % two-sided CI for $\frac{\rho_{MLP}}{\rho_{MG}}$, denoted by $\hat{\rho}$, is $(1.0443, 1.0552)$, implying that MLP produces a total weight at least 4.43 % more than that of MG under this scenario. Better performance is observed with higher T or lower d_{max}, but w_{max} does not seem to influence the algorithms' performances.

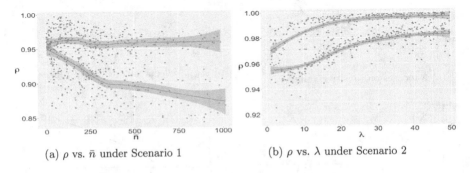

(a) ρ vs. $\bar{n}$ under Scenario 1   (b) ρ vs. λ under Scenario 2

Fig. 5. ρ_{MLP}(red) and ρ_{MG}(green) (See [16] for colored figure)

4   Comparative Analysis

In this section, we contrast the behavior of MG and MLP under a spectrum of parameter settings. We are interested in the behavior of the ratios against our parameters and expect MLP to perform better in our simulations. The general procedure for our simulations is based on sampling parameter combinations from a predefined parameter space. We impose the following parameter range restrictions: $T \in (50, 750)$, $\lambda \in (0.5, 50)$, $w_{max} \in (2, 50)$ and $d_{max} \in (1, 50)$ (Scenario 2). For each combination, we run MG, MLP (5 times each) and the offline algorithm in order to obtain values for ρ_{MG}, ρ_{MLP}, as well as $\hat{\rho} = \frac{\zeta_{MLP}}{\zeta_{MG}}$. Detailed steps for simulations are given in the full version of the paper [16].

4.1   Ratio Behavior w.r.t. Model Parameters

Comparing ρ_{MLP} and ρ_{MG} against values of λ implies that on average MLP outperforms MG (Fig. 5). As λ increases, both algorithms perform better. A 99 % one-sided CI for ρ_{MG} is $(0, 0.9734)$, implying that we are 99 % confident that ζ_{MG} is at most 97.34 % of ζ_{OFF}, while the one-sided CI for ρ_{MLP} is $(0, 0.9926)$. In fact, MG produces a wider spread of the ratios. All else constant, the performance of each algorithm improves with higher T, lower d_{max} or higher $\bar{n}$, whereas it is not influenced by the values of w_{max}. Relevant graphs are presented in [16].

Figure 6a plots $\hat{\rho}$ vs. λ, colored by T. For very small λ's, MLP and MG may perform similarly. In some cases, MG outperforms MLP, regardless of the value of T. However, for large λ, MLP tends to outperform MG. A 99 % two-sided CI for $\hat{\rho}$ is $(1.0188, 1.0216)$, implying that we are 99 % confident that ζ_{MLP} is at least 1.88 % more than ζ_{MG}. However, both algorithms have similar performance as the upper bound of the CI shows that ζ_{MLP} is at most 2.16 % more. Whether this is beneficial depends on the use case as well as time constraints (see Sect. 5). In [16], we present a brief analysis where we construct gradient booted tree predictive models on the ratios for inference purposes.

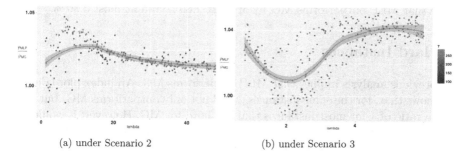

(a) under Scenario 2   (b) under Scenario 3

Fig. 6. $\hat{\rho}$ vs. λ and colored by T (See [16] for colored figure)

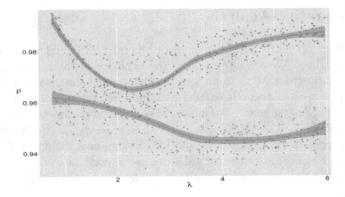

Fig. 7. ρ_{MLP}(red) and ρ_{MG}(green) vs. λ under Scenario 3 (See [16] for colored figure)

4.2 Changing the Distribution of τ

So far, we have only considered uniform distributions, however, real inputs are more complicated. Here we make one step towards modeling more realistic inputs and assume τ that follows a bimodal distribution of two distinct peaks (recall Model 2); with probability p, τ is $N(2, 0.5^2)$ and with probability $1 - p$, τ is $N(8, 0.75^2)$. We restrict our parameters to the following ranges: $T \in (100, 300)$, $\lambda \in (0.7, 6)$, $w_{max} \in (2, 7)$ and $p \in (0.75, 0.95)$ (Scenario 3). We choose a bimodal distribution because these distributions are often hard for scheduling algorithms. Indeed, we see that the results for Scenario 3 are slightly different.

While MG performs worse with increasing λ, ρ_{MLP} improves with λ and still outperforms ρ_{MG} (Fig. 7).The graph for ρ_{MLP} resembles a dip-shaped graph, yet we find this dip to be entirely above the confidence interval of ρ_{MG}. All else constant, neither algorithm is influenced greatly by any of the parameters T, d_{max} or p. Figure 6 plots $\hat{\rho}$ vs. λ, where lighter points correspond to longer T's. For very small λ, MLP and MG perform similarly. In some cases, MG outperforms MLP, regardless of the value of T. However, for large λ, a 95 % CI shows that MLP outperforms MG by at least 2.80 % and at most 3.30 %.

5 Hard Instances

The previous analysis implies that MLP outperforms MG. An index plot (Fig. 8) of ρ shows that, for the same instances, MLP not only outperforms MG, but also gives a ratio of 1 for most instances that are hard for MG. However, it would be incorrect to conclude that MLP always has a better competitive ratio than MG. In fact, we are able to create hard instances for MLP where it performs worse than MG. A small example is given in Table 1.

For $w_2 > w_1$, we can easily see that MLP is 2-competitive, while ρ_{MG} on those instances is 1. However, such worst-case instances may be rare and our results show that MLP performs better on a varied set of data. On the other hand, MG, which bases its decisions on a single comparison, is simpler than MLP which solves an LP at each time step. In our experiments, MLP was as much as 140 times slower than MG. In the next section, we consider some modifications to take advantage of the strengths of both approaches.

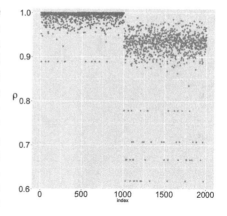

Fig. 8. Index plot of ρ_{MLP}(red) and ρ_{MG}(green) (See [16] for colored figure)

Table 1. Hard instance for MLP

Packet(r, d, w)	MLP	MG	Offline
$(1, 1, w_1)$	Assign to $t = 1$		
$(1, 2, w_2)$		Assign to $t = 1$	Assign to $t = 1$
$(2, 2, w_2)$	Assign to $t = 2$	Assign to $t = 2$	Assign to $t = 2$
Throughput	$w_1 + w_2$	$2w_2$	$2w_2$

6   Algorithm Modifications

In an attempt to find faster solutions, we introduce some possible algorithmic modifications. Preliminary results show slight performance improvement when these modifications are applied. However, more analysis is needed to verify the results and choose the best parameters for improvement.

6.1   The Mix and Match Algorithm (MM)

Algorithm. The Mix and Match Algorithm (MM) combines both MG and MLP. At each time step, MM chooses to either run MG or MLP-according to $\bar{n}$. If $\bar{n}$ is high, then by previous analysis, MG and MLP each converges to 1 (assuming Model 1), and MM runs MG, as it is faster and has a competitive ratio that is as good as that of MLP. If $\bar{n}$ is low, MM runs MLP, as it is more accurate and the running time is also small since $\bar{n}$ is low. To distinguish between "high" and "low", we define a threshold $\bar{N}$. Although MM suffers from the same limitations as MLP, it might still be preferred due to its smaller computation time.

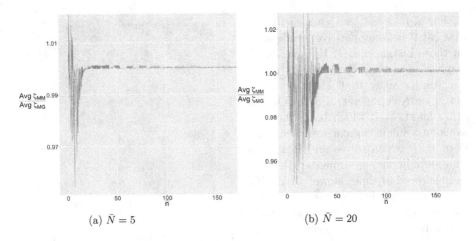

(a) $\bar{N} = 5$ (b) $\bar{N} = 20$

Fig. 9. Average ζ_{MM} over average ζ_{MG} vs. $\bar{n}$

Simulation. We set T to 200 and define ranges [0.7,15], [1,30], [1,23] and [5,20] for λ, w_{max}, d_{max} and $\bar{N}$, respectively. We compare ζ_{MM} under different values of $\bar{N}$. We also average ζ_{MM}, use the same simulations to run MG and average ζ_{MG}. We take the ratio of both averages and plot it against $\bar{n}$ (Fig. 9). Preliminary results show that for small $\bar{n}$, the algorithm, at higher $\bar{N}$, does slightly worse than at lower $\bar{N}$. However, the opposite is true for large $\bar{n}$.

Future Work. Further analysis are needed to set the optimal choice for $\bar{N}$. We may want to look at the percentage of times the algorithm chose to run MG over MLP in order to monitor time complexity. Another idea would be to take hard instances of MG into consideration and explore how to derive conditions such that the algorithm switches to MLP in such cases.

6.2 The Learning Modified Greedy Algorithm (LMG)

Algorithm. MG and MLP are memoryless [12], i.e. the assignment of packets at time t uses no information about assignments at times $t' < t$. We introduce a non-memoryless modification to MG. Recall that MG compares w_e to $\frac{w_h}{\phi}$. In the learning MG (LMG) algorithm, we try to make use of the past performance of MG, in order to replace ϕ by a more suitable divisor. If f defines the frequency of learning, then every f steps, we calculate the throughput at the current divisor, ϕ^*. Then we search for a divisor, ϕ_{better}, which lies in the vicinity of ϕ^*, but yields higher throughput using the previous data. The procedure is given below:

Step 0. Set the frequency of learning, f, i.e. the time window needed to define a learning epoch. Then run MG and use the following procedure every f steps in order to replace the divisor, ϕ, by a sequence of better divisors as follows:

1. Generate a sequence of divisors ϕ_i's starting at the current divisor ϕ^* and having jumps of ± 0.05, without going below 1 or above 2.5. For instance, if $\phi^* = 1.62$, we generate the sequence: 1.02, 1.07, ..., 1.57, 1.62, 1.67, ..., 2.47.

Table 2. Examples of choosing ϕ_{better}

Throughput/ϕ_i	1.57	$\phi^* = 1.62$	1.67	1.72	ϕ_{better}
Case 1	2	4	4	2	1.62
Case 2	2	4	2	6	1.62
Case 3	6	4	5	7	1.72
Case 4	4	4	4	4	1.62
Case 5	6	4	3	7	1.57

2. Start with the throughput associated with ϕ^* and move left in our generated sequence. At each ϕ_i, we calculate the throughput of MG on the previous data. We keep moving to subsequent divisors as long as there is an increase in throughput. Next, we do the same with divisors to the right. Given a left endpoint and a right one, we choose the divisor associated with the higher throughput and denote it by ϕ_{better}. Some toy examples are shown in Table 2. For simplicity, we only observe the weighted throughput for 4 values of ϕ.
3. The new divisor ϕ^*_{new} is given by a smoothed average of ϕ^* with ϕ_{better}, i.e. for some $\alpha \in [0,1]$,

$$\phi^*_{new} = \alpha\phi^* + (1-\alpha)\phi_{better}.$$

Simulation. We use the same parameter space as in Sect. 6.1. We set $f = \max(0.1 * T, \frac{30}{\min(1,\lambda)})$ and for simplicity, $\alpha = 0.5$. The choice for α in general must ensure that the process of finding an optimal divisor ϕ does not generate a jumpy sequence of divisors. Our analysis for 8000 sampled scenarios shows that LMG outperforms MG 83.3 % of the time. The range of the improvement is $[-0.6\%, 2.8\%]$, implying that LMG brings as much as 2.8 % increase in the ratio over MG. Performance is worse when the sequence of ϕ^*_{new}'s is around 1.618, implying that LMG is picking up on noise and should not change the divisor.

Future Work. One can avoid this noise by statistically testing and justifying the significance of changing the divisor. In terms of time complexity, LMG is not slower than MG, as it can be done while the regular process is running. Finally, no direct conclusion is made about a threshold on the number of packets beyond which LMG is particularly effective. Further analysis could yield such a conclusion, thereby indicating the instances for which LMG should be used.

6.3   The Second Max Algorithm (SMMG)

Algorithm. Inspired by the dummy packet (DP) algorithm discussed in [15] for cases of non-agreeable deadlines, we realize the importance of extending the comparison to a pool of more than two packets. The key idea in SMMG is to prevent the influence a single heavy packet may have on subsequent steps of the algorithm. We try to find an early-deadline packet that is sufficiently large compared to the heaviest packet. We set a value for $p \in (0,1)$ and the iterations are as follows:

1. If MG chooses e, send e and STOP.
2. Else find the earliest second largest packet in the buffer, denoted by s.
3. If $d_s < d_h$ and $w_s \geq \max(w_e, p * w_h)$, send s. Else send h.

The intuition here is that sending packet e is always a good choice, so we need no modification. However, we limit over-choosing packet h by finding the earliest second-largest packet s. The concern is that keeping s in the buffer may bias the choice of the packets. Hence, we send s, if its weight is significant enough, in order to eliminate its influence and keep the possibility of sending h for a subsequent iteration (as h expires after s). To evaluate that w_s is significant enough, we verify that it exceeds w_e (otherwise, we should have sent e), as well as $p * w_h$, a proportion of w_h. Note that for instance, if $p = 0.95$, it means that SMMG is very conservative allowing the fewest modifications to MG.

Simulation. We use the same parameter space as in Sect. 6.1 and try values for p as follows: 0.65, 0.75, 0.85, 0.95. Figure 10 plots the improvement of SMMG over MG (ρ_{MG} - ρ_{SMMG}) vs. $\bar{n}$, colored by p. As expected, the lower the value of p, the bigger the deviation from MG. At very low $\bar{n}$, we see that applying SMMG is not useful, however, as $\bar{n}$ increases, the improvement remains positive. At all values of p, the improvement is at its highest when $\bar{n}$ is between 8 and 12 packets. Hence, SMMG is useful when $\bar{n}$ is in the vicinity of the interval between 4 and 17. Whether this is a significant result depends on the nature of our problem. Even if $p = 0.95$, the minimum improvement within that interval is around 0.8 %. However, the maximum improvement is 1.5 % among all our simulations.

7   Model Discussion

The two-node model with no buffer limitations clearly does not capture all aspects of realistic network models. Thus, in this section, we will consider a multi-node model and also the case of finite buffer capacity.

Multi-node Model. In our analysis, we only considered the source and the target nodes. In order to understand multi-node systems, we consider first a three node system, that is a system of two tandem queues and see how the throughput behaves. Assume the arrival rate at node 1 is λ_1 and that each packet has to be sent to node 2 then reach node 3 by its deadline. Some packets are lost at node 1, as they expire before being sent. Node 2 will, hence, have less traffic. We assume as before that we can send at most one packet per node at each time step. Within this framework, we are interested in knowing whether this setup results in a deterministic online algorithm of better performance. Our simulation shows that node 2 either has the same throughput as node 1 or lower. After tracing packets, it turns out that this is a logical result because node 2 only receives

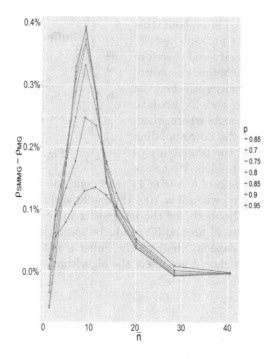

Fig. 10. (ρ_{MG} - ρ_{SMMG}) vs. $\bar{n}$, colored by p (See [16] for colored figure)

packets from node 1. The packet will either expire at stage 2 or go through to node 3. So the throughput can only be at most the same as that of node 1.

The following minor adjustment slightly improves the performance at node 2: Each arriving packet has a deadline to reach node 3, denoted by d. We introduce a temporary deadline for that packet to reach node 2 by $d - 1$. This modification guarantees that we only send packets to node 2 if after arriving at node 2 there is at least one more time unit left to its expiration in order to give the packet a chance to reach node 3. Here is a trivial example: A packet arrives with deadline 7, i.e. it should arrive at node 3 by 7. Before the adjustment it was possible for this packet at time 7 to be still at node 1 and move to node 2, then be expired at node 2 and get lost. After the adjustment, this packet will have a deadline of 6 for node 2. So if by time 6, the packet hasn't been sent yet, it gets deleted from the buffer of node 1 (one time step before its actual deadline). This adjustment improved the throughput of node 2 to be almost equal to that of node 1 because the arrival rate at node 2 is at most 1 (at most one packet is sent from node 1 at each time step). So node 2 is usually making a trivial decision of sending the only packet it has in its buffer. In conclusion, our model implicitly imposes a restriction on the maximum possible throughput at internal nodes, hence, making the multi-node model, where only one packet is sent at each time step, an uninteresting problem. In Sect. 8, we give, however, a few future directions for more interesting extensions.

Finiteness of Buffer. Throughout this paper, we chose not to put any restrictions on the buffer capacity and therefore we now look to verify whether the finiteness of the buffer has a major effect on the algorithm performance and its bounds. We run a set of experiments where, given $\lambda \in [2,100]$, we find the corresponding buffer size b, such that the probability of exceeding b is one in a million. We create

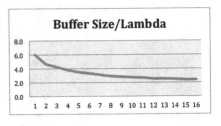

Fig. 11. b/λ vs. λ

an index plot for the ratio of b/λ (Fig. 11) and conclude that imposing a buffer size is unnecessary. Even when $\lambda = 100$, the buffer size needs to be about 1.53 as much, i.e. 153. Furthermore, for the interesting values of λ used throughout this report, a buffer size of around 30 would be more than sufficient. We conclude that imposing a capacity limit on the buffer is not necessary as choosing a reasonable buffer size should not affect the algorithm's performance.

8   Conclusion

In this paper, we consider several old and new packet scheduling algorithms. By analyzing the empirical behavior of MG, we observe that MG chooses packet h over e too frequently. We therefore develop a new algorithm, MLP, which mimics the offline algorithm and gives higher attention to early deadline packets. We then show that on a wide variety of data, including uniform and bimodal distributions, MLP is slower, but has a better empirical competitive ratio than MG (in contrast to the worst-case analysis where MG has a better competitive ratio).

We then propose three new algorithms that may offer an improvement in empirical performance, as they combine features of both algorithms. MM, at each time step, chooses between using MG or MLP in order to make a decision on the packet to send. LMG learns from past behavior to correct the divisor used in MG, while SMMG is motivated by the idea of influential packets in extending the comparison to a pool of three packets, namely e, h and s. The improvements for these algorithms are small, yet encouraging for further analysis. Moreover, it is important to consider extensions for the network model and run the algorithms on one where induced correlations are captured by more realistic distributions that are not i.i.d. Contrasting the behavior of any of the algorithms mentioned in this paper on an actual router, rather than a simulated environment, would also be important to consider.

Several interesting future directions remain. One important extension would be a multi-node model. We showed how the straightforward extension does not yield much insight but other extensions may be more interesting. For example, one could have nodes that process at different rates; this would prevent the first node from being an obvious bottleneck. Another possibility is to allow feedback, that is, if a packet expires somewhere in the multi-node system, it could return to the source to be resent. A final possibility is for each packet to have a vector of deadlines, one per node, so that different nodes could be the bottleneck at different times.

Acknowledgments. The authors would like to thank Dr. Shokri Z. Selim and Javid Ali.

References

1. Albers, S., Schmidt, M.: On the performance of greedy algorithms in packet buffering. SIAM J. Comput. (SICOMP) **35**(2), 278–304 (2005)
2. Andelman, N., Mansour, Y., Zhu, A.: Competitive queuing polices for QoS switches. In: Proceedings of the 14th Annual ACM-SIAM Symposium on Discrete Algorithms (SODA), pp. 761–770 (2003)
3. Borodin, A., El-Yaniv, R.: Online Computation and Competitive Analysis. Cambridge University Press, Cambridge (1998)
4. Bienkowski, M., Chrobak, M., Jeż, Ł.: Randomized algorithms for buffer management with 2-bounded delay. In: Bampis, E., Skutella, M. (eds.) WAOA 2008. LNCS, vol. 5426, pp. 92–104. Springer, Heidelberg (2009)
5. Chin, F.Y.L., Fung, S.P.Y.: Online scheduling with partial job values: does time-sharing or randomization help? Algorithmica **37**(3), 149–164 (2003)
6. Chin, F.Y.L., Fung, S.P.Y.: Improved competitive algorithms for online scheduling with partial job values. Theor. Comput. Sci. **325**(3), 467–478 (2004)
7. Chin, F.Y.L., Chrobak, M., Fung, S.P.Y., Jawor, W., Sgall, J., Tichy, T.: Online competitive algorithms for maximizing weighted throughput of unit jobs. J. Discrete Algorithms **4**(2), 255–276 (2006)
8. Chrobak, M., Jawor, W., Sgall, J., Tichý, T.: Improved online algorithms for buffer management in QoS switches. In: Albers, S., Radzik, T. (eds.) ESA 2004. LNCS, vol. 3221, pp. 204–215. Springer, Heidelberg (2004)
9. Englert, M., Westermann, M.: Lower and upper bounds on FIFO buffer management in QoS switches. In: Azar, Y., Erlebach, T. (eds.) ESA 2006. LNCS, vol. 4168, pp. 352–363. Springer, Heidelberg (2006)
10. Englert, M., Westermann, M.: Considering suppressed packets improves buffer management in QoS switches. In: Proceedings of 18th Annual ACM-SIAM Symposium on Discrete Algorithms (SODA), pp. 209–218 (2007)
11. Hajek, B.: On the competitiveness of online scheduling of unit-length packets with hard deadlines in slotted time. In: Proceedings of the 2001 Conference on Information Sciences and Systems (CISS), pp. 434–438 (2001)
12. Jeż, Ł., Li, F., Sethuraman, J., Stein, C.: Online scheduling of packets with agreeable deadlines. ACM Trans. Algorithms (TALG) **9**(1), 5 (2012)
13. Kesselman, A., Lotker, Z., Mansour, Y., Patt-Shamir, B., Schieber, B., Sviridenko, M.: Buffer overflow management in QoS switches. SIAM J. Comput. (SICOMP) **33**(3), 563–583 (2004)
14. Kesselman, A., Mansour, Y., van Stee, R.: Improved competitive guarantees for QoS buffering. Algorithmica **43**, 63–80 (2005)
15. Li, F., Sethuraman, J., Stein, C.: Better online buffer management. In: Proceedings of the 18th Annual ACM-SIAM Symposium on Discrete Algorithms (SODA), pp. 199–208 (2007)
16. Sakr, N., Stein, C.: An empirical study of online packet scheduling algorithms. Draft on arXiv under Computer Science: Data Structures and Algorithms. http://arxiv.org/abs/1603.07947

Advanced Multilevel Node Separator Algorithms

Peter Sanders and Christian Schulz[(✉)]

Karlsruhe Institute of Technology, Karlsruhe, Germany
{sanders,christian.schulz}@kit.edu

Abstract. A node separator of a graph is a subset S of the nodes such that removing S and its incident edges divides the graph into two disconnected components of about equal size. In this work, we introduce novel algorithms to find small node separators in large graphs. With focus on solution quality, we introduce novel flow-based local search algorithms which are integrated in a multilevel framework. In addition, we transfer techniques successfully used in the graph partitioning field. This includes the usage of edge ratings tailored to our problem to guide the graph coarsening algorithm as well as highly localized local search and iterated multilevel cycles to improve solution quality even further. Experiments indicate that flow-based local search algorithms on its own in a multilevel framework are already *highly* competitive in terms of separator quality. Adding additional local search algorithms further improves solution quality. Our strongest configuration almost always outperforms competing systems while on average computing 10 % and 62 % smaller separators than Metis and Scotch, respectively.

1 Introduction

Given a graph $G = (V, E)$, the *node separator problem* asks to find three disjoint subsets V_1, V_2 and S of the node set, such that there are no edges between V_1 and V_2 and $V = V_1 \cup V_2 \cup S$. The objective is to minimize the size of the separator S or depending on the application the weight of its nodes while V_1 and V_2 are balanced. Note that removing the set S from the graph results in at least two connected components. There are many algorithms that rely on small node separators. For example, small balanced separators are a popular tool in divide-and-conquer strategies [3,21,23], are useful to speed up the computations of shortest paths [9,11,34] or are necessary in scientific computing to compute fill reducing orderings with nested dissection algorithms [15].

Finding a balanced node separator on general graphs is NP-hard even if the maximum node degree is three [6,14]. Hence, one relies on heuristic and approximation algorithms to find small node separators in general graphs. The most commonly used method to tackle the node separator problem on large graphs in practice is the multilevel approach. During a coarsening phase, a multilevel algorithm reduces the graph size by iteratively contracting nodes and edges until the graph is small enough to compute a node separator by some other algorithm.

This paper is a short version of the TR [31].

© Springer International Publishing Switzerland 2016
A.V. Goldberg and A.S. Kulikov (Eds.): SEA 2016, LNCS 9685, pp. 294–309, 2016.
DOI: 10.1007/978-3-319-38851-9_20

A node separator of the input graph is then constructed by successively transferring the solution to the next finer graph and applying local search algorithms to improve the current solution.

Current solvers are typically more than fast enough for most applications but lack high solution quality. For example in VLSI design [3,21], node separators are computed once and even small improvements in the objective can have a large impact on production costs. Another important example in which high quality node separators are important are speedup techniques for shortest-path computations in road networks, e.g. customizable contraction hierarchies [11]. Here, smaller node separators yield better node orderings which in turn are repeatedly used to answer shortest path queries. The costs for the computation of the node separator are then amortized over a large amount of shortest path queries. Moreover, high quality node separators can be useful for benchmarking purposes and provide useful upper bounds. Hence, we focus on solution quality in this work.

The remainder of the paper is organized as follows. We begin in Sect. 2 by introducing basic concepts and by summarizing related work. Our main contributions are presented in Sect. 3 where we transfer techniques previously used for the graph partitioning problem to the node separator problem and introduce novel flow based local search algorithms for the problem that can be used in a multilevel framework. This includes edge ratings to guide a graph coarsening algorithm within a multilevel framework, highly localized local search to improve a node separator and iterated multilevel cycles to improve solution quality even further. Experiments in Sect. 4 indicate that our algorithms consistently provide excellent node separators and outperform other state-of-the-art algorithms over varying graphs. Finally, we conclude with Sect. 5. All of our algorithms have been implemented in the open source graph partitioning package KaHIP [30] and will be made available within this framework.

2   Preliminaries

2.1   Basic Concepts

In the following we consider an undirected graph $G = (V = \{0, \ldots, n-1\}, E, c, \omega)$ with node weights $c : V \to \mathbb{R}_{\geq 0}$, edge weights $\omega : E \to \mathbb{R}_{>0}$, $n = |V|$, and $m = |E|$. We extend c and ω to sets, i.e., $c(V') := \sum_{v \in V'} c(v)$ and $\omega(E') := \sum_{e \in E'} \omega(e)$. $\Gamma(v) := \{u : \{v, u\} \in E\}$ denotes the neighbors of a node v. The degree $d(v)$ of a node v is the number of its neighbors. A set $C \subset V$ of a graph is called *closed node set* if there are no connections from C to $V \setminus C$, i.e. for every node $u \in C$ an edge $(u, v) \in E$ implies that $v \in C$ as well. A graph $S = (V', E')$ is said to be a *subgraph* of $G = (V, E)$ if $V' \subseteq V$ and $E' \subseteq E \cap (V' \times V')$. We call S an *induced* subgraph when $E' = E \cap (V' \times V')$. For a set of nodes $U \subseteq V$, $G[U]$ denotes the subgraph induced by U. We define multiple partitioning problems. The *graph partitioning problem* asks for *blocks* of nodes $V_1, \ldots, V_k$ that partition V, i.e., $V_1 \cup \cdots \cup V_k = V$ and $V_i \cap V_j = \emptyset$ for $i \neq j$. A *balancing constraint* demands that $\forall i \in \{1, \ldots, k\} : c(V_i) \leq L_{\max} := (1+\epsilon)\lceil c(V)/k \rceil$ for some parameter $\epsilon \geq 0$.

In this case, the objective is often to minimize the total *cut* $\sum_{i<j} |E_{ij}|$ where $E_{ij} := \{\{u,v\} \in E : u \in V_i, v \in V_j\}$. The set of cut edges is also called *edge separator*. A node $v \in V_i$ that has a neighbor $w \in V_j, i \neq j$, is a boundary node. An abstract view of the partitioned graph is the so called *quotient graph*, where nodes represent blocks and edges are induced by connectivity between blocks. The *node separator problem* asks to find blocks, V_1, V_2 and a separator S that partition V such that there are no edges between the blocks. Again, a balancing constraint demands $c(V_i) \leq (1 + \epsilon)\lceil c(V)/2 \rceil$. However, there is no balancing constraint on the separator S. The objective is to minimize the size of the separator $c(S)$. Note that removing the set S from the graph results in at least two connected components and that the blocks V_i itself do not need to be connected components. By default, our initial inputs will have unit edge and node weights. However, the results in this paper are easily transferable to node and edge weighted problems.

A matching $M \subseteq E$ is a set of edges that do not share any common nodes, i.e. the graph (V, M) has maximum degree one. *Contracting* an edge $\{u, v\}$ means to replace the nodes u and v by a new node x connected to the former neighbors of u and v. We set $c(x) = c(u) + c(v)$. If replacing edges of the form $\{u, w\}, \{v, w\}$ would generate two parallel edges $\{x, w\}$, we insert a single edge with $\omega(\{x, w\}) = \omega(\{u, w\}) + \omega(\{v, w\})$. *Uncontracting* an edge e undos its contraction. In order to avoid tedious notation, G will denote the current state of the graph before and after a (un)contraction unless we explicitly want to refer to different states of the graph.

The multilevel approach consists of three main phases. In the *contraction* (coarsening) phase, we iteratively identify matchings $M \subseteq E$ and contract the edges in M. Contraction should quickly reduce the size of the input and each computed level should reflect the global structure of the input network. Contraction is stopped when the graph is small enough so that the problem can be *solved* by some other potentially more expensive algorithm. In the *local search* (or uncoarsening) phase, matchings are iteratively uncontracted. After uncontracting a matching, the local search algorithm moves nodes to decrease the size of the separator or to improve balance of the block while keeping the size of the separator. The succession of movements is based on priorities called *gain*, i.e., the decrease in the size of the separator. The intuition behind the approach is that a good solution at one level of the hierarchy will also be a good solution on the next finer level thus local search will quickly find a good solution.

2.2   Related Work

There has been a *huge* amount of research on graph partitioning so that we refer the reader to [4,7] for most of the material in this area. Here, we focus on issues closely related to our main contributions and previous work on the node separator problem. Lipton and Tarjan [22] provide the *planar separator theorem* stating that on planar graphs one can always find a separator S in linear time that satisfies $|S| \in O(\sqrt{|V|})$ and $|V_i| \leq 2|V|/3$ $(i = 1, 2)$. For more balanced cases, the problem remains NP-hard [13] even on planar graphs.

For general graphs there exist several heuristics to compute small node separators. A common and simple method is to derive a node separator from an edge separator [28,33] which is usually computed by a multilevel graph partitioning algorithm. Clearly, taking the boundary nodes of the edge separator in one block of the partition yields a node separator. Since one is interested in a small separator, one can use the smaller set of boundary nodes. A better method has been first described by Pothen et al. [28]. The method employs the set of cut edges of the partition and computes the smallest node separator that can be found by using a subset of the boundary nodes. The main idea is to compute a subset S of the boundary nodes such that each cut edge is incident to at least one of the nodes in S (a vertex cover). A problem of the method is that the graph partitioning problem with edge cut as objective has a different combinatorial structure compared to the node separator problem. This makes it unlikely to find high quality solutions with that approach.

Metis [19] and Scotch [26] use a multilevel approach to obtain a node separator. After contraction, both algorithms compute a node separator on the coarsest graph using a greedy algorithm. This separator is then transferred level-by-level, dropping non-needed nodes on each level and applying a Fiduccia-Mattheyses (FM) style local search. Previous versions of Metis and Scotch also included the capability to compute a node separator from an edge separator.

Recently, Hamann and Strasser [17] presented a max-flow based algorithm specialized for road networks. Their main focus is not on node separators. They focus on a different formulation of the problem, i.e. the edge-cut version graph partitioning problem. More precisely, Hamann and Strasser find Pareto solutions in terms of edge cut versus balance instead of specifying the allowed amount of imbalance in advance and finding the best solution satisfying the constraint. Their work also includes an algorithm to derive node separators, again in a different formulation of the problem, i.e. node separator size versus balance. We cannot make meaningful comparisons since the paper does not contain data on separator quality and the implementation of the algorithm is not available.

Hager et al. [16] recently proposed a multilevel approach for medium sized graphs using continuous bilinear quadratic programs and a combination of those with local search algorithms. However, a different formulation of the problem is investigated, i.e. the solver enforces upper *and* lower bounds to the block sizes which makes the results incomparable to our results.

LaSalle and Karypis [20] present a shared-memory parallel algorithm to compute node separators used to compute fill reducing orderings. Within a multilevel approach they evaluate different local search algorithms indicating that a combination of greedy local search with a segmented FM algorithm can outperform serial FM algorithms. We compare the solution quality of our algorithm against the data presented there in our experimental section (see Sect. 4).

3   Advanced Multilevel Algorithms for Node Separators

We now present our core innovations. In brevity, the novelties of our algorithm include edge ratings during coarsening to compute graph hierarchies that fulfill the needs of the node separator problem and a combination of localized local search with flow problems to improve the size of the separator. In addition, we transfer a concept called iterative multilevel scheme previously used in graph partitioning to further improve the solution quality. The description of our algorithm in this section follows the multilevel scheme. We start with the description of the edge ratings that we use during coarsening, continue with the description of the algorithm used to compute an initial node separator on the coarsest level and then describe local search algorithms as well as other techniques.

3.1   Coarsening

Before we explain the matching algorithm that we use in our system, we present the general two-phase procedure which was already used in multiple graph partitioning frameworks [18, 25, 29]. The two-phase approach makes contraction more systematic by separating two issues: A *rating function* and a *matching* algorithm. A rating function indicates how much sense it makes to contract an edge based on *local* information. A matching algorithm tries to maximize the sum of the ratings of the contracted edges looking at the *global* structure of the graph. While the rating function allows a flexible characterization of what a "good" contracted graph is, the simple, standard definition of the matching problem allows to reuse previously developed algorithms for weighted matching. Note that we can use the same edge rating functions as in the graph partitioning case but also can define new ones since the problem structure of the node separator problem is different.

We use the *Global Path Algorithm (GPA)* which runs in near linear time to compute matchings. GPA was proposed in [24] as a synthesis of the Greedy Algorithm and the Path Growing Algorithm [12]. We choose this algorithm since in [18] it gives empirically considerably better results than Sorted Heavy Edge Matching, Heavy Edge Matching or Random Matching [32]. GPA scans the edges in order of decreasing weight but rather than immediately building a matching, it first constructs a collection of paths and even length cycles. Afterwards, *optimal solutions* are computed for each of these paths and cycles using dynamic programming.

Edge Ratings for Node Separator Problems. We want to guide the contraction algorithm so that coarse levels in the graph hierarchy still contain small node separators if present in the input problem. This way we can provide a good starting point for the initial node separator routine. There are a lot of possibilities that we have tried. The most important edge rating functions for an edge $e = \{u, v\} \in E$ are the following:

$$\exp^*(e) = \omega(e)/(d(u)d(v))$$
$$\exp^{**}(e) = \omega(e)^2/(d(u)d(v))$$
$$\max(e) = 1/\max\{d(u), d(v)\}$$
$$\log(e) = 1/\log(d(u)d(v))$$

The first two ratings have already been successfully used in the graph partitioning field. To give an intuition behind these ratings, we have to characterize the properties of "good" matchings for the purpose of contraction in a multilevel algorithm for the node separator problem. Our main objective is to find a small node separator on the coarsest graph. A matching should contain a *large number of edges*, e.g. being maximal, so that there are only few levels in the hierarchy and the algorithm can converge quickly. In order to *represent the input* on the coarser levels, we want to find matchings such that the graph after contraction has somewhat *uniform node weights* and *small node degrees*. In addition, we want to keep nodes having a small degree since they are potentially good separators. Uniform node weights are also helpful to achieve a balanced node separator on coarser levels and makes local search algorithms more effective. We also included ratings that do not contain the edge weight of the graph since intuitively a matching does not have to care about large edge weights – they do not show up in the objective of the node separator problem.

3.2   Initial Node Separators

We stop coarsening as soon as the graph has less than ten thousand nodes. Our approach first computes an edge separator and then derives a node separator from that. More precisely, we partition the coarsest graph into two blocks using KaFFPa [30]. We then look at the bipartite graph induced by the set of cut edges including the given node weights. Our goal is to select a minimum weight node separator in that graph. As a side note, this corresponds to finding a minimum weight vertex cover in the bipartite graph. Also note that this is similar to the approach of Pothen et al. [28], however we integrate node weights. To solve the problem, we put all of the nodes of the bipartite graph into the initial separator S and use the *flow-based technique* defined below to select the smallest separator contained in that subgraph. Since our algorithms are randomized, we repeat the overall procedure twenty five times and pick the smallest node separator that we have found.

3.3   Local Search

Localized Local Search. In graph partitioning it has been shown that higher localization of local search can improve solution quality [25,33]. Hence, we develop a novel localized algorithm for the node separator problem that starts local search only from a couple of selected separator nodes. Our localized local search procedure is based on the FM scheme. Before we explain our approach to localization, we present a commonly used FM-variant for completeness.

For each of the two blocks V_1, V_2 under consideration, a priority queue of separator nodes eligible to move is kept. The priority is based on the *gain* concept, i.e. the decrease in the objective function value when the separator node is moved into that block. More precisely, if a node $v \in S$ would be moved to V_1, then the neighbors of v that are in V_2 have to be moved into the separator. Hence, in this case the gain of the node is the weight of v minus the weight of the nodes that have to be added to the separator. The gain value in the other case (moving v into to V_2) is similar. After the algorithm computed both gain values it chooses the largest gain value such that moving the node does not violate the balance constraint and performs the movement. Each node is moved at most once out of the separator within a single local search. The queues are initialized randomly with the separator nodes. After a node is moved, newly added separator nodes become *eligible* for movement (and hence are added to the priority queues). The moved node itself is not eligible for movement anymore and is removed from the priority queue. Note that the movement can change the gain of current separator nodes. Hence, gains of adjacent nodes are updated.

There are different possibilities to select a block to which a node shall be moved. The most common variant of the classical FM-algorithm alternates between both blocks. After a stopping criterion is applied, the smallest feasible node separator found is reconstructed (among ties choose the node separator that has better balance). This is called roll back. We have two strategies to *balance blocks*. The first strategy tries to create a balanced situation without increasing the size of the separator. It always selects the queue of the heavier block and uses the same roll back mechanism as before. The second strategy allows to increase the size of the node separator. It also selects a node from the queue of the heavier block, but the roll back mechanism recreates the node separator having the best balance (among ties we choose the smaller node separator).

Our approach to localization works as follows. Previous local search methods were initialized with *all* separator nodes, i.e. all separator nodes are *eligible* for movement at the beginning. In contrast, our method is repeatedly initialized only with a *subset* of the separator nodes (the precise amount of nodes in the subset is a tuning parameter). Intuitively, this introduces a larger amount of diversification and boosts the algorithms ability to climb out of local minima.

The algorithm is organized in rounds. One round works as follows. Instead of putting *all* separator nodes directly into the priority queues, we put the current separator nodes into a todo list T. Subsequently, we begin local search starting with a random *subset* S of the todo list T. We select the subset S by repeatedly picking a random node v from T. We add v to S if it still is a separator node and has not been moved by a previous local search in that round. Either way, v is removed from the todo list. Our localized search is restricted to the movement of nodes that have not been touched by a previous local search during the round. This assures that each node is moved at most once out of the separator during a round of the algorithm and avoids cyclic local search. By default our local search routine first uses classic local search (including balancing) to get close to a good solution and afterwards uses localization to improve the result further. We repeat this until no further improvement is found.

We now give intuition why localization of local search boosts the algorithms ability to climb out of local minima. Consider a situation in which a node separator is locally optimal in the sense that at least two node movements are necessary until moving a node out of the separator with positive gain is possible. Recall that classical local search is initialized with all separator nodes (in this case all of them have negative gain values). It then starts to move nodes with negative gain at multiple places of the graph. When it finally moves nodes with positive gain the separator is already much worse than the input node separator. Hence, the movement of these positive gain nodes does not yield an improvement with respect to the given input partition. On the other hand, a localized local search that starts close to the nodes with positive gain, can find the positive gain nodes by moving only a small number of nodes with negative gain. Since it did not move as many negative gain nodes as the classical local search, it may still find an improvement with respect to the input.

Maximum Flows as Local Search. We define the node-capacitated flow problem $\mathcal{F} = (V_{\mathcal{F}}, E_{\mathcal{F}})$ that we solve to improve a given node separator as follows. First we introduce a few notations. Given a set of nodes $A \subset V$, we define its *border* $\partial A := \{u \in A \mid \exists (u, v) \in E : v \notin A\}$. The set $\partial_1 A := \partial A \cap V_1$ is called *left border* of A and the set $\partial_2 A := \partial A \cap V_2$ is called *right border* of A. An A *induced flow problem* $\mathcal{F}$ is the node induced subgraph $G[A]$ with ∞ as edge-capacities and the node weights of the graph as node-capacities. Additionally there are two nodes s, t that are connected to the border of A. More precisely, s is connected to all left border nodes $\partial_1 A$ and all right border nodes $\partial_2 A$ are connected to t. These new edges get capacity ∞.

Note that the additional edges are directed. $\mathcal{F}$ has the *balance property* if each (s, t)-flow induces a balanced node separator in G, i.e. the blocks V_i fulfill the balancing constraint. The basic idea is to construct a flow problem $\mathcal{F}$ having the balance property.

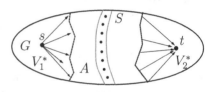

We now explain how we find such a subgraph. We start by setting A to S and extend it by performing two breadth first searches (BFS). The first BFS is initialized with the current separator nodes S and only looks at nodes in block V_1. The same is done during the second BFS with the difference that we now look at nodes from block V_2. Each node touched by any of the BFS is added to A. The first BFS is stopped as soon as the size of

Fig. 1. The construction of an A induced flow problem $\mathcal{F}$ is shown. Two BFS are started to define the area A – one into the block on the left hand side and one into the block on the right hand side. A solution of the flow problem yields the smallest node separator that can be found within the area. The area A is chosen so that each node separator being found in it yields a feasible separator for the original problem.

the newly added nodes would exceed $L_{\max} - c(V_2) - c(S)$. Similarly, the second BFS is stopped as soon as the size of the newly added nodes would exceed $L_{\max} - c(V_1) - c(S)$.

A solution of the A induced flow problem yields a valid node separator of the original graph: First, since all edges in our flow network have capacity ∞ and the separator S is contained in the problem, a maximum flow yields a separator S', $V_{\mathcal{F}} = V_1' \cup V_2' \cup S'$, in the flow network that separates $s \in V_1'$ from $t \in V_2'$. Since there is a one-to-one mapping between the nodes of our flow problem and the nodes of the input graph, we directly obtain a separator in the original network $V = V_1^* \cup V_2^* \cup S'$. Additionally, the node separator computed by our method fulfills the balance constraint – presuming that the input solution is balanced. To see this, we consider the size of V_1^*. We can bound the size of this block by assuming that all of the nodes that have been touched by the second BFS get assigned to V_1^* (including the old separator S). However, in this case the balance constraint is still fulfilled $c(V_1^*) \leq c(V_1) + c(S) + L_{\max} - c(V_1) - c(S) = L_{\max}$. The same holds for the opposite direction. Note that the separator is always smaller or equal to the input separator since S is contained in the construction.

To solve the node-capacitated flow problem $\mathcal{F}$, we transform it into a flow problem $\mathcal{H}$ without node-capacities. We use a standard technique [1]: first we insert the source and the sink into our model. Then, for each node u in our flow problem $\mathcal{F}$ that is not the source or the sink, we introduce two nodes u_1 and u_2 in $V_{\mathcal{H}}$ which are connected by a directed edge $(u_1, u_2) \in E_{\mathcal{H}}$ with an edge-capacity set to the node-capacity of the current node. For an edge $(u, v) \in E_{\mathcal{F}}$ not involving the source or the sink, we insert (u_2, v_1) into $E_{\mathcal{H}}$ with capacity ∞. If u is the source s, we insert (s, v_1) and if v is the sink, we insert (u_2, t) into $E_{\mathcal{H}}$. In both cases we use capacity ∞.

Larger Flow Problems and Better Balanced Node Separators. The definition of the flow problem to improve a node separator requires that each cut in the flow problem corresponds to a *balanced* node separator in the original graph. We now simplify this definition and stop the BFSs if the size of the touched nodes exceeds $(1 + \alpha)L_{\max} - c(V_i) - c(S)$ with $\alpha \geq 0$. We then solve the flow problem and check afterwards if the corresponding node separator is balanced. If this is the case, we accept the node separator and continue. If this is not the case, we set $\alpha := \alpha/2$ and repeat the procedure. After ten unsuccessful iterations, we set $\alpha = 0$. Additionally, we stop the process if the flow value corresponds to the separator weight of the input separator.

We apply heuristics to extract a better balanced node separator from the solved max-flow problem. Picard and Queyranne [27] made the observation that *one* (s, t)-max-flow contains information about *all* minimum (s, t)-cuts in the graph (however, finding the most balanced minimum cut is NP-hard [5]). We follow the heuristic approach of [29] and extract better balanced (s, t)-cuts from the given maximum flow in $\mathcal{H}$. This results in better balanced separators in the node-capacitated problem $\mathcal{F}$ and hence in better balanced node separators for our original problem. To be more precise, Picard and Queyranne have shown that each closed node set in the residual graph of a maximum (s, t)-flow that contains the source s but not the sink t induces a minimum s-t cut. Observe that a cycle in the residual graph cannot contain a node of both, a closed node set and its complement. Hence, Picard and Queyranne compactify the residual network by contracting all strongly connected components. Afterwards, their algorithm tries to find the most balanced minimum cut by enumeration. In [29], we find better balanced cuts heuristically. First a random topological order of the strongly connected component graph is computed. This is then scanned in reverse order. By subsequently adding strongly connected components several closed node sets are obtained, each inducing a minimum *s-t* cut. The closed node set with the best occurred balance among multiple runs of the algorithm with different topological orders is returned. An example closed node set and the scanning algorithm is shown in Fig. 2.

3.4   Miscellanea

An easy way to obtain high quality node separators is to use a multilevel algorithm multiple times using different random seeds and use the best node separator that has been found. However, instead of performing a full restart, one can use the information that has already been obtained. In the graph partitioning context, the notion of iterated multilevel schemes (V-cycles) has been introduced by Walshaw [36] and later has been augmented to more complex cycles [29]. Here, one transfers a solution of a previous multilevel cycle down the hierarchy and uses it as initial solution. More precisely, this can be done by not contracting any cut edge.

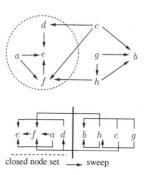

Fig. 2. Left: the set $C = \{a, d, e, f\}$ is a closed node set since no edge is starting in C and ending in $V \backslash C$. Right: using a reverse topological ordering of a DAG one can output multiple closed node sets.

We *transfer this technique* to the node separator problem as follows. One can interpret a node separator as a three way partition V_1, V_2, S. Hence, to obtain an iterated multilevel scheme for the node separator problem, our matching algorithm is not allowed to match any edge that runs between V_i and S $(i = 1, 2)$. Hence, when contraction is done, every edge leaving the separator will remain and we can transfer the node separator down in the hierarchy. Thus a given node separator can be used as initial node separator of the coarsest graph (having the same balance and size as the node separator of the finest graph). This ensures non-decreasing quality, if the local search algorithm guarantees no worsening. To increase diversification during coarsening in later V-cycles we pick a random edge rating of the ones described above.

4   Experiments

Methodology. We have implemented the algorithm described above within the KaHIP framework using C++ and compiled all algorithms using gcc 4.63 with full optimization's turned on (-O3 flag). We integrated our algorithms in KaHIP v0.71 and compare ourselves against Metis 5.1 and Scotch 6.0.4 using the quality option that has focus on solution quality instead of running time. Our new codes will be included into the KaHIP graph partitioning framework. We perform ten repetitions of each algorithm using different random seeds for initialization. When presenting the imbalance of a node separator, we report the maximum block size divided by $\lceil |V|/2 \rceil$. Not that this value can be smaller than one. Each run was made on a machine that has four Octa-Core Intel Xeon E5-4640 processors running at 2.4 GHz. It has 512 GB local memory, 20 MB L3-Cache and 8x256 KB L2-Cache. Our main objective is the cardinality of node separators $|S|$ on the input graph. In our experiments, we use $\epsilon = 20\%$ since this is the default value for node separators in Metis. We mostly present two kinds of views on the data: average values and minimum values as well as plots that show the ratios of the quality achieved by the algorithms. When further averaging over multiple instances, we use the geometric mean in order to give every instance the same influence on the final score.

Algorithm Configuration. We performed a number of experiments to evaluate the influence and choose the parameters of our algorithms. We mark the instances that have also been used for the parameter tuning in Appendix A with a * and exclude these graphs when we report average values over multiple instances in comparisons with our competitors. However, our full algorithm is not too sensitive about the precise choice with most of the parameters. In general, using more sophisticated edge ratings improves solution quality slightly and improves partitioning speed over using edge weight. We exclude further experiments from the main text and use the exp^* edge rating function as a default since it has a slight advantage in our preliminary experiments. In later iterated multilevel cycles, we pick one of the other ratings at random to introduce more diversification. Indeed, increasing the number of V-cycles reduces the objective function. We fixed the number of V-cycles to three. By default, we use the better balanced minimum cut heuristic in our node separator algorithm since it keeps the node separator cardinality and improves balance. In the localized local search algorithm, we set the size of the random subset of separator nodes from which local search is started $|S|$ to five.

Instances. We use graphs from various sources to test our algorithm. We use all 34 graphs from Chris Walshaw's benchmark archive [35]. Graphs derived from sparse matrices have been taken from the Florida Sparse Matrix Collection [8]. We also use graphs from the 10th DIMACS Implementation Challenge [2] website. Here, rggX is a *random geometric graph* with 2^X nodes where nodes represent random points in the unit square and edges connect nodes whose Euclidean distance is below $0.55\sqrt{\ln n/n}$. The graph delX is a Delaunay triangulation of 2^X random points in the unit square. The graphs af_shell9, thermal2, nlr and nlpkkt240 are from the matrix and the numeric section of the DIMACS benchmark set. The graphs europe and deu are large road networks of Europe and Germany taken from [10]. Due to large running time of our algorithm, we exclude the graph nlpkkt240 from general comparisons and only use our full algorithm to compute a result. Basic properties of the graphs under consideration can be found in Appendix A, Table 2.

4.1   Separator Quality

We now assess the size of node separators derived by our algorithms and by other state-of-the-art tools, i.e. Metis and Scotch as well as the data recently presented by LaSalle and Karypis [20]. We use multiple configurations of our algorithm to estimate the influence of the multiplicative factor α that controls the size of the flow problems solved during uncoarsening and to see the effect of adding local search. The algorithms named $Flow_\alpha$ use *only* flows during uncoarsening as local search with a multiplicative factor α. Algorithms labeled $LSFlow_\alpha$ start on each level with local search and localized local search until no improvement is found and afterwards perform flow based local search with a multiplicative factor α. Table 1 summarizes the results of the experiments. We present detailed per instances results in terms of separator size and balance as well as running times in the technical report [31].

We now summarize the results. First of all, only using flow-based local search during uncoarsening is already highly competitive, even for small flow problems with $\alpha = 0$. On average, $Flow_0$ computes 6.7% smaller separators than Metis and 57% than Scotch. It computes a smaller or equally sized separator than Metis in 89% of the cases and than Scotch in *every* case. However, it also needs more time to compute a result. This is due to the large flow problems that have to be solved. Increasing the value of α, i.e. searching for separators in larger areas around the initial separator, improves the objective further at the cost

Table 1. Avg. increase in separator size over $LSFlow_1$, avg. running times of the different algorithms and relative number of instances with a separator smaller or equal to Metis ($\# \leq_{Metis}$).

Algorithm	Avg. Inc.	t_{avg}[s]	$\# \leq_{Metis}$
Metis	10.3%	0.12	-
Scotch	62.2%	0.23	0%
$Flow_0$	3.3%	17.72	89%
$Flow_{0.5}$	0.1%	38.21	96%
$Flow_1$	0.3%	47.81	94%
$LSFlow_0$	1.5%	28.61	96%
$LSFlow_{0.5}$	−0.1%	49.08	94%
$LSFlow_1$	-	58.50	96%

of running time. For example, increasing α to 0.5 reduces the average size of the computed separator by 3.2%, but also increases the running time by more than a factor 2 on average. Using even larger values of $\alpha > 1$ did not further improve the result so that we do not include the data here. Adding non-flow-based local search also helps to improve the size of the separator. For example, it improves the separator size by 1.8% when using $\alpha = 0$. However, the impact of non-flow-based local search decreases for larger values of α.

The strongest configuration of our algorithm is $LSFlow_1$. It computes smaller or equally sized separators than Metis in all but two cases and than Scotch in every case. On average, separators are 10.3% smaller than the separators computed by Metis and 62.2% than the ones computed by Scotch. Figure 3 shows the average improvement ratios over Metis and Scotch on a per instance basis, sorted by absolute value of improvement. The largest improvement over Metis was obtained on the road network **europe** where our separator is a factor

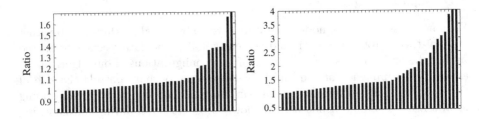

Fig. 3. Improvement of LSFlow₁ per instance over Metis (left) and Scotch (right) sorted by absolute value of ratio $\frac{\text{avg. } |S| \text{ by [Metis | Scotch]}}{\text{avg. } |S| \text{ by LSFlow}_1}$.

2.3 smaller whereas the largest improvement over Scotch is on `add32` where our separator is a factor 12 smaller. On `G2_circuit` Metis computes a 19.9 % smaller separator which is the largest improvement of Metis over our algorithm.

We now compare the size of our separators against the recently published data by LaSalle and Karypis [20]. The networks used therein that are publicly available are `auto`, `nlr`, `del24` and `nlpkkt240`. On these graphs our strongest configuration computes separators that are 10.7 %, 10.0 %, 20.1 % and 27.1 % smaller than their best configuration (Greedy+Segmented FM), respectively.

5   Conclusion

In this work, we derived algorithms to find small node separators in large graphs. We presented a multilevel algorithm that employs novel flow-based local search algorithms and transferred techniques successfully used in the graph partitioning field to the node separator problem. This includes the usage of edge ratings tailored to our problem to guide the graph coarsening algorithm as well as highly localized local search and iterated multilevel cycles to improve solution quality even further. Experiments indicate that using flow-based local search algorithms as only local search algorithm in a multilevel framework is already highly competitive in terms of separator quality.

Important future work includes shared-memory parallelization of our algorithms, e.g. currently most of the running time in our algorithm is consumed by the max-flow solver so that a parallel solver will speed up computations. In addition, it is possible to define a simple evolutionary algorithm for the node separator problem by transferring the iterated multilevel scheme to multiple input separators. This will likely result in even better solutions.

A Benchmark Set A

Table 2. Basic properties of the instances used for evaluation.

Graph	n	m	Graph	n	m
Small Walshaw Graphs			UF Graphs		
add20	2 395	7 462	cop20k_A*	99 843	1 262 244
data	2 851	15 093	2cubes_sphere*	101 492	772 886
3elt	4 720	13 722	thermomech_TC	102 158	304 700
uk	4 824	6 837	cfd2	123 440	1 482 229
add32	4 960	9 462	boneS01	127 224	3 293 964
bcsstk33	8 738	291 583	Dubcova3	146 689	1 744 980
whitaker3	9 800	28 989	bmwcra_1	148 770	5 247 616
crack	10 240	30 380	G2_circuit	150 102	288 286
wing_nodal*	10 937	75 488	c-73	169 422	554 926
fe_4elt2	11 143	32 818	shipsec5	179 860	4 966 618
vibrobox	12 328	165 250	cont-300	180 895	448 799
bcsstk29*	13 992	302 748	Large Walshaw Graphs		
4elt	15 606	45 878	598a	110 971	741 934
fe_sphere	16 386	49 152	fe_ocean	143 437	409 593
cti	16 840	48 232	144	144 649	1 074 393
memplus	17 758	54 196	wave	156 317	1 059 331
cs4	22 499	43 858	m14b	214 765	1 679 018
bcsstk30	28 924	1 007 284	auto	448 695	3 314 611
bcsstk31	35 588	572 914	Large Other Graphs		
fe_pwt	36 519	144 794	del23	≈8.4M	≈25.2M
bcsstk32	44 609	985 046	del24	≈16.7M	≈50.3M
fe_body	45 087	163 734	rgg23	≈8.4M	≈63.5M
t60k*	60 005	89 440	rgg24	≈16.7M	≈132.6M
wing	62 032	121 544	deu	≈4.4M	≈5.5M
brack2	62 631	366 559	eur	≈18.0M	≈22.2M
finan512*	74 752	261 120	af_shell9	≈504K	≈8.5M
fe_tooth	78 136	452 591	thermal2	≈1.2M	≈3.7M
fe_rotor	99 617	662 431	nlr	≈4.2M	≈12.5M
			nlpkkt240	≈27.9M	≈373M

References

1. Ahuja, R.K., Magnanti, T.L., Orlin, J.B.: Network flows: theory, algorithms, and applications (1993)
2. Bader, D., Kappes, A., Meyerhenke, H., Sanders, P., Schulz, C., Wagner, D.: Benchmarking for graph clustering and partitioning. In: Alhajj, R., Rokne, J. (eds.) Encyclopedia of Social Network Analysis and Mining. Springer, New York (2014)
3. Bhatt, S.N., Leighton, F.T.: A framework for solving vlsi graph layout problems. J. Comput. Syst. Sci. **28**(2), 300–343 (1984)
4. Bichot, C., Siarry, P. (eds.): Graph Partitioning. Wiley, New York (2011)
5. Bonsma, P.: Most balanced minimum cuts. Discrete Appl. Math. **158**(4), 261–276 (2010)
6. Bui, T.N., Jones, C.: Finding good approximate vertex and edge partitions is NP-hard. Inf. Process. Lett. **42**(3), 153–159 (1992)
7. Buluç, A., Meyerhenke, H., Safro, I., Sanders, P., Schulz, C.: Recent advances in graph partitioning. In: Algorithm Engineering – Selected Topics, to app., arXiv:1311.3144 (2014)
8. Davis, T.: The University of Florida Sparse Matrix Collection
9. Delling, D., Holzer, M., Müller, K., Schulz, F., Wagner, D.: High-performance multi-level routing. In: The Shortest Path Problem: Ninth DIMACS Implementation Challenge, vol. 74, pp. 73–92 (2009)
10. Delling, D., Sanders, P., Schultes, D., Wagner, D.: Engineering route planning algorithms. In: Lerner, J., Wagner, D., Zweig, K.A. (eds.) Algorithmics. LNCS, vol. 5515, pp. 117–139. Springer, Heidelberg (2009)
11. Dibbelt, J., Strasser, B., Wagner, D.: Customizable contraction hierarchies. In: Gudmundsson, J., Katajainen, J. (eds.) SEA 2014. LNCS, vol. 8504, pp. 271–282. Springer, Heidelberg (2014)
12. Drake, D., Hougardy, S.: A simple approximation algorithm for the weighted matching problem. Inf. Process. Lett. **85**, 211–213 (2003)
13. Fukuyama, J.: NP-completeness of the planar separator problems. J. Graph Algorithms Appl. **10**(2), 317–328 (2006)
14. Garey, M.R., Johnson, D.S.: Computers and Intractability, vol. 29. WH Freeman & Co., San Francisco (2002)
15. George, A.: Nested dissection of a regular finite element mesh. SIAM J. Numer. Anal. **10**(2), 345–363 (1973)
16. Hager, W.W., Hungerford, J.T., Safro, I.: A multilevel bilinear programming algorithm for the vertex separator problem. Technical report (2014)
17. Hamann, M., Strasser, B.: Graph bisection with pareto-optimization. In: Proceedings of the Eighteenth Workshop on Algorithm Engineering and Experiments, ALENEX 2016, pp. 90–102. SIAM (2016)
18. Holtgrewe, M., Sanders, P., Schulz, C.: Engineering a scalable high quality graph partitioner. In: Proceedings of the 24th International Parallal and Distributed Processing Symposium, pp. 1–12 (2010)
19. Karypis, G., Kumar, V.: A fast and high quality multilevel scheme for partitioning irregular graphs. SIAM J. Sci. Comput. **20**(1), 359–392 (1998)
20. LaSalle, D., Karypis, G.: Efficient nested dissection for multicore architectures. In: Träff, J.L., Hunold, S., Versaci, F. (eds.) Euro-Par 2015. LNCS, vol. 9233, pp. 467–478. Springer, Heidelberg (2015)
21. Leiserson, C.E.: Area-efficient graph layouts. In: 21st Symposium on Foundations of Computer Science, pp. 270–281. IEEE (1980)

22. Lipton, R.J., Tarjan, R.E.: A separator theorem for planar graphs. SIAM J. Appl. Math. **36**(2), 177–189 (1979)
23. Lipton, R.J., Tarjan, R.E.: Applications of a planar separator theorem. SIAM J. Comput. **9**(3), 615–627 (1980)
24. Maue, J., Sanders, P.: Engineering algorithms for approximate weighted matching. In: Demetrescu, C. (ed.) WEA 2007. LNCS, vol. 4525, pp. 242–255. Springer, Heidelberg (2007)
25. Osipov, V., Sanders, P.: n-Level graph partitioning. In: Berg, M., Meyer, U. (eds.) ESA 2010, Part I. LNCS, vol. 6346, pp. 278–289. Springer, Heidelberg (2010)
26. Pellegrini, F.: Scotch Home Page. http://www.labri.fr/pelegrin/scotch
27. Picard, J.C., Queyranne, M.: On the structure of all minimum cuts in a network and applications. Math. Program. Stud. **13**, 8–16 (1980)
28. Pothen, A., Simon, H.D., Liou, K.P.: Partitioning sparse matrices with eigenvectors of graphs. SIAM J. Matrix Anal. Appl. **11**(3), 430–452 (1990)
29. Sanders, P., Schulz, C.: Engineering multilevel graph partitioning algorithms. In: Demetrescu, C., Halldórsson, M.M. (eds.) ESA 2011. LNCS, vol. 6942, pp. 469–480. Springer, Heidelberg (2011)
30. Sanders, P., Schulz, C.: Think locally, act globally: highly balanced graph partitioning. In: Bonifaci, V., Demetrescu, C., Marchetti-Spaccamela, A. (eds.) SEA 2013. LNCS, vol. 7933, pp. 164–175. Springer, Heidelberg (2013)
31. Sanders, P., Schulz, C.: Advanced Multilevel Node Separator Algorithms. Technical report. arXiv:1509.01190 (2016)
32. Schloegel, K., Karypis, G., Kumar, V.: Graph partitioning for high performance scientific simulations. In: Dongarra, J., et al. (eds.) CRPC Parallel Computing Handbook. Morgan Kaufmann, San Francisco (2000)
33. C. Schulz. High Quality Graph Partititioning. Ph.D. thesis, Karlsruhe Institute of Technology (2013)
34. Schulz, F., Wagner, D., Zaroliagis, C.D.: Using multi-level graphs for timetable information in railway systems. In: Mount, D.M., Stein, C. (eds.) ALENEX 2002. LNCS, vol. 2409, pp. 43–59. Springer, Heidelberg (2002)
35. Soper, A.J., Walshaw, C., Cross, M.: A combined evolutionary search and multi-level optimisation approach to graph-partitioning. J. Global Optim. **29**(2), 225–241 (2004)
36. Walshaw, C.: Multilevel refinement for combinatorial optimisation problems. Ann. Oper. Res. **131**(1), 325–372 (2004)

A Merging Heuristic for the Rectangle Decomposition of Binary Matrices

Julien Subercaze[⊠], Christophe Gravier, and Pierre-Olivier Rocher

Université de Lyon, UJM-Saint-Etienne CNRS, UMR5516,
Laboratoire Hubert Curien, 42023 Saint-Etienne, France
{julien.subercaze,christophe.gravier,
pierreolivier.rocher}@univ-st-etienne.fr

Abstract. In this paper we present a linear-time and linear-space algorithm for the decomposition of binary images into rectangles. Our contribution is a two-stage algorithm. In the first stage we compute a $\frac{1}{min(h,w)}$-approximation for the largest rectangle starting at each point of the matrix. In the second stage the algorithm walks through the matrix, alternatively stacking, merging or removing encountered rectangles. Through an experimental evaluation, we show that our algorithm outperforms state-of-the-art linear-time algorithms for small to medium-sized rectilinear polygons.

1 Introduction

Rectangle Decomposition of Binary Matrices (RDBM) is a family of problems that has practical applications in various domains such as pattern recognition, video encoding, image processin and VLSI design. The common goal is these problems is to find the best decomposition of a binary matrix intro a set of rectangles that forms a partition of its 1-entries. Different criteria may be subject to optimization in this family of problems, the two most representative problems being DBMR-MINLINES [4,6,7,9,12] and DBMR-MINRECT.

In this paper we discuss the Rectangle Decomposition of Binary Matrices problem (DBMR-MINRECT). This problem is of the utmost practical interest for moment computation, compression, pattern recognition [14] or multimodal video encoding [11]. The optimal solution has been independently proved by [2,3,8,10] with a time complexity of $\mathcal{O}(n^{2.5})$. It was later shown [1] that this solution can be computed in $\mathcal{O}(n^{\frac{3}{2}}logn)$. However due to the complexity of the required datastructures, existing implementations uses the alternative solution in $\mathcal{O}(n^{2.5})$ [14]. Unfortunately, it is hardly suitable for near-real time applications like streaming of dynamic video content, since computation time goes over a second for a single 4 K picture.

The main contributions of the paper are twofold: *First*, we present a linear time and space approximation of the largest rectangle starting in each point, using dynamic programming. *Second*, we present WSRM, a linear time and space heuristic for the DBMR-MINRECT problem using the previous approximation

© Springer International Publishing Switzerland 2016
A.V. Goldberg and A.S. Kulikov (Eds.): SEA 2016, LNCS 9685, pp. 310–325, 2016.
DOI: 10.1007/978-3-319-38851-9_21

as input. We rely on a linked matrix structure to ensure the linearity of the algorithm. We evaluate WSRM through extensive experiments against the state-of-the-art linear time algorithm.

Our experimental comparison shows that WSRM performs better decomposition than the state of the art linear heuristic for rectilinear with a small to medium number of vertices. We also show that both heuristics have different strengths and weaknesses and that their complementary use is a suitable alternative to the expensive optimal algorithm. An open-source implementation of our algorithm is available under the Apache 2 license.

The paper is organized as follows. Section 2 presents the problem definition as well as the related work. In Sect. 3 we describe the first stage of the algorithm that computes an approximated value of the largest rectangle starting in every point of the binary matrix. Section 4 presents the second stage of the algorithm that makes a linear walk in the matrix to stack, merge and/or remove the processed rectangles. Section 5 gives implementation details and presents the complexity analysis. Section 6 presents the experimental evaluation. Section 7 concludes.

2   DBRM-MinRect

In this section, we present the problem definition, its optimal solution, and finally heuristics of the DBRM-MINRECT problem. Let M be a $(0,1)$-matrix of size $m \times n$. Let $\mathcal{H} = \{1, 2, \ldots, n-1, n\}$ and $\mathcal{W} = \{1, 2, \ldots, m-1, m\}$. A non-empty subset $R_{t,l,b,r}$ of M is said to be a rectangle (a set of $(i,j) \in \mathcal{W} \times \mathcal{H}$) if there are $l, r \in \mathcal{W}$ and $b, t \in \mathcal{H}$ such as:

$$\begin{cases} R_{t,l,b,r} = \{(i,j) : b \leq i \leq t, l \leq j \leq r\} \\ \sum_{i=l}^{r} \sum_{j=b}^{t} R[i][j] = (|r-l|+1) \times (|b-t|+1) \end{cases} \quad (1)$$

For each rectangle $R_{t,l,b,r}$, $R[t][l]$ (respectively $R[b][r]$) is the upper-left (respectively bottom-right) corner of the rectangle.

The DBMR-MINRECT problem is therefore to *find a **minimal** set of rectangles* $\mathcal{R}$*, such as* $\mathcal{R}$ *is a partition of the binary matrix* M, wich can be formulated as follows:

$$argmin(|\mathcal{R}|) : \begin{cases} \mathcal{R} = \bigcup_{k=1}^{K} R^k, \text{ with } R^k \text{ a rectangle of M} \\ R_i \cap R_j = \emptyset \text{ for } i \neq j \\ \sum_{k=1}^{K} \sum_{i=R_l^k l}^{R_r^k} \sum_{j=R_b^k}^{R_t^k} R^k[i][j] = \sum_{i=0}^{m} \sum_{j=0}^{n} M[i][j] \end{cases} \quad (2)$$

Figure 1 provides an illustration of the DBMR-MINRECT problem. It is important to note that this problem is focused on finding a minimal partition of

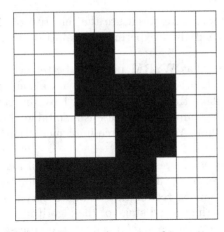

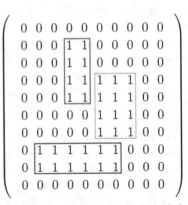

(a) A rectilinear polygon in a binary matrix.

(b) A minimal decomposition of the binary matrix representing the image.

Fig. 1. Illustration of a solution (b) to the DBMR-MINRECT problem applied to the binary matrix (a).

the binary matrix into rectangles. It should not be confused with the rectangle coverage problem [5], in which rectangles in the decomposition can overlap.

2.1 Optimal Solution to the RDBM-MinRect Problem

The optimal solution for rectilinear polygons with holes has been independently proven by [2,3,8,10]. This optimal solution is in Theorem 1 and proofs can be found in the aforementioned works.

Theorem 1. Optimal solution for the RDBM-MINRECT problem. *An orthogonal polygon can be minimally partitioned into $N - L - H + 1$ rectangles, where N is the number of vertices with interior angle greater than 180 degrees. H is the number of holes and L is the maximum number of non-intersecting chords that can be drawn either horizontally or vertically between reflex vertices. (Theorem rephrased from [3]).*

Corollary 1. *From Eq. (2) and Theroem (1), $argmin(|\mathcal{R}|) = N - L - H + 1$.*

This result is not trivial and was found by investigating a dual problem of the RDBM-MINRECT problem in the field of computational geometry [1]. However, [1] provides a proof that an optimal algorithm is in the PTIME complexity class. [2,8,10] exhibited independently an optimal polynomial-time algorithm running in $\mathcal{O}(n^{\frac{3}{2}} \times log(n))$. We refer to this algorithm leading to the optimal solution as the Bipartite Graph-based Decomposition (GBD). The outline of the GBD algorithm is as follows:

1. List all chords for concave vertices, and create their bipartite graphs.
2. The maximum independent set of nodes gives the extremity of chords to keep for decomposing the polygon.
3. For remaining subpolygons, choose a chord of arbitrary direction.

Unfortunately, like in the RDBM-MinLines problem, this complexity prevents scalability for near real-time applications. This primarily encompasses image processing, and especially video processing applications, in which the binary matrix dimensions are not negligible and where the running time must be limited to a dozen of milliseconds. Thus, subsequent efforts focused on finding heuristics for this problem.

2.2   Heuristics for the RDBM-MinRect Problem

From 1985, most of the efforts on developing heuristics have been made by the image processing community. The major application was compacity of binary images: by encoding the set $\mathcal{R}$ instead of the $m \times n$-dimension bits sequence, new algorithms to the RDBM-MinRect problem have been developed. In 2012, [14] provided an exhaustive study in the light of image processing applications. It presents four heuristics known so far, which are outlined below.

Image Block Representation (IBR). The IBR method [13] is an improvement over the trivial row segmentation [14]. The principle follows the RS method, with the extension that it merges identical adjacent lines. If two consecutive lines contain intervals with the same bounds, they are merged. While keeping a linear complexity and being an improvement over the RS method, the IBR still generates an unecessary amount of rectangles in most cases. IBR was identified as the best heuristic by [14] for tasks where a lower complexity prevails over an optimal decomposition.

Quad Tree Decomposition (QTD). QTD is a hierarchical decomposition in which the output of the decomposition is a quad tree data structure. The algorithm divides the matrix recursively in four quadrants until each quadrant contains only 1-entries. The decomposition only contains squares, and often lead to absurd results.

Morphological Decomposition (MED). MED relies on a mask, usually of size 3×3, applied on all pixels. If the 3×3 neighborhood centered on a given pixel contains only 1-entries, the pixel stay 1, and is set to 0 otherwise. The algorithm erodes a rectilinear polygon until all pixels are set to 0 in the binary matrix. This process presents a higher bound and a higher complexity than the previous algorithms. The Distance-Transform Decomposition (DTD) [14] enhances slightly the running time but presents similar complexity.

In order to give an insight on the complexity, advantages and disadvantages of each approach, we provide a synoptic view of all methods in Table 1.

Table 1. Synoptic view of existing methods for the RDBM-MINRECT problem (GBD is an optimal solution, while other methods are approximations.)

Method	Complexity	Analysis	
		Cons	Pros
RS	$\mathcal{O}(n)$	max rectangle height is 1 pixel	Fastest
IBR	$\mathcal{O}(n)$	sub-optimal	Near fastest
QTD	$\mathcal{O}(n \times log_2(n))$	squares only, possibly absurd results	The highest compression ratio
MED	$\mathcal{O}(n^2)$	sub-optimal, slowest	-
DTD	$\mathcal{O}(n^2)$	sub-optimal, slow	Results similar to MED
BGD	$\mathcal{O}(n^{\frac{3}{2}} \times log(n))$	Complexity	Optimal solution

Assuming M a $w \times l$ $(0-1)$-matrix whose decomposition results in k rectangles, hence $\mathcal{O}(n)$ means $(w \times l)$-time.

3   Approximating Largest Area

The underlying idea of our algorithm is to perform a greedy walk in the matrix by successively removing the largest encountered rectangles. The walker considers rectangles starting from their upper left corners. It validates the current rectangle by visiting all its point. If a larger rectangle is encountered the current one is stacked and the walker validates the largest one. For this purpose, this greedy walker requires as input a matrix containing the largest rectangle starting (upper left) in each point. In this section, we present an existing solution and its limitations and we devise an adapted algorithm that matches our requirements. Considering each point of the binary matrix as the upper left corner of a rectangle, our goal is to compute the size of the largest rectangle for each point. Figure 2 depicts the expected result on the example matrix.

A solution for computing the largest rectangles in linear time has been proposed by M. Gallotta[1], based on the observation that the largest rectangle will always touch a 0 on all four sides. The proposed solution is implemented using three Dynamic Programming relations (DP later in this document). Unfortunately this method only computes the largest rectangles to which each point belongs and does not output the largest rectangle starting at each point in the left-to-right and top-to-bottom directions.

Inspired by the approach of Gallotta, we devise a linear-time algorithm that computes an approximated largest rectangle for each point in the matrix by using four DPs. The idea is to maximize one dimension of the rectangle, i.e. height or width and find the corresponding value for the other dimension. For a point of coordinates (x, y), $h_1(x, y) \times w_2(x, y)$ is the area of the rectangle for which the height is maximized, whereas $h_2(x, y) \times w_1(x, y)$ is the area of the rectangle where the width is maximized. We detail here first the DPs maximizing height

---

[1] https://olympiad.cs.uct.ac.za/old/camp1_2009/day2_camp1_2009_solutions.pdf.

$$\begin{vmatrix}
0 & 0 & 0 & 0 & 0 & 0 & 0 & 0 & 0 & 0 \\
0 & 0 & 0 & 8 & 4 & 0 & 0 & 0 & 0 & 0 \\
0 & 0 & 0 & 6 & 3 & 0 & 0 & 0 & 0 & 0 \\
0 & 0 & 0 & 10 & 8 & 12 & 8 & 4 & 0 & 0 \\
0 & 0 & 0 & 5 & 4 & 10 & 6 & 3 & 0 & 0 \\
0 & 0 & 0 & 0 & 0 & 8 & 4 & 2 & 0 & 0 \\
0 & 0 & 0 & 0 & 0 & 6 & 3 & 1 & 0 & 0 \\
0 & 12 & 10 & 8 & 6 & 4 & 2 & 0 & 0 & 0 \\
0 & 6 & 5 & 4 & 3 & 2 & 1 & 0 & 0 & 0 \\
0 & 0 & 0 & 0 & 0 & 0 & 0 & 0 & 0 & 0
\end{vmatrix}.$$

(a) A rectilinear polygon in a binary image.

(b) Area of the largest rectangle considering each point as the upper left corner of a rectangle

Fig. 2. Illustration of the largest area starting in each point

and width, given an $n \times m$ matrix:

$$h_1(x,y) = \begin{cases} 0 & if \ x = 0 \ or \ matrix(x,y) = 0 \\ h_1(x-1,y)+1 & otherwise \end{cases}$$

$$w_1(x,y) = \begin{cases} 0 & if \ y = 0 \ or \ matrix(x,y) = 0 \\ w_1(x,y-1)+1 & otherwise \end{cases}$$

We compute the allowed width (w_2) for a rectangle maximizing the height ($h = h_1$). The idea is that for a point (x,y), a lower bound of the width of the rectangle of height $h1(x,y)$ is given by $argmin_{j=y...y_1}[w_1(x,j)]$. y_1 is the coordinate of the last 1-entry, starting at y and going down.

Conversely, the same applies for the computation of the height h_2 corresponding to the maximized width w_1.

$$area(x,y) = max(h_1(x,y) \times w_2(x,y), h_2(x,y) \times w_1(x,y)) \quad (3)$$

The area computed in Eq. 3 is a $\frac{1}{min(h,w)}$-approximation, where h and w are the height and width of the optimal rectangle. By using four DPs, each requiring one pass over the matrix and one additional matrix for storage. Complexity of this stage is linear, for both time and space.

4 Walk, Stack, Remove and Merge (WSRM)

The principle of the second stage of algorithm is to walk through the matrix and to successively remove the encountered rectangles in a greedy manner. The walk starts from the upper left of the matrix. Whenever a rectangle is encountered (due to the preprocessing described in stage one), the walker walks through this rectangle to check if all its points are present in the matrix. If a larger rectangle is encountered, the previous rectangle is stacked, as well as the position of the walker. If the rectangle is fully validated, i.e. the walker walks through all its points, these latters are removed from the matrix. The walker restarts at the position stored before stacking the rectangle. The process stops when the matrix is empty. The algorithm has four main operations:

Walk. The walker goes from left to right, top to bottom. The walker ensures the presence of every point of the current rectangle. If a point is missing, dimensions of the current rectangle are updated accordingly.

Stack. If the walker reaches a point that is the beginning of a rectangle larger than the current one, the current one is stacked and its size is updated.

Remove. When the current rectangle has been validated by the walker, it is removed from the matrix.

Merge. The merge occurs when the last two removed rectangles are suitable for merge. In order to ensure linearity, candidates for merge are not looked for in the whole set of removed rectangles. Only the last two removed rectangles are considered for merging.

Algorithm 1 presents the main loops of the WSRM algorithm. In the following, we first detail a run of the algorithm on the example presented along this paper (Fig. 2) to introduce the global principles of the algorithm. We detail in another subsection particular cases and behaviour of the algorithm. The function `resizeOrRemove` has been omitted from Algorithm 1 and will be presented in Sect. 4.2.

4.1 Example

Figures 4 and 5 depicts the key steps of the algorithm on a sample matrix. The dotted green boundaries indicate the position of the walker. The rectangle delimited with the dashed blue line is the rectangle on top of the stack, i.e. the one being validated by the walker. Light greyed-out parts have been validated by the walker. Dark greyed indicates parts of stacked rectangles that have not yet been validated. Red arrows show the movement of the walker.

The algorithm starts at step (a) with the first non zero point which is $(2, 4)$. At this point, the maximal area computed in the previous step is equal to 8. The rectangle has a width of 2 and a height of 4. The walker goes rightward through the current rectangle until it reaches the third line of the rectangle at step (b).

The walker is at $(4, 4)$ where the area is equal to 10, greater than the one of the current rectangle in the stack. Since the last line is reached (see more details in Sect. 4.2), the height of the rectangle on the top of the stack is resized to 2 (2 lines have been validated by the walker). The previous rectangle is resized and a new rectangle of area 10 is stacked. The walker continues walking until reaching position $(4, 6)$.

Algorithm 1. Outline of the WSRM algorithm

Data: Binary Matrix M

Result: List L of non overlapping rectangles r_i such as $\cup_i r_i = M$

$stack \leftarrow \emptyset$, $rectangles \leftarrow \emptyset$, $area \leftarrow 0$

while $M \neq \emptyset$ **do**

  if $stack = \emptyset$ **then**

    | stackAndWalk()

  else

    if $currentpoint_{area} \leq area$ **then**

      if *current point is continuous with previous point* **then**

        decrease counter value

        if *current rectangle is validated* **then**

          | remove rectangle

        end

      else

        | resizeOrRemove()

      end

    else

      | stackAndWalk()

    end

  end

end

Function stackAndWalk()

  Stack new rectangle starting at the current point

  $area \leftarrow currentarea$

  Walk to next element

At step (c) the walker encounters a point having a larger associated area. The previously applied process is again executed and a new rectangle of width 3 and height 4 is stacked.

The walker goes through the whole rectangle until reaching coordinates $(7, 8)$ at step (d). Since the rectangle has been validated by the walker, it is safely removed from the matrix and added into the list of removed rectangles. Now the rectangle starting in $(4, 4)$ is on top of the stack, its width had been updated to 2. The last position of the walker was kept in the column P of the stack and was in $(4, 5)$. As the current rectangle starts in $(4, 4)$, the first line has been validated by the walker, its current position is then updated to the beginning of the next line in $(5, 4)$.

The walker validates the last two points $(5, 4)$ and $(5, 5)$ of the current rectangle which is removed at step (e) and added in the list of removed rectangles.

The current rectangle is then the first stacked rectangle, starting in $(2, 4)$, which has been resized at the end of step (a) to a width of 2. Since the last position of the walker is at the end of the rectangle, this latter is fully validated and can be removed at step (f). The two last removed rectangles are starting on the same column, they have same width and are next to each other (i.e. no gap), therefore they are valid candidates for the merging that takes place at step (g). The walker is now set to the next non zero element in the matrix, which is at coordinates $(8, 2)$ and a new rectangle is stacked. The walker then validates the current rectangle when it reaches $(9, 7)$ at step (h), this rectangle is then removed. The matrix is empty, the algorithm returns the list of removed rectangles.

4.2 Algorithm Details

As we discussed in the previous example, different situations may occur whenever a new rectangle is stacked or when a rectangle is removed and poped from the stack. In this subsection we detail the different cases encountered in the `resizeOrRemove` method introduced in Algorithm 1.

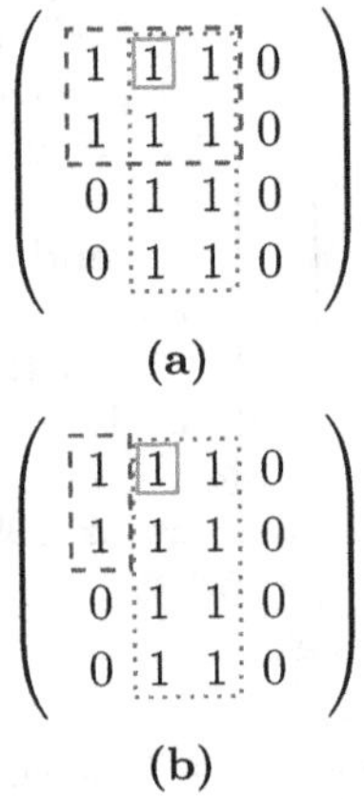

$$\begin{pmatrix} 1 & 1 & 1 & 0 \\ 1 & 1 & 1 & 0 \\ 0 & 1 & 1 & 0 \\ 0 & 1 & 1 & 0 \end{pmatrix}$$

(a)

Stack on First Line. The case where a potential largest rectangle is encountered by the walker on the first line of the current rectangle is the simplest case of stacking in our algorithm. Figure 3 presents an example of such a situation. In this figure, as well as in all this whole subsection, the dashed blue rectangle represents the current rectangle, the full green line delimits the current position of the walker and finally the dotted red line shows the next largest rectangle.

$$\begin{pmatrix} 1 & 1 & 1 & 0 \\ 1 & 1 & 1 & 0 \\ 0 & 1 & 1 & 0 \\ 0 & 1 & 1 & 0 \end{pmatrix}$$

(b)

Fig. 3. The width of the current rectangle in (a) is resized when a largest rectangle is encountered (b) (Color figure online)

In this case, WSRM resizes the current rectangle to the validated width. The current rectangle will not reach its full size except if the part only in red (coordinates $(3, 2)$ to $(4, 2)$) has already been removed from the matrix. However this situation cannot be known in advance when the largest rectangle is encountered. Therefore the decision to resize the current rectangle width to the currently validated width is safe and allows a later merge with the new largest rectangle. When the current rectangle (dashed, blue) appears again on top the stack, after other rectangles have been popped, the rectangle is simply removed and can be merged with the previous removed rectangle, when possible.

Stack at the Beginning of a Line. The same principle applies when the largest rectangle is encountered at the beginning of a line of the current rectangle. In this case, the height of the rectangle is resized to the size that has been validated by the walker. Figure 6 depicts such a situation.

General Case. If a larger rectangle is found in the middle of a line of the current rectangle, one cannot resize the rectangle, since the validated width is greater on the above line. Figure 7(a) depicts such an example.

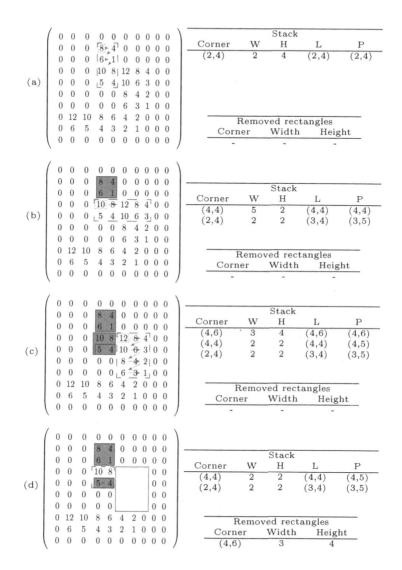

Fig. 4. First four key steps of WSRM on the example figure. L and P columns are detailed in Sect. 5. (Color figure online)

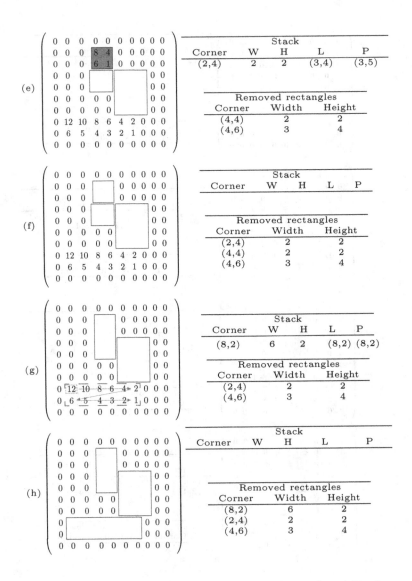

Fig. 5. Laste four key steps of WSRM on the example figure. L and P columns are detailed in Sect. 5 (Color figure online)

When stacking the new rectangle, no resizing takes place. After removal of the new largest rectangle (and others if required), the validated part of the dashed blue rectangle may be larger or smaller than the resized non validated part. For instance in Fig. 7(b), the gray part has been validated by the walker and does not form

(a)   (b)

Fig. 6. Resizing current rectangle

a rectangle. Figure 7(c) shows the final decomposition. At this point, the algorithm will cut the validated part into two rectangles depending on their ratio. The cut may be taken either in a horizontal or vertical way. In Fig. 7(b) one can choose to cut horizontally under the first line or vertically after the first point. In both cases, two rectangles of area 2 and 3 would be obtained. The largest rectangle being already validated with the horizontal cut, this latter will be chosen by the algorithm. The motivation is to favor the cut where the largest rectangle has been validated by the walker.

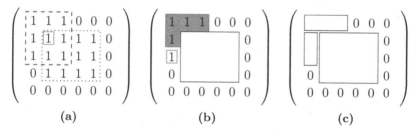

(a)   (b)   (c)

Fig. 7. General case, the new largest rectangle starts in the middle of a line. (Color figure online)

5   Implementation

WSRM is a two-stage algorithm. Implementing the first stage in linear time does not present any noticeable difficulty. However the implementation of the second stage in linear time requires dedicated data structures that are presented in this section. We also present a time and space complexity analysis of the algorithm. Deeper implementation details and can be found in the source code[2].

5.1   Data Structure

The main idea of the second stage of WSRM is to visit each point of the matrix only once in order to ensure linearity. For the walking operation, the walker must

---

[2] http://www.github.com/jsubercaze/wsrm.

reach its next element in $O(1)$. Using a flag to set points to *visited* would result in a $O(n)$ algorithm. Figure 7(b) shows such an example. The walker (in green) has no element after him, however the naive implementation would revisit points from the already removed rectangle. Moreover, storing 0's is of no use and results in a waste of memory. We therefore introduce a data structure inspired by the two dimensional doubly linked list representation for sparse matrices. Our data structure is as follows. Each nonzero is represented by:

- Its coordinates x, y in the original matrix.
- The dimension $width, height, area$ of the rectangle starting in this point. These values have been computed during the first stage.
- Four pointers to link to previous and next elements in the horizontal and vertical directions.

Hence, walk and remove are then constant-time operations. The `Root` element, depicted in Fig. 8 has its next horizontal pointer linking to the first nonzero in the matrix. When this pointer is `null`, the algorithm terminates.

The second subtlety in the implementation is to ensure the linearity after the removal of a rectangle or when reaching the end of the line of the current rectangle. Both are linked to the same issue. For instance when validating a rectangle, for example at the step (c) of Fig. 4, the walker is first

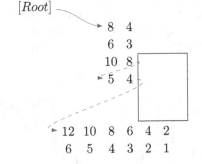

Fig. 8. `Root` element points to the next available value. Dashed arrows points to next elements.

at the coordinate $(4, 6)$ of area 12, it then walks two steps to the right to $(4, 8)$ and then should go to the next line on coordinates $(5, 6)$, of area 10. The point $(4, 8)$ has the following pointers:

- Horizontal next : $(5, 4)$ of area 5
- Horizontal previous $(4, 7)$ of area 8
- Vertical next : $(4, 8)$ of area 3
- Vertical previous : `null`

The beginning of the next line is directly reachable from the point at the beginning of the current line with its downward pointer. Consequently, a pointer to the starting point of the current line is maintained for the rectangle on top of the stack. This pointer is denoted L in Figs. 4 and 5. In a similar manner, the pointer P denotes the current position of the walker. This pointer is used when a rectangle is removed to restore the walker position and to possibly decide the cut as described in Sect. 4.2.

We show that the two stages of the algorithm exhibit linear complexity for both time and space. The case of the first stage is trivial. In the second stage, the

Table 2. Overview of the complexity analysis of the algorithm.

Operation	Time	Space
Stage one		
Computing 4 DPs	2-pass of the matrix: $O(n)$	Store two matrices: $O(n)$
Stage two		
Sparse Matrix	construction in one pass	$n_{\neq 0} \times (4\ ptr + 5\ int)$
Walk	visit each nonzero: $O(n)$	-
Stack	at most $n/2$ elts: $O(n)$	$\leq n/2$ elements: $O(n)$
Merge	at most $n/2$ elts: $O(n)$	-
Remove	each nonzero is removed: $O(n)$	-
Overall	$O(n)$	$O(n)$

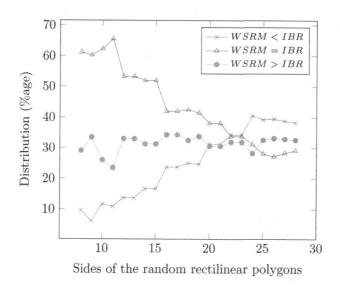

Sides of the random rectilinear polygons

Fig. 9. Decomposition quality comparison between WSRM and IBR, depending on the number of poylgons' sides. The sign $<$ indicates a better decomposition, i.e. a lower number of rectangles.

walker goes once through every point. Each operation has a linear complexity. Considering the space complexity, WSMR requires a matrix in which each point has four pointers to its neighbours. For the worst case (chessboard), the size of the stack reaches its maximum with $n/2$ rectangles. Table 2 gives an overview of the complexity for each stage and operation.

6   Evaluation

We evaluated our algorithm against its direct competitor, *Image Block Representation (IBR)*, and against the optimal algorithm *GBD*. For a global comparison of the different algorithms, we once again refer the reader to the well-documented state-of-the-art by [14].

In order to evaluate the performance of the algorithms, we generated random rectilinear polygons of arbitrary edges. [15] developed an INFLATE-CUT approach that generates random rectilinear polygons of a fixed number of edges. We generated random rectilinear polygons for a growing number of sides, from 8 to 25. For each number of sides, 50 000 random polygons have been generated.

In a first experiment we compared the decomposition quality of the two approaches, i.e. the number of rectangles outputted by both algorithms. Figure 9 shows the decomposition quality depending on the number of sides of the random polygons. When the number of sides is smaller than 22, WSRM outperforms IBR in 30 % of the cases. Both algorithms perform similarly, 60 % of the time for small rectilinear polygons. This value decreases to 30 % when the number of sides grows, whereas the IBR's performance increases with the number of sides. An equal distribution is reached for 22 sided rectilinear polygons. Afterwards, IBR presents a better decomposition. Taking the best decomposition of both heuristics leads to the optimal decomposition for over 90 % of the pictures.

In a second experiment, we compared the ratio of the rectangles' largest sides to their smallest for the two algorithms. The IBR algorithm proceeds in row scanning manner that favors a large ratio. Again, for this experiment, 50 000 random rectilinear polygons have been generated for each number of sides. As expected, IBR outputs rectangles with a large ratio (See Fig. 10). The ratio starts over 2 for 8 sided rectilinear polygons and is increasing over 4 for 25 sided rectilinear polygons. On the other side, WSRM outputs decompositions whose rectangles' sides ratio is lower, starting at 1.9 and slowly reaching 2.5 for 25 sided polygons. WSRM performs for this criterion a better decomposition than IBR, regardless of the number of sides.

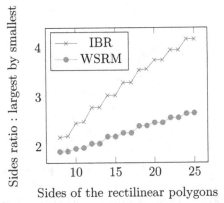

Fig. 10. Sides ratio of the decomposed rectangles, largest side by smallest.

7   Conclusion

In this paper we presented WSRM, a two-stage algorithm that performs rectangle decomposition of rectilinear polygons. Both space and time complexities of the algorithm are linear in the number of elements of the binary input matrix. We also presented implementation details regarding the required data structures to maintain linearity. Evaluation showed that WSRM outperforms IBR for decomposing rectilinear polygons with less than 25 sides. The evaluation also highlighed that the sides of the rectangles obtained using WSRM are in the much lower ratio than the ones outputted by IBR.

References

1. Eppstein, D.: Graph-theoretic solutions to computational geometry problems. In: Paul, C., Habib, M. (eds.) WG 2009. LNCS, vol. 5911, pp. 1–16. Springer, Heidelberg (2010)
2. Ferrari, L., Sankar, P.V., Sklansky, J.: Minimal rectangular partitions of digitized blobs. Comput. Graph. Image Process. **28**(1), 58–71 (1984)
3. Gao, D., Wang, Y.: Decomposing document images by heuristic search. In: Yuille, A.L., Zhu, S.-C., Cremers, D., Wang, Y. (eds.) EMMCVPR 2007. LNCS, vol. 4679, pp. 97–111. Springer, Heidelberg (2007)
4. Gonzalez, T., Zheng, S.-Q.: Bounds for partitioning rectilinear polygons. In: 1st Symposium on Computational Geometry, pp. 281–287. ACM (1985)
5. Levcopoulos, C.: Improved bounds for covering general polygons with rectangles. In: Nori, K.V. (ed.) Foundations of Software Technology and Theoretical Computer Science. LNCS, vol. 287, pp. 95–102. Springer, Heidelberg (1987)
6. Lingas, A., Pinter, R.Y., Rivest, R.L., Shamir, A.: Minimum edge length partitioning of rectilinear polygons. In: Proceeding of 20th Allerton Conference Communication Control and Computing, pp. 53–63 (1982)
7. Liou, W.T., Tan, J.J., Lee, R.C.: Minimum partitioning simple rectilinear polygons in o (n log log n)-time. In: Proceedings of the Fifth Annual Symposium on Computational Geometry, pp. 344–353. ACM (1989)
8. Lipski, W., Lodi, E., Luccio, F., Mugnai, C., Pagli, L.: On two dimensional data organization ii. Fundamenta Informaticae **2**(3), 245–260 (1979)
9. Nahar, S., Sahni, S.: Fast algorithm for polygon decomposition. IEEE Trans. Comput. Aided Des. Integr. Circuits Syst. **7**(4), 473–483 (1988)
10. Ohtsuki, T.: Minimum dissection of rectilinear regions. In: Proceeding IEEE Symposium on Circuits and Systems, Rome, pp. 1210–1213 (1982)
11. Rocher, P.-O., Gravier, C., Subercaze, J., Preda, M.: Video stream transmodality. In: Cordeiro, J., Hammoudi, S., Maciaszek, L., Camp, O., Filipe, J. (eds.) ICEIS 2014. LNBIP, vol. 227, pp. 361–378. Springer, Heidelberg (2015)
12. Soltan, V., Gorpinevich, A.: Minimum dissection of a rectilinear polygon with arbitrary holes into rectangles. Discrete Comput. Geom. **9**(1), 57–79 (1993)
13. Spiliotis, I.M., Mertzios, B.G.: Real-time computation of two-dimensional moments on binary images using image block representation. IEEE Trans. Image Process. **7**(11), 1609–1615 (1998)
14. Suk, T., Höschl IV, C., Flusser, J.: Decomposition of binary images a survey and comparison. Pattern Recogn. **45**(12), 4279–4291 (2012)
15. Tomás, A.P., Bajuelos, A.L.: Generating Random Orthogonal Polygons. In: Conejo, R., Urretavizcaya, M., Pérez-de-la-Cruz, J.-L. (eds.) CAEPIA/TTIA 2003. LNCS (LNAI), vol. 3040, pp. 364–373. Springer, Heidelberg (2004)

CHICO: A Compressed Hybrid Index for Repetitive Collections

Daniel Valenzuela[✉]

Department of Computer Science,
Helsinki Institute for Information Technology HIIT,
University of Helsinki, Helsinki, Finland
dvalenzu@cs.helsinki.fi

Abstract. Indexing text collections to support pattern matching queries is a fundamental problem in computer science. New challenges keep arising as databases grow, and for repetitive collections, compressed indexes become relevant. To successfully exploit the regularities of repetitive collections different approaches have been proposed. Some of these are Compressed Suffix Array, Lempel-Ziv, and Grammar based indexes.

In this paper, we present an implementation of an hybrid index that combines the effectiveness of Lempel-Ziv factorization with a modular design. This allows to easily substitute some components of the index, such as the Lempel-Ziv factorization algorithm, or the pattern matching machinery.

Our implementation reduces the size up to a 50 % over its predecessor, while improving query times up to a 15 %. Also, it is able to successfully index thousands of genomes in a commodity desktop, and it scales up to multi-terabyte collections, provided there is enough secondary memory. As a byproduct, we developed a parallel version of Relative Lempel-Ziv compression algorithm.

1 Introduction

In 1977 Abraham Lempel and Jacob Ziv developed powerful compression algorithms, namely LZ77 and LZ78 [29,30]. Almost forty years from their conception, they remain central to the data compression community. LZ77 is among the most effective compressors which also offers extremely good decompression speed. Those attributes have made it the algorithm of choice for many compression utilities like zip, gzip, 7zip, lzma, and the GIF image format.

These algorithms are still being actively researched [1,3,10,16], and with the increasing need to handle ever-growing large databases the Lempel-Ziv family of algorithms still has a lot to offer.

Repetitive datasets and the challenges on how to index them are actively researched at least since 2009 [21,22,27]. Canonical examples of such datasets are biological databases, such as the 1000 Genomes projects, UK10K, 1001 plant genomes, etc.

Among the different approaches to index repetitive collections, we will focus in one of the Lempel-Ziv based indexes, the Hybrid Index of Ferrada et al. [7].

© Springer International Publishing Switzerland 2016
A.V. Goldberg and A.S. Kulikov (Eds.): SEA 2016, LNCS 9685, pp. 326–338, 2016.
DOI: 10.1007/978-3-319-38851-9_22

Their approach is very promising, specially to be used in biological databases. Among the features that make the Hybrid Index attractive, is that it offers a framework to answer approximate pattern matching queries. Also, by design it does not stick to any specific (approximate) pattern matching technique. For instance, if the index was built to align genomic reads to a reference, it could use a tool like BWA that is highly specialized for this kind of queries (by taking care of the reverse complements, considering quality scores, specificities of the sequencing technologies, etc.).

In this paper we present an improved version of that index that achieves up to a 50 % reduction in its space usage, while maintaining or improving query times up to a 15 %. We achieve faster building time than other indexes for repetitive collections. For collections in the tens of gigabytes, our index can still be built in a commodity machine. Using a more powerful server, we successfully indexed a collection of 201 GB in about an hour, and a collection of 2.4 TB in about 12 h.

A key element to achieve those results is that our index does not stick to a particular Lempel-Ziv factorization. Instead, it accepts the LZ77 parsing as input, and also a variety of LZ77-like parsings. That allows our index to offer appealing trade-offs between indexing time, index size, and resource usage during construction time. As a byproduct, we developed a parallel Relative Lempel-Ziv parser that can parse 2.4 TB in about 10 h using 24 cores.

The structure of this paper is as follows. In Sect. 2 we briefly discuss previous work. In particular we discuss relevant aspects and variants of the Lempel-Ziv parsing. In Sect. 3 we review the fundamental ideas behind the Hybrid Index of Ferrada et al. In Sect. 4 we present our improvements on the Hybrid Index and we discuss how it compares with the original proposal. In Sect. 5 we discuss the practicalities of our implementation. In Sect. 6 we show the experimental behavior of our index. Finally in Sect. 7 we discuss our findings and future research.

2   Related Work

The idea of exploiting repetitiveness to decrease the space requirement of indexing data structures can be traced back at least to Kärkkäinen and Ukkonen [17].

In the string processing community, a full-text index that uses space proportional to that of the compressed text is called a *compressed index* [25]. The main strategies are based on the compression of the suffix array, the compression of the Burrows Wheeler transform, or the use of the Lempel-Ziv parsing. For further reading, we refer the reader to a survey covering the theoretical aspects of compressed full-text indexes [25] and to a recent experimental study [12].

2.1   Repetitive Collections

A pioneer work that formally introduced the challenges of indexing repetitive collection and that offered a successful solution was the Run Length Compressed Suffix Array [21,22]. In parallel, in the context of plant genomics, Schneeberger et al. [27] faced similar challenges, and they solved them by using a q-gram index.

Full-Text Indexing. Since the publication of the Run Length Compressed Suffix Array (RLCSA) [21,22], many techniques have been developed to tackle this problem [24]. Some of them are mainly theoretical works [2,6,11], and some have been implemented [4,7,19,26]. However, to the best of our knowledge, the only one that successfully scales to collections in the tens of gigabytes is the RLCSA.

Bioinformatics Tools. Some indexes are specially tailored for biological databases. One of the foundational works handling large repetitive datasets is GenomeMapper [27]. Some recent works, like MuGI [5], also take as an input a *reference* sequence and a file that informs about the variations present in a set of sequences corresponding to individuals. Other tools [23,28] consider the input to be a (multiple) sequence alignment. This assumption, even though natural in some contexts, restricts the usability of those solutions.

2.2   LZ77

Lempel-Ziv is a family of very powerful compression algorithms that achieve compression by using a dictionary to encode substrings of the text. Here we focus on the LZ77 algorithm, in which the dictionary (implicitly) contains all the substrings of the part of the text that has already been processed.

The compression algorithm consists on two phases: *parsing* (also called *factorization*) and *encoding*. Given a text $T[1, N]$, a *LZ77 valid parsing* is a partition of T into z substrings T^i (often called *phrases* or *factors*) such that $T = T^1 T^2 \ldots T^z$, and for all $i \in [1, z]$ either there is at least one occurrence of T^i with starting position strictly smaller than $|T^1 T^2 \ldots T^{i-1}|$, or T^i is the first occurrence of a single character.

The *encoding* process represents each phrase T^i using a pair (p_i, l_i), where p_i is the position of the previous occurrence of T_i and $l_i = |T^i|$, for phrases that are not single characters. When $T^i = \alpha \in \Sigma$, then it is encoded with the pair $(\alpha, 0)$. We call the latter *literal phrases* and the former *copying phrases*.

Decoding LZ77 compressed text is particularly simple and fast: the pairs (p_i, l_i) are read from left to right, if $l_i = 0$, then p_i is interpreted as a char and it is appended to the output, if $l_i \neq 0$, then l_i characters are copied from the position p_i to $p_i + l_i - 1$ and are appended to the current output.

Greedy Parsing. So far we have defined what is a *LZ77 valid parsing*, but we have not discussed how to compute such a parsing. Indeed, there are different possibilities. It is a common agreement in the literature to reserve the name LZ77 for the case when the parsing is a *greedy parsing*. That is, if we assume that we have already parsed a prefix of T, $T' = T^1 T^2 \ldots T^p$ then the phrase T^{p+1} must be the longest substring starting at position $|T^1 T^2 \ldots T^p| + 1$ such that there is a previous occurrence of T^{p+1} in T starting at some position smaller than $|T^1 T^2 \ldots T^p|$. There are numerous reasons to choose the *greedy parsing*. One of them is that it can be computed in linear time [15]. Another reason is that the parsing it produces is optimal in the number of phrases [8]. Therefore, if the pairs are encoded using any fixed-length encoder, the *greedy parsing* always produces the smaller representation. Moreover, various authors [18,29] proved that greedy parsing achieves asymptotically the (empirical) entropy of the source generating the input string.

3   Hybrid Index

The Hybrid Index of Ferrada et al. [7] extends the ideas of Kärkkäinen and Ukkonen [17] to use the Lempel-Ziv parsing as a basis to capture repetitiveness. This Hybrid Index combines the efficiency of LZ77 with any other index in a way that can support not only exact matches but also approximate matching.

Given the text to be indexed T, a LZ77 parsing of T consisting of z phrases, and also the maximum length M of a query pattern and the maximum number of mismatches K, the Hybrid Index is a data structure using space proportional to z and to M that supports approximate string matching queries. That is, it is able to find all the positions i in T such that $ed(T[i, |P| - 1], P[1, |P| - 1]) \leq K$ for a given query pattern P, where $ed(x, y)$ stands for the edit distance between strings x and y.

Let us adopt the following definitions [7,17]: A *primary occurrence* is an (exact or approximate) occurrence of P in T that spans two or more phrases. A *secondary match* is an (exact or approximate) occurrence of P in T that is entirely contained in one phrase. Kärkkäinen and Ukkonen [17] noted that every secondary match is an exact copy of a previous (secondary or primary) match. Therefore, the pattern matching procedure can be done in two stages: first all the primary occurrences are identified, then, using the structure of the LZ77 parse, all the secondary matches can be discovered.

3.1   Kernelization to Find Primary Occurrences

Conceptually, the kernel string aims to get rid of large repetitions in the input text, and extract only the non-repetitive areas. To do that, it extracts the characters in the neighborhoods of the phrase boundaries, while discarding most of the content of large phrases.

More formally, given the LZ77 parsing of T, the kernel text $\mathcal{K}_{M,K}$ is defined as the concatenation of characters within distance $M + K - 1$ from their nearest phrase boundaries. Characters not contiguous in T are separated in $\mathcal{K}_{M,K}$ by $K + 1$ copies of a special separator #. It is important to note that for any substring of T with length at most $M + K$ that crosses a phrase boundary in the LZ77 parse of T, there is a corresponding and equal substring in $\mathcal{K}_{M,K}$.

To be able to map the positions from the kernel text to the original text the Hybrid Index uses two sorted lists with the phrase boundaries. L_T stores the phrase boundaries in T and $L_{\mathcal{K}_{M,K}}$ stores the list of phrase boundaries in $\mathcal{K}_{M,K}$.

The kernel text does not need to be stored explicitly; what is required is the ability to query it and, for that reason, the Hybrid Index stores an index $I_{M,K}$

that supports the desired queries on $\mathcal{K}_{M,K}$ (e.g. exact and approximate pattern matching).

By construction of $\mathcal{K}_{M,K}$, it is guaranteed that all the primary matches of P occur in $\mathcal{K}_{M,K}$. However, there are also some secondary matches that may appear on $\mathcal{K}_{M,K}$. When a query pattern P, $|P| \leq M$ is given the first step is to query the index $I_{M,K}$ to find all the matches of P in $\mathcal{K}_{M,K}$. These are all the potential primary occurrences. They are mapped to T using L and L_M. Those matches that do not overlap a phrase boundary are discarded. For queries of length one, special care is taken with matches that corresponds to the first occurrence of the character [7].

3.2 Reporting Secondary Occurrences

The secondary occurrences are reported using *2-sided range reporting* [7]. The idea is that, once the positions of the primary occurrences are known, the parsing information is used to discover the secondary matches. Instead of looking for theoretically optimal data structures to solve this problem, the Hybrid Index proposes a simple and practical way to solve it.

Each phrase (pos, len) of the LZ77 parsing can be expressed as triplets $(x, y) \rightarrow w$ where $(x = pos, y = pos + len)$ is said to be the source, and w is the position in the text that is encoded with such phrase. The sources are sorted by the x coordinate, and the sorted x values are stored in an array $\mathcal{X}$. The corresponding w positions are stored in an array $\mathcal{W}$. The values of the y coordinates are not explicitly stored. However, a position-only Range Maximum Query [9] data structure is stored for the y values.

The 2-sided recursive reporting procedure works as follows: For a given primary occurrence in position pos, the goal is to find all the phrases whose source entirely contains $(pos, pos + |P| - 1)$. To do that the first step is to look for the position of the predecessor of pos in $\mathcal{X}$. The second step is to do a range maximum query of the y values in that range. Even though y is not stored explicitly, it can be easily computed [7]. If that value is smaller than $pos + |P| - 1$ the search stops. If the y value is equal or larger than $pos + |P| - 1$, it corresponds to a phrase that contains the primary position. Then, the procedure recurses on the two intervals that are induced by it. For further details we refer the reader to the Hybrid Index paper [7].

4   CHICO: Beyond Greedy LZ77

When we described the LZ77 algorithm in Sect. 2.2, we first used a general description of the LZ parsing. We noted that in the description of the Hybrid Index the parsing strategy is not specified, and indeed, it does not affect the working of the index. Therefore, the index works the same way with any *LZ77 valid parsing* where the phrases are either single characters (*literal phrases*, or pairs (pos, len) representing a reference to a previously occurring substring (*copying phrases*).

The greedy parsing is usually the chosen scheme because it produces the minimum number of phrases. This produces the smallest possible output if a plain encoding is used. If other encoding schemes are allowed, it has been proven that the greedy parsing does not imply the smallest output anymore. Moreover, different parsing algorithms exists that provide bit-optimality for a wide variety of encoders [8].

4.1   Reducing the Number of Phrases

It is useful to note that all the phrases shorter than $2M$ are entirely contained in the kernel text. Occurrences that are entirely contained in such phrases are found using the index of the kernel string $I_{M,K}$. Then they are discarded, just to be rediscovered later by the recursive reporting procedure.

To avoid this redundancy we modify the parsing in a way that phrases smaller than $2M$ are avoided. First we need to accept that *literal phrases* (those pairs $(\alpha, 0)$ such that $\alpha \in \Sigma$) can be used not only to represent characters but also to hold longer strings $(s, 0)$, $s \in \Sigma^*$.

To actually reduce the number of phrases in the recursive data structure we designed a **phrase merging** procedure. The phrase merging procedures receives a LZ77 parsing of a text, and produces a parsing with less phrases, using *literal phrases* longer than one character. The procedure works as follows. It reads the phrases in order, and when it finds a *copying phrase* (pos, len) such that $len < 2M$, it is transformed into a *literal phrase* $(T[pos, pos + len - 1], 0)$. That is, a *literal phrase* that decodes to the same string. If two *literal phrases* $(s_1, 0)$ and $(s_2, 0)$ are consecutive, they are merged into $(s_1 \circ s_2, 0)$ where $\circ$ denotes string concatenation. It is clear that the output parsing decodes to the same text as the input parsing. Moreover, the output parsing produce the same kernel text $\mathcal{K}_{M,K}$ as the input parsing.

Because the number of phrases after the phrase merging procedure is strictly smaller, the space needed for the recursive reporting data structure also decreases. In addition, the search space for the recursive reporting queries shrinks.

4.2   Finding the Occurrences

We extend the definition of primary occurrences as follows: A *primary occurrence* is an (exact or approximate) occurrence of P in T that either crosses one or more phrase boundary, or it lies entirely within a *literal phrase*. To ensure that every occurrence is reported once and only once we store a bitmap $F[1..z]$ such that $F[i] = 1$ if the i-th phrase is a *literal phrase* and $F[i] = 0$ otherwise.

Hence, when a query pattern P is processed, first $I_{M,K}$ is used to find potential primary occurrence. Then the only occurrences to be discarded are those that lie entirely within *copying phrases*. Using this approach, there is also no need to handle a special case for queries of length one.

4.3 RLZ: A Faster Building Algorithm

A classical way to reduce the parsing time of the LZ77 algorithm is to use a smaller dictionary. For instance, the *sliding window* approach constrains the parsing algorithm to look for matches only in the ω last positions of the processed text. This approach is used in the popular gzip program.

A recent approach that presents a different modification to the LZ77 algorithm is the Relative Lempel-Ziv algorithm (RLZ) [20]. Here, the dictionary is not a prefix of the text. Instead, the dictionary is a different text that is provided separately. We note that while the sliding window approach still generates a *LZ77 valid parsing*, the RLZ algorithm does not. Therefore, the former one is a valid input for the Hybrid Index, while the second is not.

To make the RLZ algorithm compatible with the Hybrid Index, a natural way would be to prepend the reference to the input text. In that way, the first phrase would be an exact copy of the reference. We mark this phrase as a *literal phrase* to ensure all its contents goes to the kernel text. One caveat of this approach is that if the reference is compressible we would not exploit it. An alternative way to modify the RLZ algorithm without having this trouble is as follows: We parse the input text using the traditional RLZ algorithm, to obtain a parse $\mathcal{P}_T$. Then, we parse the reference using the LZ77 greedy parsing, to obtain a parse $\mathcal{P}_R$. It is easy to see that $\mathcal{P}_R \circ \mathcal{P}_T$ is a LZ77 *valid parsing* of $R \circ T$.

5 Implementation

We implemented the index in C++, relying on the Succinct Data Structure Library 2.0.3 (SDSL) [12] for most succinct data structures, such as the RMQ [9] data structure on the Y array.

We encoded the phrase boundaries L_T, $L_{K,M}$ and the x-values of the sources in the $\mathcal{X}$ array using SDSL elias delta codes implementation. Following the ideas of the original Hybrid Index we did not implement specialized data structure for predecessor queries on arrays $\mathcal{X}$ and $L_{K,M}$. Instead, we sampled these arrays and perform binary searches to find the predecessors.

For the index of the kernel text $I_{K,M}$ we used a SDSL implementation of the FM-Index. As that FM-Index does not support approximate searches natively, we decided to exclude those queries in our experimental study.

For further details, the source code of our implementation is publicly available at https://www.cs.helsinki.fi/u/dvalenzu/software/.

5.1 LZ77

To offer different trade-offs we included alternative modules for LZ77 factorization, all of them specialized in repetitive collections. The default construction algorithm of the index uses **LZscan** [14], an in-memory algorithm that computes the LZ77 greedy parsing.

To handle larger inputs, we also included an external memory algorithm called **EM-LZscan** [16]. An important feature is that it allows the user to specify the amount of memory that the algorithm is allowed to use.

5.2   Relative Lempel-Ziv

The third parser that we included is our own version of Relative Lempel-Ziv **RLZ**. Our implementation is based on the Relative Lempel-Ziv of Hoobin et al. [13]. This algorithm uses the suffix array of the reference and then it uses it to compute the parsing of the input text.

Our implementation differs in two aspects with the original algorithm. The first (explained in Sect. 4.3) is that the reference itself is also represented using LZ77. To compute the LZ77 parsing of the reference we use the KKP algorithm [15]. The second difference is that instead of using an arbitrary reference, we restrict ourselves to use a prefix of the input text. In that way there is no need to modify the input text by prepending the reference (see Sect. 4.3).

5.3   Parallel Relative Lempel-Ziv

We implemented a fourth parser, **PRLZ**, which is a parallel version of RLZ. This is a simple yet effective parallelization of our RLZ implementation. The first step of the algorithm is to build the suffix array of the reference. This is done sequentially. Then, instead of processing the text sequentially, we split it into chunks and process them in parallel. Each chunk is assigned to a different thread and the thread computes the RLZ parsing of its assigned chunk using the reference. This is easily implemented using OpenMP. Using a moderate number of chunks (e.g. the number of available processors) we expect similar compression ratios to those achieved by a sequential RLZ implementation.

6   Experimental Results

We used collections of different kinds and sizes to evaluate our index in practice. In the first round of experiments we used some repetitive collections from the pizzachilli repetitive corpus[1].

Einstein: All versions of Wikipedia page of Albert Einstein up to November 10, 2006. Total size is 446 MB.
Cere: 37 sequences of Saccharomyces Cerevisiae. Total size is 440 MB.
Para: 36 sequences of Saccharomyces Paradoxus. Total size is 410 MB.
Influenza: 78, 041 sequences of *Haemophilus Influenzae*. Total size is 148 MB.
Coreutils: Source code from all 5.x versions of the *coreutils* package. Total size is 196 MB.

We extracted 1000 random patterns of size 50 and 100 from each collection to evaluate the indexes. In addition, we generated two large collections using data from the 1000 genomes project:

CHR$_{21}$: 2000 versions of Human Chromosome 21. Total size 90 GB.

---

[1] http://pizzachili.dcc.uchile.cl/.

CHR$_{14}$: 2000 versions of Human Chromosome 14. Total size 201 GB.
CHR$_{1...5}$: 2000 versions of Human Chromosomes $1, 2, 3, 4$ and 5. Total size 2.4 TB.

To demonstrate the efficiency of our index, we ran most of our experiments in a commodity computer. This machine has 16 GB of RAM and 500 GB of hard drive. The operative system is Ubuntu 14.04.3. The code was compiled using gcc 4.8.4 with full optimization.

In our first round of experiments we compared CHICO with the original Hybrid Index on some collections of the pizzachilli corpus. As the size of those collections is moderate, we ran our index using the in-memory version of LZscan. We also compared with the LZ77 Index of Kreft and Navarro [19], and with RLCSA [21,22], with sampling parameters 128 and 1024. To illustrate how small can be the *RLCSA*, we also show the space usage of the *RLCSA* without the samples. This version of the RLCSA can only count the number of occurrences but cannot find them.

The results are presented in Tables 1 and 2. First we observe that the construction time of our index dominates every other alternative. This is possible as we use a highly engineered LZ parsing algorithm [14]. As expected, our index is consistently smaller and faster than the original Hybrid Index. Also it is competitive with other alternatives for repetitive collections.

6.1   Larger Collections

The next experiment considered the 90 GB collection CHR$_{21}$. None of the competitors were able to index the collection in the machine we were using. As the input text greatly exceeded the available RAM, we studied different parsing strategies using external memory.

Table 1. Construction time in seconds and size of the resulting index in bytes per character for moderate-sized collections.

	HI		CHICO		LZ77-Index		$RLCSA_{128}$		$RLCSA_{1024}$		$RLCSA_{min}$
	Time	Size	Time	Size	Time	Size	Time	Size	Time	Size	Size
Influenza	204.1	0.0795	30.1	0.0572	40.8	0.0458	64.3	0.0959	63.9	0.0462	0.039
Coreutils	393.1	0.1136	32.9	0.0508	49.9	0.0792	139.2	0.1234	134.1	0.0737	0.066
Einstein	98.9	0.0033	63.5	0.0019	95.3	0.0019	389.0	0.0613	347.5	0.0097	0.002
Para	1065.7	0.0991	33.8	0.0577	157.3	0.0539	232.0	0.1598	217.3	0.1082	0.100
Cere	620.3	0.0767	42.6	0.0517	175.1	0.0376	264.9	0.1366	268.0	0.0850	0.077

Table 2. Time in milliseconds to find all the occurrences of a query pattern of length 50 and 100. Times were computed as average of 1000 query patterns randomly extracted from the collection.

	HI		CHICO		LZ77-Index		$RLCSA_{128}$		$RLCSA_{1024}$	
	$\lvert P \rvert = 50$	$\lvert P \rvert = 100$	$\lvert P \rvert = 50$	$\lvert P \rvert = 100$	$\lvert P \rvert = 50$	$\lvert P \rvert = 100$	$\lvert P \rvert = 50$	$\lvert P \rvert = 100$	$\lvert P \rvert = 50$	$\lvert P \rvert = 100$
Influenza	43.51	7.63	42.39	7.20	20.01	48.72	2.54	1.01	57.21	25.16
Coreutils	28.17	2.24	26.92	1.79	10.43	20.07	0.86	0.14	16.08	0.78
Einstein	18.65	11.41	16.43	9.45	23.03	34.90	3.28	2.55	30.94	22.66
Para	2.56	1.55	2.38	1.34	14.37	32.31	0.14	0.15	1.54	1.23
Cere	3.07	1.80	2.88	1.58	13.81	33.31	0.15	0.17	1.95	1.68

Table 3. Different parsing algorithms to index collection CHR$_{21}$, 90 GB. The first row shows the results for EM-LZ, which computes the LZ77 greedy parsing. The next rows show the results for the RLZ parsing using prefixes of size 500 MB and 1 GB. The size of the index is expressed in bytes per char, the building times are expressed in minutes, and the query times are expressed in milliseconds. Query times were computed as average of 1000 query patterns randomly extracted from the collection.

	Size(bpc)	Build time(min)	Query time (ms)	
			$|P| = 50$	$|P| = 70$
EM-LZ	0.00126	4647	18.16	14.90
RLZ$_{0.5\,GB}$	0.0060	143	55.28	46.67
RLZ$_{1\,GB}$	0.0047	65	50.30	40.72

The first setting tested was using the EM-LZscan algorithm to compute the greedy parsing. We ran it allowing a maximum usage of 10 GB of RAM. We also tried the RLZ parser using as reference prefixes of sizes 500 MB and 1 GB. For each of the resulting indexes, we measure the query time as the average over 1000 query patterns. The results are presented in Table 3.

Table 3 shows the impact of choosing different parsing algorithms to build the index. We can see different trade-offs between building time and the resulting index: EM-LZ generates the greedy parsing and is indeed the one that generates the smaller index: 0.00126 bytes per character. The building time to achieve that is more than 70 h. On the other hand, the RLZ parser is able to compute the index in about 10 h. As expected, using RLZ instead of the *greedy parsing* results in a larger index. However, the compression ratio is still good enough and the resulting index fits comfortably in RAM.

Table 4. Results using RLZ and PRLZ to parse CHR$_{14}$, a 201 GB collection. Query times were computed as average of 1000 query patterns randomly extracted from the collection.

	Size (bpc)	Build time (min)		Query Time (ms)	
		LZ parsing	Others	P=50	P=70
RLZ$_{1\,GB}$	0.00656	308	51	225.43	178.47
PRLZ$_{1\,GB}$	0.00658	22	55	224.04	181.21

Table 5. Results using PRLZ to parse CHR$_{1...5}$, a collection of 2.4 TB of data. Query times were computed as average of 1000 query patterns randomly extracted from the collection.

	Size (bpc)	Build time (min)		Query Time	
		LZ parsing	Others	P=50 (ms)	P=70 (ms)
PRLZ$_{10\,GB}$	0.000233	600	191	61.20	36.38

The next test was on CHR_{14}, a 201 GB collection that we could no longer process in the same commodity computer. We indexed it in a large server, equipped with 48 cores, 12 TB of hard disk, and 1.5 TB of RAM. We compared the RLZ and PRLZ parsers. The results are shown in Table 4.

We can see that the parallelization had almost no impact in the size of the index, but that the indexing time decreased considerably. In the parallel version about 20 min where spent parsing the input, and 55 min building the index. In all the previous settings, the building time was largely dominated by the parsing procedure.

Finally, to demonstrate the scalability of our approach, we indexed $CHR_{1...5}$, a 2.4 TB collection. For this collection we only run the fastest parsing, and the results can be seen in Table 5.

7   Conclusions

We have presented an improved version of the Hybrid Index of Ferrada et al., that achieves up to a 50 % reduction in its space usage, while also improving the query times. By using state of the art Lempel-Ziv parsing algorithms we achieved different trade-offs between building time and space usage: When the collections size is moderate, we could compare to available implementations, and ours achieved the fastest building time. For collections in the tens of gigabytes, our index can still be built in a commodity machine. Finally, we developed a parallel Relative Lempel-Ziv parser to be run in a more powerful machine. In that setting, we indexed a 201 GB collection in about an hour and a 2.4 TB collection in about 12 h.

Some of our parsing schemes worked effectively in the genomic collections, because a prefix of the collection is a natural reference for the RLZ algorithm. For future developments, we will study alternatives such as artificial references [13], so that the index can be equally effective in different contexts.

We also plan to build a version specialized for read alignment. To that end, it is not enough to replace the kernel index by an approximate pattern matching index: Read aligners must consider different factors, such as base quality scores, reverse complements, among other aspects that are relevant to manage genomic data.

Acknowledgments. Many thanks to Travis Gagie, Simon Puglisi, Veli Mäkinen, Dominik Kempa and Juha Kärkkäinen for insightful discussions. The author is funded by Academy of Finland grant 284598 (CoECGR).

References

1. Al-Hafeedh, A., Crochemore, M., Ilie, L., Kopylova, E., Smyth, W.F., Tischler, G., Yusufu, M.: A comparison of index-based Lempel-Ziv LZ77 factorization algorithms. ACM Comput. Surv. (CSUR) **45**(1), 5 (2012)

2. Belazzougui, D., Cunial, F., Gagie, T., Prezza, N., Raffinot, M.: Composite repetition-aware data structures. In: Cicalese, F., Porat, E., Vaccaro, U. (eds.) CPM 2015. LNCS, vol. 9133, pp. 26–39. Springer, Heidelberg (2015)
3. Belazzougui, D., Puglisi, S.J.: Range predecessor and Lempel-Ziv parsing. In: Proceedings of the Twenty-Seventh Annual ACM-SIAM Symposium on Discrete Algorithms. SIAM (2016) (to appear)
4. Claude, F., Fariña, A., Martínez-Prieto, M., Navarro, G.: Indexes for highly repetitive document collections. In: Proceedings of the 20th ACM International Conference on Information and Knowledge Management (CIKM), pp. 463–468. ACM (2011)
5. Danek, A., Deorowicz, S., Grabowski, S.: Indexing large genome collections on a PC. PLoS ONE 9(10), e109384 (2014)
6. Do, H.H., Jansson, J., Sadakane, K., Sung, W.K.: Fast relative Lempel-Ziv self-index for similar sequences. Theor. Comput. Sci. 532, 14–30 (2014)
7. Ferrada, H., Gagie, T., Hirvola, T., Puglisi, S.J.: Hybrid indexes for repetitive datasets. Philos. Trans. R. Soc. A 372, 20130137 (2014)
8. Ferragina, P., Nitto, I., Venturini, R.: On the bit-complexity of Lempel-Ziv compression. In: Proceedings of the Twentieth Annual ACM-SIAM Symposium on Discrete Algorithms, pp. 768–777. Society for Industrial and Applied Mathematics (2009)
9. Fischer, J.: Optimal succinctness for range minimum queries. In: López-Ortiz, A. (ed.) LATIN 2010. LNCS, vol. 6034, pp. 158–169. Springer, Heidelberg (2010)
10. Fischer, J., Gagie, T., Gawrychowski, P., Kociumaka, T.: Approximating LZ77 via small-space multiple-pattern matching. In: Bansal, N., Finocchi, I. (eds.) Algorithms - ESA 2015. LNCS, vol. 9294, pp. 533–544. Springer, Heidelberg (2015)
11. Gagie, T., Puglisi, S.J.: Searching and indexing genomic databases via kernelization. Front. Bioeng. Biotechnol. 3(12) (2015)
12. Gog, S., Beller, T., Moffat, A., Petri, M.: From theory to practice: plug and play with succinct data structures. In: Gudmundsson, J., Katajainen, J. (eds.) SEA 2014. LNCS, vol. 8504, pp. 326–337. Springer, Heidelberg (2014)
13. Hoobin, C., Puglisi, S.J., Zobel, J.: Relative Lempel-Ziv factorization for efficient storage and retrieval of web collections. Proc. VLDB Endow. 5(3), 265–273 (2011)
14. Kärkkäinen, J., Kempa, D., Puglisi, S.J.: Lightweight Lempel-Ziv parsing. In: Bonifaci, V., Demetrescu, C., Marchetti-Spaccamela, A. (eds.) SEA 2013. LNCS, vol. 7933, pp. 139–150. Springer, Heidelberg (2013)
15. Kärkkäinen, J., Kempa, D., Puglisi, S.J.: Linear time Lempel-Ziv factorization: simple, fast, small. In: Fischer, J., Sanders, P. (eds.) CPM 2013. LNCS, vol. 7922, pp. 189–200. Springer, Heidelberg (2013)
16. Karkkainen, J., Kempa, D., Puglisi, S.J.: Lempel-Ziv parsing in external memory. In: Data Compression Conference (DCC), pp. 153–162. IEEE (2014)
17. Kärkkäinen, J., Ukkonen, E.: Lempel-Ziv parsing and sublinear-size index structures for string matching. In: Proceedings of the 3rd South American Workshop on String Processing (WSP 1996). Citeseer (1996)
18. Kosaraju, S.R., Manzini, G.: Compression of low entropy strings with Lempel-Ziv algorithms. SIAM J. Comput. 29(3), 893–911 (2000)
19. Kreft, S., Navarro, G.: On compressing and indexing repetitive sequences. Theor. Comput. Sci. 483, 115–133 (2013)
20. Kuruppu, S., Puglisi, S.J., Zobel, J.: Relative Lempel-Ziv compression of genomes for large-scale storage and retrieval. In: Chavez, E., Lonardi, S. (eds.) SPIRE 2010. LNCS, vol. 6393, pp. 201–206. Springer, Heidelberg (2010)

21. Mäkinen, V., Navarro, G., Sirén, J., Välimäki, N.: Storage and retrieval of individual genomes. In: Batzoglou, S. (ed.) RECOMB 2009. LNCS, vol. 5541, pp. 121–137. Springer, Heidelberg (2009)

22. Mäkinen, V., Navarro, G., Sirén, J., Välimäki, N.: Storage and retrieval of highly repetitive sequence collections. J. Comput. Biol. **17**(3), 281–308 (2010)

23. Na, J.C., Park, H., Crochemore, M., Holub, J., Iliopoulos, C.S., Mouchard, L., Park, K.: Suffix tree of alignment: an efficient index for similar data. In: Lecroq, T., Mouchard, L. (eds.) IWOCA 2013. LNCS, vol. 8288, pp. 337–348. Springer, Heidelberg (2013)

24. Navarro, G.: Indexing highly repetitive collections. In: Arumugam, S., Smyth, W.F. (eds.) IWOCA 2012. LNCS, vol. 7643, pp. 274–279. Springer, Heidelberg (2012)

25. Navarro, G., Mäkinen, V.: Compressed full-text indexes. ACM Comput. Surv. **39**(1), article 2 (2007)

26. Navarro, G., Ordóñez, A.: Faster compressed suffix trees for repetitive collections. ACM J. Exp. Alg. **21**(1), article 1.8 (2016)

27. Schneeberger, K., Hagmann, J., Ossowski, S., Warthmann, N., Gesing, S., Kohlbacher, O., Weigel, D.: Simultaneous alignment of short reads against multiple genomes. Genome Biol. **10**, R98 (2009)

28. Sirén, J., Välimäki, N., Mäkinen, V.: Indexing graphs for path queries with applications in genome research. IEEE/ACM Trans. Comput. Biol. Bioinf. **11**(2), 375–388 (2014)

29. Ziv, J., Lempel, A.: A universal algorithm for sequential data compression. IEEE Trans. Inf. Theory **23**(3), 337–343 (1977)

30. Ziv, J., Lempel, A.: Compression of individual sequences via variable-rate coding. IEEE Trans. Inf. Theory **24**(5), 530–536 (1978)

Fast Scalable Construction
of (Minimal Perfect Hash) Functions

Marco Genuzio[1], Giuseppe Ottaviano[2], and Sebastiano Vigna[1]([✉])

[1] Dipartimento di Informatica, Università degli Studi di Milano, Milan, Italy
vigna@di.unimi.it
[2] Facebook, Menlo Park, USA

Abstract. Recent advances in random linear systems on finite fields
have paved the way for the construction of constant-time data structures
representing static functions and minimal perfect hash functions using
less space with respect to existing techniques. The main obstruction for
any practical application of these results is the cubic-time Gaussian elimination required to solve these linear systems: despite they can be made
very small, the computation is still too slow to be feasible.

In this paper we describe in detail a number of heuristics and programming techniques to speed up the resolution of these systems by several orders of magnitude, making the overall construction competitive
with the standard and widely used MWHC technique, which is based on
hypergraph peeling. In particular, we introduce *broadword programming*
techniques for fast equation manipulation and a *lazy Gaussian elimination* algorithm. We also describe a number of technical improvements to
the data structure which further reduce space usage and improve lookup
speed.

Our implementation of these techniques yields a minimal perfect hash
function data structure occupying 2.24 bits per element, compared to
2.68 for MWHC-based ones, and a static function data structure which
reduces the multiplicative overhead from 1.23 to 1.03.

1 Introduction

Static functions are data structures designed to store arbitrary mappings from
finite sets to integers; that is, given a set of n pairs (k_i, v_i) where $k_i \in S \subseteq
U, |S| = n$ and $v_i \in 2^b$, a static function will retrieve v_i given k_i in constant time.
Closely related are *minimal perfect hash functions (MPHFs)*, where only the set
S of k_i's is given, and the data structure produces an injective numbering $S \to n$.
While these tasks can be easily implemented using hash tables, static functions
and MPHFs are allowed to return *any* value if the queried key is not in the
original set S; this relaxation enables to break the information-theoretical lower
bound of storing the set S. In fact, constructions for static functions achieve
just $O(nb)$ bits space and MPHFs $O(n)$ bits space, regardless of the size of
the keys. This makes static functions and MPHFs powerful techniques when
handling, for instance, large sets of strings, and they are important building

© Springer International Publishing Switzerland 2016
A.V. Goldberg and A.S. Kulikov (Eds.): SEA 2016, LNCS 9685, pp. 339–352, 2016.
DOI: 10.1007/978-3-319-38851-9_23

blocks of space-efficient data structures such as (compressed) full-text indexes [7], monotone MPHFs [3,5], Bloom filter-like data structures [8], and prefix-search data structures [4].

An important line of research, both theoretical and practical, involves lowering the multiplicative constants in the big-O space bounds, while keeping feasible construction times. In this paper we build on recent advances in random linear systems theory, and in perfect hash data structures [14,22], to achieve practical static functions with the lowest space bounds so far, and construction time comparable with widely used techniques. The new results, however, require solving linear systems rather than a simple depth-first visit of a hypergraph, as it happens in current state-of-the-art solutions.

Since we aim at structures that can manage billions of keys, the main challenge in making such structures usable is taming the cubic running time of Gaussian elimination at construction time. To this purpose, we introduce novel techniques based on *broadword programming* [18] and a lazy version of *structured Gaussian elimination*. We obtain data structures that are significantly smaller than widely used hypergraph-based constructions, while maintaining or improving the lookup times and providing still feasible construction time.

All implementations discussed in this paper are distributed as free software as part of the Sux4J project (http://sux4j.di.unimi.it/).

2   Notation and Tools

We use von Neumann's definition and notation for natural numbers, identifying n with $\{0, 1, \ldots, n-1\}$, so $2 = \{0, 1\}$ and 2^b is the set of b-bit numbers.

Model and Assumptions. Our model of computation is a unit-cost word RAM with word size w. We assume that $n = |S| = O(2^{cw})$ for some constant c, so that constant-time static data structures depending on $|S|$ can be used.

Hypergraphs. An r-hypergraph on a vertex set V is a subset E of $\binom{V}{r}$, the set of subsets of V of cardinality r. An element of E is called an *edge*. The k-core of a hypergraph is its maximal induced subgraph having degree at least k.

A hypergraph is *peelable* if it is possible to sort its edges in a list so that for each edge there is a vertex that does not appear in following elements of the list. A hypergraph is peelable if and only if it has an empty 2-core. It is *orientable* if it is possible to associate with each hyperedge a distinct vertex. Clearly, a peelable hypergraph is orientable, but the converse is not necessarily true.

3   Background and Related Work

Linear Functions and MWHC. Most static function constructions work by finding a *linear function* that satisfies the requirements. For simplicity start with the special case of functions with binary values, that is $v_i \in \mathbf{F}_2$ (the field with two elements); the task is to find a vector $w \in \mathbf{F}_2^m$ such that for each i

$$h_\theta(k_i)^T w = v_i \tag{1}$$

where h_θ is a function $U \to \mathbf{F}_2^m$ from a suitable family $\mathcal{H}$ indexed by θ. To make the lookup constant-time, we add the additional constraint that $h_\theta(k)$ has a constant number r of ones, and that the positions of these ones can be computed in constant time. Then, with a slight abuse of notation, we can write $h_{\theta,j}$ to be the *position* of the j-th nonzero element, and hence the lookup just becomes

$$w_{h_{\theta,0}(k_i)} + \cdots + w_{h_{\theta,r-1}(k_i)} = v_i. \tag{2}$$

It is clear that, if such a function exists, the data structure just requires to store w and θ. Note that if h_θ is fixed, just writing down the n equations above yields a linear system: stacking the row vectors $h_\theta(k_i)^T$ into a matrix H and the values v_i into the vector v, we are looking to solve the equation

$$Hw = v. \tag{3}$$

A sufficient condition for the solution w to exist is that the matrix H has full rank. To generalize to the case where $v_i \in \mathbf{F}_2^b$ is a b-bit integer, just replace v with the $n \times b$ matrix V obtained by stacking the v_i's as rows, and w by a $m \times b$ matrix. Full rank of H is still a sufficient condition for the solvability of $HW = V$. It remains to show how to pick the number of variables m, and the functions h_θ, so that H has full rank.

In their seminal paper [20], Majewski, Wormald, Havas and Czech (MWHC hereinafter) introduced the first static function construction that can be described using the framework above. They pick as $\mathcal{H}$ the set of functions $U \to \mathbf{F}_2^m$ whose values have exactly r ones, that is, $h_\theta(k)$ is the vector with r ones in positions $h_{\theta,j}(k)$ for $j \in r$, using the same notation above. If the functions are picked uniformly at random, the r-uples $\big(h_{\theta,0}(k), \ldots, h_{\theta,r-1}(k)\big)$ can be seen as edges of a random hypergraph with m nodes. When $m > c_r n$ for a suitable c_r, with high probability the hypergraph is peelable, and the peeling process *triangulates* the associated linear system; in other words, we have both a probabilistic guarantee that the system is solvable, and that the solution can be found in linear time. The constant c_r depends on the degree r, which attains its minimum at $r = 3$, $c_3 \approx 1.23$. The family $\mathcal{H}$ can be substituted with a smaller set where the parameter θ can be represented with a sublinear number of bits, so the overall space is $1.23bn + o(n)$ bits. In practice, $h_{\theta,j}(k)$ will be simply a hash function with random seed θ, which can be represented in $O(1)$ bits.

MPHFs. Chazelle et al. [12], unaware of the MWHC construction, proposed it independently, but also noted that as a side-effect of the peeling process each hyperedge can be assigned an unique node; that is, each key can be assigned injectively an integer in m. We just need to store which of the r nodes of the hyperedge is the assigned one to obtain a perfect hash function $S \to m$, and this can be done in $c_r \lceil \log r \rceil n + o(n)$ bits. To make it perfect, that is, $S \to n$, it is possible to add a ranking structure. Again, the best r is 3, which yields theoretically a $2.46n + o(n)$ data structure [10].

HEM. Botelho et al. [10] introduced a practical external-memory algorithm called Heuristic External Memory (HEM) to construct MPHFs for sets that are too large to store their hypergraph in memory. They replace each key with a *signature* of $\Theta(\log n)$ bits computed with a random hash function, and check that no collision occurs. The signatures are then sorted and divided into small chunks based on their most significant bits, and a separate function is computed for each chunk with the approach described above (using a local seed). The representations of the chunk functions are then concatenated into a single array and their offsets (i.e., for each chunk, the position of the start of the chunk in the global array) are stored separately.

Cache-Oblivious Constructions. As an alternative to HEM, in [2] the authors propose *cache-oblivious* algorithms that use only scanning and sorting to peel hypergraphs and compute the corresponding structures. The main advantage is that of avoiding the cost of accessing the offset array of HEM without sacrificing scalability.

CHD. Finally, specifically for the purpose of computing MPFHs Belazzougui et al. [6] introduced a completely different construction, called CHD (compressed hash-and-displace), which, at the price of increasing the expected construction time makes it possible, in theory, to reach the information-theoretical lower bound of ≈1.44 bits per key.

Beyond Hypergraphs. The MWHC construction for static functions can be improved: Dietzfelbinger and Pagh [14] introduced a new construction that allows to make the constant in front of the nb space bound for static functions *arbitrarily small*; by Calkin's theorem, a constant β_r exists such that if $m > \beta_r n$ and the rows of the matrix H are just drawn at random from vectors of weight r then H has full rank with high probability. Contrary to c_r which has a finite minimum, β_r vanishes quickly as r increases, thus the denser the rows, the closer m can be to n. For example, if $r = 3$, $\beta_3 \approx 1.12 < c_3 \approx 1.23$. Unlike MWHC's linear-time peeling algorithm, general matrix inversion requires super*quadratic* time ($O(n^3)$ with Gaussian elimination); to obtain a linear-time algorithm, they shard the set S into small sets using a hash function, and compute the static functions on each subset independently; the actual construction is rather involved, to account for some corner cases (note that the HEM algorithm described above is essentially a practical simplified version of this scheme).

The authors also notice that solvability of the system implies that the corresponding hypergraph is orientable, thus making it possible to construct minimal perfect hash functions. Later works [13,15,16] further improve the thresholds for solvability and orientability: less than 1.09 for $r = 3$, and less than 1.03 for $r = 4$.

4   Squeezing Space

In this paper, we combine a number of new results and techniques to provide improved constructions. Our data structure is based on the HEM construction [10]: the key set is randomly sharded into chunks of expected constant size, and then the (minimal perfect hash) function is computed independently on each chunk. Instead of using a vertex/edge ratio that guarantees peelability, however, we choose a lower one that still guarantees orientability and solvability of the associated linear system (with high probability). Losing peelability implies that we have to use Gaussian elimination to solve the linear system, but since the chunks have constant size the overall construction is linear-time (plus an $O(n \log n)$ step to sort the signatures, which is actually a small part of the execution time in practice). We also describe improvements to the HEM data structure in Sect. 7.

First of all, we use the orientability thresholds in [13], which are shown to be the same as those of XORSAT solvability; for example, when a random 3-hypergraph has a vertex/edge ratio $c > 1.09$, it contains a nonempty 2-core (i.e., a maximal subgraph all whose vertices have degree at least 2), but the hypergraph is orientable and the incidence matrix has full rank. We can thus extend the MWHC technique to 3-hypergraphs with a nonempty 2-core: after the peeling procedure, we simply solve the equations specified by the 2-core. The main obstacle to this approach, before the results described in this paper, was that construction time was two orders of magnitude slower than that of the MWHC construction [1], making the whole construction unusable in practice. In Michael Rink's Ph.D. thesis [22] these considerations are described in some detail.

Moreover, since recently Goerdt and Falke have proved a result analogous to XORSAT for modulo-3 systems [17],[1] we can also obtain an orientation of a random 3-hypergraph using the *generalized selfless algorithm* [13], and then solve the modulo-3 linear system induced by the orientation to obtain a perfect hash function. Both procedures have some controlled probability of failure. In case such a failure occurs, we generate a new hypergraph. We then show how to manage the ranking part essentially with no space cost.

5   Broadword Programming for Row Operations

Our first step towards a practical solution by Gaussian elimination is *broadword programming* [18] (a.k.a. SWAR—"SIMD in A Register"), a set of techniques to process simultaneously multiple values by packing them into machine words of w bits and performing the computations on the whole words. In theoretical succinct data structures it is common to assume that $w = \Theta(\log n)$ and reduce to subproblems of size $O(w)$, whose results can be precomputed into sublinear-sized tables and looked up in constant time. For practical values of n, however, these tables are far from negligible; in this case broadword algorithms are usually sufficient to compute the same functions in constant or near-constant time without having to store a lookup table.

For our problem, the inner loop of the Gaussian elimination is entirely composed of row operations: given vectors x and y, and a scalar α, compute $x + \alpha y$.

---

[1] Technically, the proof in the paper is for $k > 15$, but the author claim that the result can be proved for $k \geq 3$ with the same techniques, and in practice we never needed more than two attempts to generate a solvable system.

It is trivial to perform this operation w elements at a time when the field is $\mathbf{F}_2$, which is the case for static functions computation: we can just pack one element per bit, and since the scalar can be only 1 the sum is just a bitwise XOR x ^ y, using the C notation. For MPHFs, instead, the field is $\mathbf{F}_3$, which requires more sophisticated algorithms. First, we can encode each element $\{0, 1, 2\}$ into 2 bits, thus fitting $w/2$ elements into a word. The scalar α can be only 1 or -1, so we can treat the cases $x + y$ and $x - y$ separately.

For the addition, we can start by simply adding x and y. When elements on both sides are smaller than 2, there's nothing to do: the result will be smaller than 3. When however at least one of the two is 2 and the other one is not 0, we need to subtract 3 from the result to bring it back to the canonical representation in $[0 .. 3)$. Note that when the two sides are both 2 the result overflows its 2 bits $(10_2 + 10_2 = 100_2)$, but since addition and subtraction modulo 2^w are associative we can imagine that the operation is performed independently on each 2-bit element, as long as the final result fits into 2 bits. Thus we need to compute a mask that is 3 wherever the results is at least 3, and then subtract it from $x + y$.

```c
uint64_t add_mod3_step2(uint64_t x, uint64_t y) {
    uint64_t xy = x | y;
    // Set MSB if (x or y == 2) and (x or y == 1).
    uint64_t mask = (xy << 1) & xy;
    // Set MSB if (x == 2) and (y == 2).
    mask |= x & y;
    // The MSB of each 2-bit element is now set
    // iff the result is >= 3. Clear the LSBs.
    mask &= 0x5555555555555555 << 1;
    // Now turn the elements with MSB set into 3.
    mask |= mask >> 1;
    return x + y - mask;
}
```

Subtraction is very similar. We begin by subtracting elementwise y from 3, which does not cause any carry since all the elements are strictly smaller than 3. The resulting elements are thus at least 1. We can now proceed to compute $x + y$ with the same case analysis as before, except now the right-hand elements are in $[1 .. 3]$ so the conditions for the mask are slightly different.

```c
uint64_t sub_mod3_step2(uint64_t x, uint64_t y) {
    // y = 3 - y.
    y = 0xFFFFFFFFFFFFFFFF - y;
    // Now y > 0
    // Set MSB if x == 2.
    uint64_t mask = x;
    // Set MSB if (x == 2 and y >= 2) or (y == 3).
    mask |= ((x | y) << 1) & y;
    mask &= 0x5555555555555555 << 1;
    mask |= mask >> 1;
    return x + y - mask;
}
```

Both addition and subtraction take just 10 arithmetic operations, and on modern 64-bit CPUs they can process vectors of 32 elements at a time.

Finally, when performing back substitution we will need to compute row-matrix multiplications, where a row is given by the coefficients of an equation and the matrix contains the solutions computed so far.

In the field $\mathbf{F}_2$, this can be achieved by iterating on the ones of the row, and adding up the corresponding b-bit rows in the right-hand matrix. The ones can iterate by finding the LSB of the current row word, and deleting it with the standard broadword trick x = x & - x.

For MPHFs, instead, the field is $\mathbf{F}_3$ but the matrix of solutions is a vector, so the product is just a scalar product. To compute it, we use the following broadword algorithm that computes the scalar product of two vectors represented as 64-bit words.

```
uint64_t prod_mod3_step2(uint64_t x, uint64_t y) {
    uint64_t high = x & 0xAAAAAAAAAAAAAAAA;
    uint64_t low = x & 0x5555555555555555;
    // Make every 10 into a 11 and zero everything else.
    uint64_t high_shift = high >> 1;
    // Exchange ones with twos, and make 00 into 11.
    uint64_t t = (y ^ (high | high_shift))
        & (x | high_shift | low << 1);
    return popcount(t & 0xAAAAAAAAAAAAAAAA) * 2
      + popcount(t & 0x5555555555555555);
}
```

The expression computing t takes care of placing in a given position a value equivalent to the product of the associated positions in x and y (this can be easily check with a case-by-case analysis). We remark that in some cases we actually use 3 as equivalent to zero. At that point, the last lines compute the contribution of each product (popcount() returns the number of bit in a word that are set). Note that the results has still to be reduced modulo 3.

6   Lazy Gaussian Elimination

Even if armed with broadword algorithms, solving by Gaussian elimination systems of the size of a HEM chunk (thousands of equations and variables) would be prohibitively slow, making construction of our data structures an order of magnitude slower than the standard MWHC technique.

Structured Gaussian elimination aims at reducing the number of operations in the solution of a linear system by trying to isolate a number of variables appearing in a large number of equations, and then rewrite the rest of the system using just those variables. It is a heuristics developed in the context of computations of discrete logarithms, which require the solution of large sparse systems [19,21]. The standard formulation requires the selection of a fraction (chosen arbitrarily) of variables that appear in a large number of equations, and then a number of loosely defined refinement steps.

We describe here a new parameterless version of structured Gauss elimination, which we call *lazy Gaussian elimination*. This heuristics turned out to be extremely effective on our systems, reducing the size of the system to be solved by standard elimination to around 4 % of the original one.

Consider a system of equations on some field. At any time a variable can be *active*, *idle*, or *solved* and an equation can be *sparse* or *dense*. Initially, all equations are sparse and all variables are idle. We will modify the system maintaining the following invariants:

- dense equations do not contain idle variables;
- an equation can contain at most one solved variable;
- a solved variable appears in exactly one dense equation.

Our purpose is to modify the system so that all equations are dense, trying to minimize the number of active variables (or, equivalently, maximize the number of solved variables). At that point, values for the active variables can be computed by standard Gaussian elimination on the dense equations that do not contain solved variables, and solved variables can be computed easily from the values assigned to active variables.

The *weight* of a variable is the number of sparse equations in which it appears. The *priority* of a sparse equation is the number of idle variables in the equation. Lazy Gaussian elimination keeps equations in a min-priority queue, and performs the following actions:

1. If there is a sparse equation of priority zero that contains some variables, it is made dense. If there are no variables, the equation is either an identity, in which case it is discarded, or it is impossible, in which case the system is unsolvable and the procedure stops.
2. If there is a sparse equation of priority one, the only idle variable in the equation becomes solved, and the equation becomes dense. The equation is then used to eliminate the solved variable from all other equations.
3. Otherwise, the idle variable appearing in the largest number of sparse equations becomes active.

Note that if the system is solvable the procedure always completes—in the worst case, by making all idle variables active (and thus all equations dense).

Two observations are in order:

- The weight of an idle variable never changes, as in step 2 we eliminate the solved variable and modify the coefficients of active variables only. This means that we can simply sort initially (e.g., by countsort) the variables by the number of equations in which they appear, and pick idle variables in that order at step 3.
- We do not actually need a priority queue for equations: simply, when an equation becomes of priority zero or one, it is moved to the left or right side, respectively, of a deque that we check in the first step.

Thus, the only operations requiring superlinear time are the eliminations performed in step 2, and the final Gaussian elimination on the dense equations, which we compute, however, using broadword programming.

7   Data Structure Improvements

Improving HEM. Our HEM version uses *on-disk bucket sorting* to speed up construction: keys are first divided into 256 on-disk *physical* chunks, depending on the highest bits of their hash value (we use Jenkins's SpookyHash). The on-disk chunks are then loaded in memory and sorted, and virtual chunks of the desired size are computed either splitting or merging physical chunks. Since we store a 192-bit hash plus a 64-bit value for each key, we can guarantee that the amount of memory used that depends on the number of keys cannot exceed one bit per key (beside the structure to be computed).

Eliminating the Ranking Structure. In the case of minimal perfect hashing, we can further speed up the structure and reduce space by getting rid of the ranking structure that is necessary to make minimal the perfect hashing computed by the system of equations.

In the standard HEM construction, the number of vertices associated to a chunk of size s is given by $\lceil cs \rceil$, where c is a suitable constant, and the offset information contains the partial sums of such numbers.

We will use a different approach: the number of vertices associated with the chunk will be $\lceil c(S + s) \rceil - \lceil cS \rceil$, where S is the number of elements stored in previous chunks. The difference to $\lceil cs \rceil$ is at most one, but using our approach we can compute, given S and s, the number of vertices associated with the chunk.

Thus, instead of storing the offset information, we will store for each chunk the number S of elements stored in previous chunks. The value can be used as a base for the ranking inside the chunk: this way, the ranking structure is no longer necessary, reducing space and the number of memory accesses. When $r = 3$, as it is customary we can use two bits for each value, taking care of using the value 3, instead of 0, for the vertex associated to a hyperedge. As a result, ranking requires just counting the number of nonzero pairs in the values associated with a chunk, which can be performed again by broadword programming:

```
int count_nonzero_pairs (uint64_t x) {
    return popcount((x | x >> 1) & 0x5555555555555555);
}
```

Compacting Offsets and Seeds. After removing the ranking structure, it is only left to store the partial sums of the number of keys per chunk, and the seed used for the chunk hash function. This is the totality of the overhead imposed by the HEM data structure with respect to constructing the function over the whole input set at once.

Instead of storing these two numbers separately, we combine them into a single 64-bit integer. The main observation that allows us to do so is that due to the extremely high probability of finding a good seed for each chunk, few random bits are necessary to store it: we can just use the same sequence of seeds for each chunk, and store the number of failed attempts before the successful one. In our experiments this number is distributed geometrically and never greater than 24. If we are storing n keys, $64 - \lceil \log n \rceil$ bits are available for the seed, which are more than sufficient for any realistic n.

8   Experimental Results

We performed experiments in Java using two datasets derived from the eu-2015 crawls gathered by BUbiNG [9] on an Intel® Core™ i7-4770 CPU @3.40 GHz (Haswell). The smaller dataset is the list of hosts (11 264 052 keys, ≈22 B/key), while the larger dataset is the list of pages (1 070 557 254 keys, ≈80 B/key). The crawl data is publicly available at the LAW website.[2]

Besides the final performance figures (which depends on the chosen chunk size), it is interesting to see how the measures of interest vary with the chunk size. In Fig. 1 we show how the number of bits per element, construction time and lookup time vary with the chunk size for $r = 3$. Note that in the case of minimal perfect hash functions we show the actual number of bits per key. In the case of general static function, we build a function mapping each key to its ordinal position and report the number of additional bits per key used by the algorithm.

As chunks gets larger, the number of bits per key slightly decreases (as the impact of the offset structure is better amortized); at the same time:

- construction time increases because the Gaussian elimination process is super-linear (very sharply after chunk size 2^{11});
- in the case of minimal perfect hash functions, larger chunks cause the rank function to do more work linearly with the chunk size, and indeed lookup time increases sharply in this case;
- in the case of static functions, chunks larger than 2^{10} yield a slightly improved lookup time as the offset array becomes small enough to fit in the L3 cache.

In Table 1, we show the lookup and construction time of our "best choice" chunk size, 2^{10}, with respect to the data reported in [1] for the same space usage (i.e., additional 1.10 b/key), and to the C code for the CHD technique made available by the authors (http://cmph.sourceforge.net/) when $\lambda = 3$, in which case the number of bits per key is almost identical to ours. We remark that in the case of CHD for the larger dataset we had to use different hardware, as the memory available (16 GB) was not sufficient to complete the construction, in spite of the final result being just 3 GB.

Table 1. A comparison of per-key construction and evaluation time, $r = 3$. CHD is from [6], ADR is from [1].

	eu-2015-host			eu-2015			ADR
	MPHF	SF	CHD	MPHF	SF	CHD	SF
Lookup (ns)	186	210	408	499	438	1030	?
Construction (μs)	1.61	1.12	0.98	2.45	1.73	3.53	270

---

[2] http://law.di.unimi.it/.

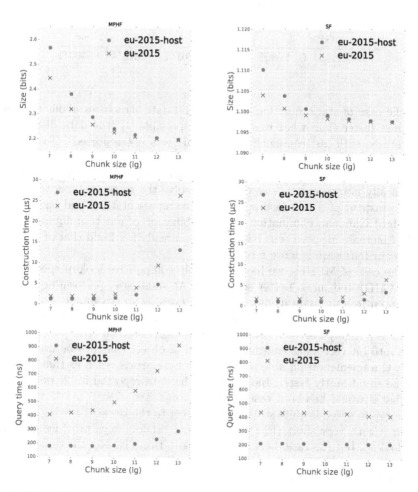

Fig. 1. Size in bits per element, and construction and lookup time in microseconds for the `eu-2015` and `eu-2015-host` datasets when $r = 3$.

Table 2. Per-key construction and evaluation time of static functions, $r = 4$.

	eu-2015-host	eu-2015	ADR
Lookup (ns)	236	466	?
Construction (µs)	1.75	2.6	≈ 2000

Table 3. Increase in construction time for $r = 3$ using just pre-peeling (P), broadword computation (B), lazy Gaussian elimination (G) or a combination.

BG	GP	G	BP	B	P	None
+13 %	+57 %	+98 %	+296 %	+701 %	+2218 %	+5490 %

In the case of static function, we can build data structures about two hundred times faster than what was previously possible [1] (the data displayed is on a dataset with 10^7 elements; lookup time was not reported). To give our reader an idea of the contribution of each technique we use, Table 3 shows the increase in construction time using any combination of the peeling phase (which is technically not necessary—we could just solve the system), broadword computation instead of a standard sparse system representation, and lazy instead of standard Gaussian elimination. The combination of our techniques brings a *fifty-fold* increase in speed (our basic speed is already fourfould that of [1], likely because our hardware is more recent).

In the case of MPHFs, we have extremely competitive lookup speed (twice that of CHD) and much better scalability. At small size, performing the construction entirely in main memory, as CHD does, is an advantage, but as soon as the dataset gets large our approach scales much better. We also remark that our code is a highly abstract Java implementation based on *strategies* that turn objects into bit vectors at runtime: any kind of object can thus be used as key. A tight C implementation able to hash only byte arrays, such as that of CHD, would be significantly faster. Indeed, from the data reported in [2] we can estimate that it would be about twice as fast.

The gap in speed is quite stable with respect to the key size: testing the same structures with very short (less than 8 bytes) random keys provides of course faster lookup, but the ratio between the lookup times remain the same.

Finally, one must consider that CHD, at the price of a much greater construction time, can further decrease its space usage, but just a 9 % decrease in space increases construction time by an order of magnitude, which makes the tradeoff unattractive for large datasets.

With respect to our previous peeling-based implementations, we increase construction time by $\approx 50\%$ (SF) and $\approx 100\%$ (MPHF), at the same time decreasing lookup time.

In Table 2 we report timings for the case $r = 4$ (the construction time for [1] has been extrapolated, as the authors do not provide timings for this case). Additional space required now is just $\approx 3\%$ (as opposed to $\approx 10\%$ when $r = 3$). The main drawbacks are the slower construction time (as the system becomes denser) and the slower lookup time (as more memory has to be accessed). Larger values of r are not interesting as the marginal gain in space becomes negligible.

9   Further Applications

Static functions are a basic building block of *monotone* minimal perfect hash functions [5], data structures for weak prefix search [4], and so on. Replacing the common MWHC implementation of these building blocks with our improved construction will automatically decrease the space used and the lookup time in these data structures.

We remark that an interesting application of static functions is the almost optimal storage of *static approximate dictionaries*. By encoding as a static function the mapping from a key to a b-bit signature generated by a random hash function, one can answer to the question "$x \in X$?" in constant time, with false positive rate 2^{-b}, using (when $r = 4$) just 1.03 nb bits; the lower bound is nb [11].

10   Conclusions

We have discussed new practical data structures for static functions and minimal perfect hash functions. Both scale to billion keys, and both improve significantly lookup speed with respect to previous constructions. In particular, we can build static functions based on Gaussian elimination two orders of magnitude faster than previous approaches, thanks to a combination of broadword programming and a new, parameterless lazy approach to the solution of sparse system. We expect that these structure will eventually replace the venerable MWHC approach as a scalable method with high-performance lookup.

References

1. Aumüller, M., Dietzfelbinger, M., Rink, M.: Experimental variations of a theoretically good retrieval data structure. In: Fiat, A., Sanders, P. (eds.) ESA 2009. LNCS, vol. 5757, pp. 742–751. Springer, Heidelberg (2009)
2. Belazzougui, D., Boldi, P., Ottaviano, G., Venturini, R., Vigna, S.: Cache-oblivious peeling of random hypergraphs. In: 2014 Data Compression Conference (DCC 2014), pp. 352–361. IEEE (2014)
3. Belazzougui, D., Boldi, P., Pagh, R., Vigna, S.: Monotone minimal perfect hashing: searching a sorted table with $O(1)$ accesses. In: Proceedings of the 20th Annual ACM-SIAM Symposium on Discrete Mathematics (SODA), pp. 785–794. ACM, New York (2009)
4. Belazzougui, D., Boldi, P., Pagh, R., Vigna, S.: Fast prefix search in little space, with applications. In: de Berg, M., Meyer, U. (eds.) ESA 2010, Part I. LNCS, vol. 6346, pp. 427–438. Springer, Heidelberg (2010)
5. Belazzougui, D., Boldi, P., Pagh, R., Vigna, S.: Theory and practice of monotone minimal perfect hashing. ACM J. Exp. Algorithm. **16**(3), 3.2:1–3.2:26 (2011)
6. Belazzougui, D., Botelho, F.C., Dietzfelbinger, M.: Hash, displace, and compress. In: Fiat, A., Sanders, P. (eds.) ESA 2009. LNCS, vol. 5757, pp. 682–693. Springer, Heidelberg (2009)
7. Belazzougui, D., Navarro, G.: Alphabet-independent compressed text indexing. In: Demetrescu, C., Halldórsson, M.M. (eds.) ESA 2011. LNCS, vol. 6942, pp. 748–759. Springer, Heidelberg (2011)

8. Belazzougui, D., Venturini, R.: Compressed static functions with applications. In: SODA, pp. 229–240 (2013)

9. Boldi, P., Marino, A., Santini, M., Vigna, S.: BUbiNG: massive crawling for the masses. In: Proceedings of the Companion Publication of the 23rd International Conference on World Wide Web Companion, WWW Companion 2014, pp. 227–228. International World Wide Web Conferences Steering Committee (2014)

10. Botelho, F.C., Pagh, R., Ziviani, N.: Practical perfect hashing in nearly optimal space. Inf. Syst. **38**(1), 108–131 (2013)

11. Carter, L., Floyd, R., Gill, J., Markowsky, G., Wegman, M.: Exact and approximate membership testers. In: Proceedings of Symposium on Theory of Computation (STOC 1978), pp. 59–65. ACM Press (1978)

12. Chazelle, B., Kilian, J., Rubinfeld, R., Tal, A.: The Bloomier filter: an efficient data structure for static support lookup tables. In: Munro, J.I. (ed.) Proceedings of the Fifteenth Annual ACM-SIAM Symposium on Discrete Algorithms, SODA 2004, pp. 30–39. SIAM (2004)

13. Dietzfelbinger, M., Goerdt, A., Mitzenmacher, M., Montanari, A., Pagh, R., Rink, M.: Tight thresholds for cuckoo hashing via XORSAT. In: Abramsky, S., Gavoille, C., Kirchner, C., Meyer auf der Heide, F., Spirakis, P.G. (eds.) ICALP 2010. LNCS, vol. 6198, pp. 213–225. Springer, Heidelberg (2010)

14. Dietzfelbinger, M., Pagh, R.: Succinct data structures for retrieval and approximate membership (extended abstract). In: Aceto, L., Damgård, I., Goldberg, L.A., Halldórsson, M.M., Ingólfsdóttir, A., Walukiewicz, I. (eds.) ICALP 2008, Part I. LNCS, vol. 5125, pp. 385–396. Springer, Heidelberg (2008)

15. Fountoulakis, N., Panagiotou, K.: Sharp load thresholds for cuckoo hashing. Random Struct. Algorithms **41**(3), 306–333 (2012)

16. Frieze, A.M., Melsted, P.: Maximum matchings in random bipartite graphs and the space utilization of cuckoo hash tables. Random Struct. Algorithms **41**(3), 334–364 (2012)

17. Goerdt, A., Falke, L.: Satisfiability thresholds beyond k−XORSAT. In: Hirsch, E.A., Karhumäki, J., Lepistö, A., Prilutskii, M. (eds.) CSR 2012. LNCS, vol. 7353, pp. 148–159. Springer, Heidelberg (2012)

18. Knuth, D.E.: The Art of Computer Programming. Pre-Fascicle 1A. Draft of Section 7.1.3: Bitwise Tricks and Techniques (2007)

19. LaMacchia, B.A., Odlyzko, A.M.: Solving large sparse linear systems over finite fields. In: Menezes, A., Vanstone, S.A. (eds.) CRYPTO 1990. LNCS, vol. 537, pp. 109–133. Springer, Heidelberg (1991)

20. Majewski, B.S., Wormald, N.C., Havas, G., Czech, Z.J.: A family of perfect hashing methods. Comput. J. **39**(6), 547–554 (1996)

21. Odlyzko, A.M.: Discrete logarithms in finite fields and their cryptographic significance. In: Beth, T., Cot, N., Ingemarsson, I. (eds.) EUROCRYPT 1984. LNCS, vol. 209, pp. 224–314. Springer, Heidelberg (1985)

22. Rink, M.: Thresholds for matchings in random bipartite graphs with applications to hashing-based data structures. Ph.D. thesis, Technische Universität Ilmenau (2015)

Better Partitions of Protein Graphs
for Subsystem Quantum Chemistry

Moritz von Looz[1][(✉)], Mario Wolter[2], Christoph R. Jacob[2],
and Henning Meyerhenke[1]

[1] Institute of Theoretical Informatics Karlsruhe Institute of Technology (KIT),
Karlsruhe, Germany
{moritz.looz-corswarem,meyerhenke}@kit.edu
[2] Institute of Physical and Theoretical Chemistry,
TU Braunschweig, Braunschweig, Germany
{m.wolter,c.jacob}@tu-braunschweig.de

Abstract. Determining the interaction strength between proteins and
small molecules is key to analyzing their biological function. Quantum-
mechanical calculations such as *Density Functional Theory* (DFT) give
accurate and theoretically well-founded results. With common implementations the running time of DFT calculations increases quadrati-
cally with molecule size. Thus, numerous subsystem-based approaches
have been developed to accelerate quantum-chemical calculations. These
approaches partition the protein into different fragments, which are
treated separately. Interactions between different fragments are approximated and introduce inaccuracies in the calculated interaction energies.

To minimize these inaccuracies, we represent the amino acids and
their interactions as a weighted graph in order to apply graph partitioning. None of the existing graph partitioning work can be directly
used, though, due to the unique constraints in partitioning such protein
graphs. We therefore present and evaluate several algorithms, partially
building upon established concepts, but adapted to handle the new constraints. For the special case of partitioning a protein along the main
chain, we also present an efficient dynamic programming algorithm that
yields provably optimal results. In the general scenario our algorithms
usually improve the previous approach significantly and take at most a
few seconds.

1 Introduction

Context. The biological role of proteins is largely determined by their interactions with other proteins and small molecules. Quantum-chemical methods,
such as *Density Functional Theory* (DFT), provide an accurate description of
these interactions based on quantum mechanics. A major drawback of DFT is its
time complexity, which has been shown to be cubic with respect to the protein
size in the worst case [4,17]. For special cases this complexity can be reduced
to being linear [9,20]. DFT implementations used for calculations on proteins

© Springer International Publishing Switzerland 2016
A.V. Goldberg and A.S. Kulikov (Eds.): SEA 2016, LNCS 9685, pp. 353–368, 2016.
DOI: 10.1007/978-3-319-38851-9_24

are in between these bounds and typically show quadratic behavior with significant constant factors, rendering proteins bigger than a few hundred amino acids prohibitively expensive to compute [4,17].

To mitigate the computational cost, quantum-chemical subsystem methods have been developed [12,15]. In such approaches, large molecules are separated into fragments (= subsystems) which are then treated individually. A common way to deal with individual fragments is to assume that they do not interact with each other. The error this introduces for protein–protein or protein–molecule interaction energies (or for other local molecular properties of interest) depends on the size and location of fragments: A partition that cuts right through the strongest interaction in a molecule will give worse results than one that carefully avoids this. It should also be considered that a protein consists of a *main chain* (also called *backbone*) of amino acids. This main chain folds into 3D-secondary-structures, stabilized by non-bonding interactions (those not on the backbone) between the individual amino acids. These different connection types (backbone vs non-backbone) have different influence on the interaction energies.

Motivation. Subsystem methods are very powerful in quantum chemistry [12,15] but so far require manual cuts with chemical insight to achieve good partitions [18]. Currently, when automating the process, domain scientists typically cut every X amino acids along the main chain (which we will call the *naive approach* in the following). This gives in general suboptimal and unpredictable results.

By considering amino acids as nodes connected by edges weighted with the expected error in the interaction energies, one can construct (dense) graphs representing the proteins. Graph partitions with a light cut, i.e. partitions of the vertex set whose inter-fragment edges have low total weight, should then correspond to a low error for interaction energies. A general solution to this problem has high significance, since it is applicable to any subsystem-based method and since it will enable such calculations on larger systems with controlled accuracy. Yet, while several established graph partitioning algorithms exist, none of them is directly applicable to our problem scenarios due to additional domain-specific optimization constraints (which are outlined in Sect. 2).

Contributions. For the first of two problem scenarios, the special case of continuous fragments along the main chain, we provide in Sect. 4 a dynamic programming (DP) algorithm. We prove that it yields an optimal solution with a worst-case time complexity of $\mathcal{O}(n^2 \cdot maxSize)$.

For the general protein partitioning problem, we provide three algorithms using established partitioning concepts, now equipped with techniques for adhering to the new constraints (see Sect. 5): (i) a greedy agglomerative method, (ii) a multilevel algorithm with Fiduccia-Mattheyses [8] refinement, and (iii) a simple postprocessing step that "repairs" traditional graph partitions.

Our experiments (Sect. 6) use several protein graphs representative for DFT calculations. Their number of nodes is rather small (up to 357), but they are complete graphs. The results show that our algorithms are usually better in quality than the naive approach. While none of the new algorithms is consistently the best one, the DP algorithm can be called most robust since it is always better in quality than the naive approach. A meta algorithm that runs all single algorithms and picks the best solution would still take only about ten seconds per instance and improve the naive approach on average by 13.5 % to 20 %, depending on the imbalance. In the whole quantum-chemical workflow the total partitioning time of this meta algorithm is still small. Further experiments and visualizations omitted due to space constraints can be found in the full version [28].

2   Problem Description

Given an undirected connected graph $G = (V, E)$ with n nodes and m edges, a set of k disjoint non-empty node subsets $V_1, V_2, ...V_k$ is called a k-partition of G if the union of the subsets yields V ($V = \bigcup_{1 \leq i \leq k} V_i$). We denote partitions with the letter Π and call the subsets *fragments* in this paper.

Let $w(u, v)$ be the weight of edge $\{u, v\} \in E$, or 1 in an unweighted graph. Then, the *cut weight* of a graph partition is the sum of the weights of edges with endpoints in different subsets: $\text{cutweight}(\Pi, G) = \sum_{u \in V_i, v \in V_j, i \neq j, V_i, V_j \in \Pi} w(u, v)$. The largest fragment's size should not exceed $\text{maxSize} := (1 + \epsilon) \cdot \lceil n/k \rceil$, where ϵ is the so-called *imbalance* parameter. A partition is balanced iff $\epsilon = 0$.

Given a graph $G = (V, E)$ and $k \in \mathbb{N}_{\geq 2}$, *graph partitioning* is often defined as the problem of finding a k-partition with minimum cut weight while respecting the constraint of maximum imbalance ϵ. This problem is $\mathcal{NP}$-hard [10] for general graphs and values of ϵ. For the case of $\epsilon = 0$, no polynomial time algorithm can deliver a constant factor approximation guarantee unless $\mathcal{P}$ equals $\mathcal{NP}$ [1].

2.1   Protein Partitioning

We represent a protein as a weighted undirected graph. Nodes represent amino acids, edges represent bonds or other interactions. (Note that our graphs are different from protein interaction networks [23].) Edge weights are determined both by the strength of the bond or interaction and the importance of this edge to the protein function. Such a graph can be constructed from the geometrical structure of the protein using chemical heuristics whose detailed discussion is beyond our scope. Partitioning into fragments yields faster running time for DFT since the time required for a fragment is quadratic in its size. The cut weight of a partition corresponds to the total error caused by dividing this protein into fragments. A balanced partition is desirable as it maximizes this acceleration effect. However, relaxing the constraint with a small $\epsilon > 0$ makes sense as this usually helps in obtaining solutions with a lower error.

Note that the positions on the main chain define an ordering of the nodes. From now on we assume the nodes to be numbered along the chain.

New Constraints. Established graph partitioning tools using the model of the previous section cannot be applied directly to our problem since protein partitioning introduces additional constraints and an incompatible scenario due to chemical idiosyncrasies:

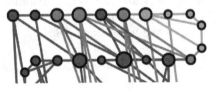 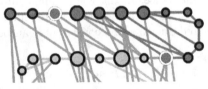

(a) Excerpt from a partition where the gap constraint is violated, since nodes 4 and 6 (counting clockwise from the upper left) are in the green fragment, but node 5 is in the blue fragment.

(b) Excerpt from a partition where the charge constraint is violated. Nodes 3 and 13 are charged, indicated by the white circles, but are both in the blue fragment.

Fig. 1. Examples of violated gap and charge constraints, with fragments represented by colors. (Color figure online)

- The first constraint is caused by so-called *cap molecules* added for the subsystem calculation. These cap molecules are added at fragment boundaries (only in the DFT, not in our graph) to obtain chemically meaningful fragments. This means for the graph that if node i and node $i + 2$ belong to the same fragment, node $i + 1$ must also belong to that fragment. Otherwise the introduced cap molecules will overlap spatially and therefore not represent a chemically meaningful structure. We call this the *gap* constraint. Figure 1a shows an example where the gap constraint is violated.
- More importantly, some graph nodes can have a charge. It is difficult to obtain robust convergence in quantum-mechanical calculations for fragments with more than one charge. Therefore, together with the graph a (possibly empty) list of charged nodes is given and two charged nodes must not be in the same fragment. This is called the *charge* constraint. Figure 1b shows an example where the charge constraint is violated.

We consider here **two problem scenarios** (with different chemical interpretations) in the context of protein partitioning:

- **Partitioning along the main chain:** The main chain of a protein gives a natural structure to it. We thus consider a scenario where partition fragments are forced to be continuous on the main chain. This minimizes the number of cap molecules necessary for the simulation and has the additional advantage of better comparability with the naive partition.

  Formally, the problem can be stated like this: Given a graph $G = (V, E)$ with ascending node IDs according to the node's main chain position, an integer

k and a maximum imbalance ϵ, find a k-partition with minimum cut weight such that $v_j \in V_i \wedge v_j + l \in V_i \rightarrow v_j + 1 \in V_i, 1 \leq j \leq n, l \in \mathbb{N}^+, 1 \leq i \leq k$ and which respects the balance, gap, and charge constraints.

- **General protein partitioning:** The general problem does not require continuous fragments on the main chain, but also minimizes the cut weight while adhering to the balance, gap, and charge constraints.

3 Related Work

3.1 General-Purpose Graph Partitioning

General-purpose graph partitioning tools only require the adjacency information of the graph and no additional problem-related information. For special inputs (very small n or $k = 2$ and small cuts) sophisticated methods from mathematical programming [11] or using branch-and-bound [5] are feasible – and give provably optimal results. To be of general practical use, in particular for larger instances, most widely used tools employ local heuristics within a multilevel approach, though (see the survey by Buluc et al. [2]).

The multilevel metaheuristic, popularized for graph partitioning in the mid-1990s [14], is a powerful technique and consists of three phases: First, one computes a hierarchy of graphs $G_0, \ldots, G_l$ by recursive coarsening in the first phase. G_l ought to be small in size, but topologically similar to the input graph G_0. A very good initial solution for G_l is computed in the second phase. After that, the recursive coarsening is undone and the solution prolongated to the next-finer level. In this final phase, in successive steps, the respective prolongated solution on each level is improved using local search.

A popular local search algorithm for the third phase of the multilevel process is based on the method by Fiduccia and Mattheyses (FM) [8] (many others exist, see [2]). The main idea of FM is to exchange nodes between blocks in the order of the cost reductions possible, while maintaining a balanced partition. After every node has been moved once, the solution with the best cost improvement is chosen. Such a phase is repeated several times, each running in time $\mathcal{O}(m)$.

3.2 Methods for Subsystem Quantum Chemistry

While this work is based on the *molecular fractionation with conjugate cap* (MFCC) scheme [13,30], several more sophisticated approaches have been developed which allow to decrease the size of the error in subsystem quantum-mechanical calculations [6,7,15]. The general idea is to reintroduce the interactions missed by the fragmentation of the supermolecule. A prominent example is the *frozen density embedding* (FDE) approach [15,16,29]. All these methods strongly depend on the underlying fragmentation of the supermolecule and it is therefore desirable to minimize the error in the form of the cut weight itself. Thus, the implementation shown in this paper is applicable to all quantum-chemical subsystem methods needing molecule fragments as an input.

4   Solving Main Chain Partitioning Optimally

As discussed in the introduction, a protein consists of a main chain, which is folded to yield its characteristic spatial structure. Aligning a partition along the main chain uses the locality information in the node order and minimizes the number of cap molecules necessary for a given number of fragments. The problem description from Sect. 2 – finding fragments with continuous node IDs – is equivalent to finding a set of $k-1$ *delimiter nodes* $v_{d_1}, v_{d_2}, ... v_{d_{k-1}}$ that separate the fragments. Note that this is not a vertex separator, instead the delimiter nodes induce a set of cut edges due to the continuous node IDs. More precisely, delimiter node v_{d_j} belongs to fragment j, $1 \leq j \leq k-1$.

Consider the delimiter nodes in ascending order. Given the node v_{d_2}, the optimal placement of node v_{d_1} only depends on edges among nodes $u < v_{d_2}$, since all edges $\{u, v\}$ from nodes $u < v_{d_2}$ to nodes $v > v_{d_2}$ are cut no matter where v_{d_1} is placed. Placing node v_{d_2} thus induces an optimal placement for v_{d_1}, using only information from edges to nodes $u < v_{d_2}$. With this dependency of the positions of v_{d_1} and v_{d_2}, placing node v_{d_3} similarly induces an optimal choice for v_{d_2} and v_{d_1}, using only information from nodes smaller than v_{d_3}. The same argument can be continued inductively for nodes $v_{d_4} \ldots v_{d_k}$.

Algorithm 1 is our dynamic-programming-based solution to the main chain partitioning problem. It uses the property stated above to iteratively compute the optimal placement of $v_{d_{j-1}}$ for all possible values of v_{d_j}. Finding the optimal placements of $v_{d_1}, \ldots v_{d_{j-1}}$ given a delimiter v_{d_j} at node i is equivalent to the subproblem of partitioning the first i nodes into j fragments, for increasing values of i and j. If n nodes and k fragments are reached, the desired global solution is found. We allocate (Line 3) and fill an $n \times k$ table partCut with the optimal values for the subproblems. More precisely, the table entry partCut$[i][j]$ denotes the minimum cut weight of a j-partition of the first i nodes:

Lemma 1. *After the execution of Algorithm 1,* partCut$[i][j]$ *contains the minimum cut value for a continuous j-partition of the first i nodes. If such a partition is impossible,* partCut$[i][j]$ *contains* ∞.

We prove the lemma after describing the algorithm. After the initialization of data structures in Lines 2 and 3, the initial values are set in Line 4: A partition consisting of only one fragment has a cut weight of zero.

All further partitions are built from a *predecessor partition* and a new fragment. A j-partition $\Pi_{i,j}$ of the first i nodes consists of the jth fragment and a $(j-1)$-partition with fewer than i nodes. A valid predecessor partition of $\Pi_{i,j}$ is a partition $\Pi_{l,j-1}$ of the first l nodes, with l between $i - $ maxSize and $i - 1$. Node charges have to be taken into account when compiling the set of valid predecessors. If a backwards search for $\Pi_{i,j}$ from node i encounters two charged nodes a and b with $a < b$, all valid predecessors of $\Pi_{i,j}$ contain at least node a (Line 7).

The additional cut weight induced by adding a fragment containing the nodes $[l+1, i]$ to a predecessor partition $\Pi_{l,j-1}$ is the weight sum of edges connecting

---

Algorithm 1. Main Chain Partitioning with Dynamic Programming

---

Input: Graph $G = (V, E)$, fragment count k, bool list $isCharged$, imbalance ϵ
Output: partition Π

1 maxSize= $\lceil |V|/k \rceil \cdot (1 + \epsilon)$;
2 allocate empty partition Π;
3 partCut[i][j] $= \infty, \forall i \in [1, n], \forall j \in [1, k]$;
   /* initialize empty table partCut with n rows and k columns   */
4 partCut[i][1] $= 0, \forall i \in [1, \text{maxSize}]$;
5 **for** $1 \leq i \leq n$ **do**
6 | windowStart $= \max(i - \text{maxSize}, 1)$;
7 | if necessary, increase windowStart so that [windowStart, i] contains at most
   | one charged node;
8 | compute column i of cut cost table c;
9 | **for** $2 \leq j \leq k$ **do**
10 | | partCut[i][j] $= \min_{l \in [windowStart, i]} \text{partCut}[l][j-1] + c[l][i]$;
11 | | pred[i][j] $= \text{argmin}_{l \in [windowStart, i]} \text{partCut}[l][j-1] + c[l][i]$;
12 | **end**
13 **end**
14 $i = n$;
15 **for** $j = k; j \geq 2; j- = 1$ **do**
16 | $nextI = \text{pred}[i][j]$;
17 | assign nodes between $nextI$ and i to fragment Π_j;
18 | $i = nextI$;
19 **end**
20 **return** Π

---

nodes in $[1, l]$ to nodes in $[l+1, i]$: $c[l][i] = \sum_{\{u,v\} \in E, u \in [1,l], v \in [l+1,i]} w(u, v)$. Line 8 computes this weight difference for the current node i and all valid predecessors l.

For each i and j, the partition $\Pi_{i,j}$ with the minimum cut weight is then found in Line 10 by iterating backwards over all valid predecessor partitions and selecting the one leading to the minimum cut. To reconstruct the partition, we store the predecessor in each step (Line 11). If no partition with the given values is possible, the corresponding entry in partCut remains at ∞.

After the table is filled, the resulting minimum cut weight is at partCut[n][k], the corresponding partition is found by following the predecessors (Line 16).

We are now ready to prove Lemma 1 and the algorithm's correctness and time complexity.

Proof (of Lemma 1). By induction over the number of partitions j.

Base Case: $j = 1, \forall i$. A 1-partition is a continuous block of nodes. The cut value is zero exactly if the first i nodes contain at most one charge and i is not larger than maxSize. This cut value is written into partCut in Lines 3 and 4 and not changed afterwards.

Inductive Step: $j - 1 \rightarrow j$. Let i be the current node: A cut-minimal j-partition $\Pi_{i,j}$ for the first i nodes contains a cut-minimal $(j-1)$-partition $\Pi_{i',j-1}$ with continuous node blocks. If $\Pi_{i',j-1}$ were not minimum, we could find a better partition $\Pi'_{i',j-1}$ and use it to improve $\Pi_{i,j}$, a contradiction to $\Pi_{i,j}$ being cut-minimal. Due to the induction hypothesis, partCut$[l][j-1]$ contains the minimum cut value for all node indices l, which includes i'. The loop in Line 10 iterates over possible predecessor partitions $\Pi_{l,j-1}$ and selects the one leading to the minimum cut after node i. Given that partitions for $j-1$ are cut-minimal, the partition whose weight is stored in partCut$[i][j]$ is cut-minimal as well.

If no allowed predecessor partition with a finite weight exists, partCut$[i][j]$ remains at infinity. $\qquad\square$

Theorem 1. *Algorithm 1 computes the optimal main chain partition in time* $\mathcal{O}(n^2 \cdot \text{maxSize})$.

Proof. The correctness in terms of optimality follows directly from Lemma 1. We thus continue with establishing the time complexity. The nested loops in Lines 5 and 9 require $\mathcal{O}(n \cdot k)$ iterations in total. Line 7 is executed n times and has a complexity of maxSize. At Line 10 in the inner loop, up to maxSize predecessor partitions need to be evaluated, each with two constant time table accesses. Computing the cut weight column $c[\cdot][i]$ for fragments ending at node i (Line 8) involves summing over the edges of $\mathcal{O}(\text{maxSize})$ predecessors, each having at most $\mathcal{O}(n)$ neighbors. Since the cut weights constitute a reverse prefix sum, the column $c[\cdot][i]$ can be computed in $\mathcal{O}(n \cdot \text{maxSize})$ time by iterating backwards. Line 8 is executed n times, leading to a total complexity of $\mathcal{O}(n^2 \cdot \text{maxSize})$. Following the predecessors and assigning nodes to fragments is possible in linear time, thus the $\mathcal{O}(n^2 \cdot \text{maxSize})$ to compile the cut cost table dominates the running time. $\qquad\square$

5   Algorithms for General Protein Partitioning

As discussed in Sect. 2, one cannot use general-purpose graph partitioning programs due to the new constraints required by the DFT calculations. Moreover, if the constraint of the previous section is dropped, the DP-based algorithm is not optimal in general any more. Thus, we propose three algorithms for the general problem in this section: The first two, a greedy agglomerative method and Multilevel-FM, build on existing graph partitioning knowledge but incorporate the new constraints directly into the optimization process. The third one is a simple postprocessing repair procedure that works in many cases. It takes the output of a traditional graph partitioner and fixes it so as to fulfill the constraints.

5.1   Greedy Agglomerative Algorithm

The greedy agglomerative approach, shown in Algorithm 2, is similar in spirit to Kruskal's MST algorithm and to approaches proposed for clustering graphs with respect to the objective function modularity [3]. It initially sorts edges by weight and puts each node into a singleton fragment. Edges are then considered

---

Algorithm 2. Greedy Agglomerative Algorithm

Input: Graph $G = (V, E)$, fragment count k, list *charged*, imbalance ϵ
Output: partition Π

1 sort edges by weight, descending;
2 Π = create one singleton partition for each node;
3 chargedPartitions = partitions containing a charged node;
4 maxSize= $\lceil |V|/k \rceil \cdot (1 + \epsilon)$;
5 **for** *edge* $\{u, v\}$ **do**
6 $\quad$ allowed = True;
7 $\quad$ **if** $\Pi[u] \in$ chargedPartitions *and* $\Pi[v] \in$ chargedPartitions **then**
8 $\quad\quad$ | allowed = False;
9 $\quad$ **end**
10 $\quad$ **if** $|\Pi[u]| + |\Pi[v]| >$ maxSize **then**
11 $\quad\quad$ | allowed = False;
12 $\quad$ **end**
13 $\quad$ **for** *node* $x \in \Pi[u] \cup \Pi[v]$ **do**
14 $\quad\quad$ **if** $x + 2 \in \Pi[u] \cup \Pi[v]$ *and* $x + 1 \notin \Pi[u] \cup \Pi[v]$ **then**
15 $\quad\quad\quad$ | allowed = False;
16 $\quad\quad$ **end**
17 $\quad$ **end**
18 $\quad$ **if** *allowed* **then**
19 $\quad\quad$ | merge $\Pi[u]$ and $\Pi[v]$;
20 $\quad\quad$ | update chargedPartitions;
21 $\quad$ **end**
22 $\quad$ **if** *number of fragments in Π equals k* **then**
23 $\quad\quad$ | break;
24 $\quad$ **end**
25 **end**
26 **return** Π

---

iteratively with the heaviest first; the fragments belonging to the incident nodes are merged if no constraints are violated. This is repeated until no edges are left or the desired fragment count is achieved.

The initial edge sorting takes $\mathcal{O}(m \log m)$ time. Initializing the data structures is possible in linear time. The main loop (Line 5) has at most m iterations. Checking the size and charge constraints is possible in constant time by keeping arrays of fragment sizes and charge states. The time needed for checking the gaps and merging is linear in the fragment size and thus at most $\mathcal{O}(\text{maxSize})$.

The total time complexity of the greedy algorithm is thus:

$$T(\text{Greedy}) \in \mathcal{O}(m \cdot \max\{\text{maxSize}, \log m\}).$$

5.2 Multilevel Algorithm with Fiduccia-Mattheyses Local Search

Algorithm 3 is similar to existing multilevel partitioners using non-binary (i. e. $k > 2$) Fiduccia-Mattheyses (FM) local search. Our adaptation incorporates the

---

Algorithm 3. Multilevel-FM

Input: Graph $G = (V, E)$, fragment count k, list *charged*, imbalance ϵ, $[\Pi']$
Output: partition Π

1   $G_0, \ldots, G_l$ = hierarchy of coarsened Graphs, $G_0 = G$;
2   Π_l = partition G_l with region growing or recursive bisection;
3   **for** $0 \leq i < l$ **do**
4   uncoarsen G_i from G_{i+1};
5   Π_i = projected partition from Π_{i+1};
6   rebalance Π_i, possibly worsen cut weight;
     /* Local improvements   */
7   gain = NaN;
8   **repeat**
9   oldcut = cut(Π_i', G);
10   Π_i' = Fiduccia-Mattheyses-Step of Π_i with constraints;
11   gain = cut(Π_i', G) - oldcut;
12   **until** *gain == 0*;
13 **end**

---

constraints throughout the whole partitioning process, though. First a hierarchy of graphs $G_0, G_1, \ldots G_l$ is created by recursive coarsening (Line 1). The edges contracted during coarsening are chosen with a local matching strategy. An edge connecting two charged nodes stays uncontracted, thus ensuring that a fragment contains at most one charged node even in the coarsest partitioning phase. The coarsest graph is then partitioned into Π_l using region growing or recursive bisection. If an optional input partition Π' is given, it is used as a guideline during coarsening and replaces Π_l if it yields a better cut. We execute both our greedy and DP algorithm and use the partition with the better cut as input partition Π' for the multilevel algorithm.

After obtaining a partition for the coarsest graph, the graph is iteratively uncoarsened and the partition projected to the next finer level. We add a rebalancing step at each level (Line 6), since a non-binary FM step does not guarantee balanced partitions if the input is imbalanced. A Fiduccia-Mattheyses step is then performed to yield local improvements (Line 10): For a partition with k fragments, this non-binary FM step consists of one priority queue for each fragment. Each node v is inserted into the priority queue of its current fragment, the maximum gain (i.e. reduction in cut weight when v is moved to another fragment) is used as key. While at least one queue is non-empty, the highest vertex of the largest queue is moved if the constraints are still fulfilled, and the movement recorded. After all nodes have been moved, the partition yielding the minimum cut is taken. In our variant, nodes are only moved if the charge constraint stays fulfilled.

5.3   Repair Procedure

As already mentioned, traditional graph partitioners produce in general solutions that do not adhere to the constraints for protein partitioning. To be able to use existing tools, however, we propose a simple repair procedure for an existing partition which possibly does not fulfill the charge, gap, or balance constraints. To this end, Algorithm 4 performs one sweep over all nodes (Line 6) and checks for every node v whether the constraints are violated at this point. If they are and v has to be moved, an FM step is performed: Among all fragments that could possibly receive v, the one minimizing the cut weight is selected. If no suitable target fragment exists, a new singleton fragment is created. Note that due to the local search, this step can lead to more than k fragments, even if a partition with k fragments is possible.

The cut weight table allocated in Line 1 takes $\mathcal{O}(n \cdot k + m)$ time to create. Whether a constraint is violated can be checked in constant time per node by counting the number of nodes and charges observed for each fragment. A node needs to be moved when at least one charge or at least maxSize nodes have already been encountered in the same fragment. Finding the best target partition (Line 13) takes $\mathcal{O}(k)$ iterations, updating the cut weight table after moving a node v is linear in the degree $\deg(v)$ of v. The total time complexity of a repair step is thus: $\mathcal{O}(n \cdot k + m + n \cdot k + \sum_v \deg(v)) = \mathcal{O}(n \cdot k + m)$.

6   Experiments

6.1   Settings

We evaluate our algorithms on graphs derived from several proteins and compare the resulting cut weight. As main chain partitioning is a special case of general protein partitioning, the solutions generated by our dynamic programming algorithm are valid solutions of the general problem, though perhaps not optimal. Other algorithms evaluated are Algorithms 2 (Greedy), 3 (Multilevel), and the external partitioner KaHiP [25], used with the repair step discussed in Sect. 5.3. The algorithms are implemented in C++ and Python using the NetworKit tool suite [26], the source code is available from a hg repository[1].

We use graphs derived from five common proteins, covering the most frequent structural properties. Ubiquitin [24] and the Bubble Protein [21] are rather small proteins with 76 and 64 amino acids, respectively. Due to their biological functions, their overall size and their diversity in the contained structural features, they are commonly used as test cases for quantum-chemical subsystem methods [18]. The Green Fluorescent Protein (GFP) [22] plays a crucial role in the bioluminescence of marine organisms and is widely expressed in other organisms as a fluorescent label for microscopic techniques. Like the latter one, Bacteriorhodopsin (bR) [19] and the Fenna-Matthews-Olson protein (FMO) [27] are large enough to render quantum-chemical calculations on the whole proteins

---

[1] https://algohub.iti.kit.edu/parco/NetworKit/NetworKit-chemfork/.

Algorithm 4. Repairing a partition

Input: Graph $G = (V, E)$, k-partition Π, list *charged*, imbalance ϵ
Output: partition Π'

1 cutWeight$[i][j] = 0, 1 \leq i \leq n, 1 \leq j \leq k$;
2 **for** *edge* $\{u, v\}$ *in* E **do**
3 $\quad$ cutWeight$[u][\Pi(u)] += w(u, v)$;
4 $\quad$ cutWeight$[v][\Pi(v)] += w(u, v)$;
5 **end**
6 **for** *node* v *in* V **do**
   $\quad$ /* Check whether node can stay   */
7 $\quad$ **if** *charge violated* or *size violated* or *gap of size 1* **then**
8 $\quad\quad$ $\Psi =$ set of allowed target fragments;
9 $\quad\quad$ **if** Ψ *is empty* **then**
10 $\quad\quad\quad$ create new fragment for v;
11 $\quad\quad$ **end**
12 $\quad\quad$ **else**
      $\quad\quad\quad$ /* Fiduccia-Mattheyses-step: To minimize the cut weight,
      $\quad\quad\quad$ move the node to the fragment to which it has the
      $\quad\quad\quad$ strongest connection   */
13 $\quad\quad\quad$ target $= \text{argmax}_{i \in \Psi}\{\text{cutWeight}[v][i]\}$;
14 $\quad\quad\quad$ move v to target;
15 $\quad\quad$ **end**
16 $\quad\quad$ update charge counter, size counter and cutWeight;
17 $\quad$ **end**
18 **end**

---

practically infeasible. Yet, investigating them with quantum-chemical methods is key to understanding the photochemical processes they are involved in. The graphs derived from the latter three proteins have 225, 226 and 357 nodes, respectively. They are complete graphs with weighted $n(n-1)/2$ edges. All instances can be found in the mentioned hg repository in folder `input/`.

In our experiments we partition the graphs into fragments of different sizes (i.e. we vary the fragment number k). The small proteins ubiquitin and bubble are partitioned into 2, 4, 6 and 8 fragments, leading to fragments of average size 8–38. The other proteins are partitioned into 8, 12, 16, 20 and 24 fragments, yielding average sizes between 10 and 45. As maximum imbalance, we use values for ϵ of 0.1 and 0.2. While this may be larger than usual values of ϵ in graph partitioning, fragment sizes in our case are comparably small and an imbalance of 0.1 is possibly reached with the movement of a single node.

On these proteins, the running time of all partitioning implementations is on the order of a few seconds on a commodity laptop, we therefore omit detailed time measurements.

Charged Nodes. Depending on the environment, some of the amino acids are charged. As discussed in Sect. 2, at most one charge is allowed per fragment. We repeatedly sample $\lfloor 0.8 \cdot k \rfloor$ random charged nodes among the potentially charged, under the constraint that a valid main chain partition is still possible. To smooth out random effects, we perform 20 runs with different random nodes charged. Introducing charged nodes may cause the naive partition to become invalid. In these cases, we use the repair procedure on the invalid naive partition and compare the cut weights of other algorithms with the cut weight of the repaired naive partition.

6.2 Results

For the uncharged scenario, Fig. 2 shows a comparison of cut weights for different numbers of fragments and a maximum imbalance of 0.1. The cut weight is up to 34.5 % smaller than with the naive approach (or 42.8 % with $\epsilon = 0.2$). The best algorithm choice depends on the protein: For ubiquitin, green fluorescent protein, and Fenna-Matthew-Olson protein, the external partitioner KaHiP in combination with the repair step described in Sect. 5.3 gives the lowest cut weight when averaged over different fragment sizes. For the bubble protein, the multilevel algorithm from Sect. 5.2 gives on average the best result, while for bacteriorhodopsin, the best cut weight is achieved by the dynamic programming (DP) algorithm. The DP algorithm is always as least as good as the naive approach. This already follows from Theorem 1, as the naive partition is aligned along the main chain and thus found by DP in case it is optimal. DP is the only algorithm with this property, all others perform worse than the naive approach for at least one combination of parameters.

The general intuition that smaller fragment sizes leave less room for improvements compared to the naive solution is confirmed by our experimental results. While the general trend is similar and the best choice of algorithm depends on the protein, the cut weight is usually more clearly improved. Moreover, a meta algorithm that executes all single algorithms and picks their best solution yields average improvements (geometric mean) of 13.5 %, 16 %, and 20 % for $\epsilon = 0.1, 0.2$, and 0.3, respectively, compared to the naive reference. Such a meta algorithm requires only about ten seconds per instance, negligible in the whole DFT workflow.

Randomly charging nodes changes the results only insignificantly. The necessary increase in cut weight for the algorithm's solutions is likely compensated by a similar increase in the naive partition due to the necessary repairs. Further experimental results can be found in the full version [28].

7 Conclusions

Partitioning protein graphs for subsystem quantum-chemistry is a new problem with unique constraints which general-purpose graph partitioning algorithms were unable to handle. We have provided several algorithms for this problem and proved the optimality of one in the special case of partitioning along the main chain. With our algorithms chemists are now able to address larger problems in an automated manner with smaller error. Larger proteins, in turn, in connection

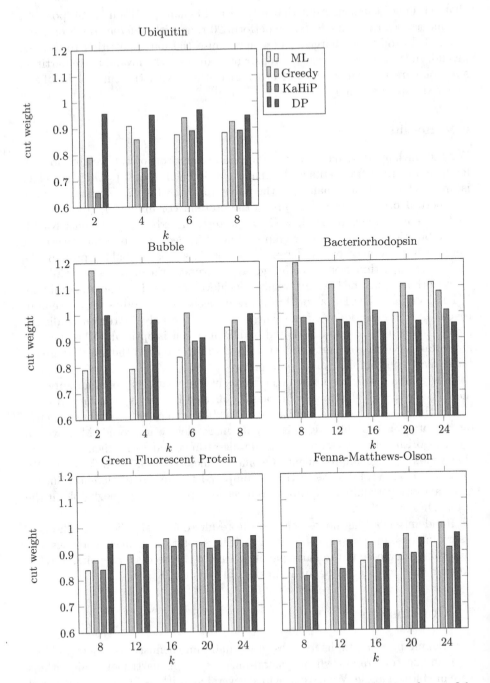

Fig. 2. Comparison of partitions given by several algorithms and proteins, for $\epsilon = 0.1$. The partition quality is measured by the cut weight in comparison to the naive solution.

with a reasonable imbalance, may provide more opportunities for improving the quality of the naive solution further.

References

1. Andreev, K., Racke, H.: Balanced graph partitioning. Theor. Comput. Syst. **39**(6), 929–939 (2006)
2. Buluç, A., Meyerhenke, H., Safro, I., Sanders, P., Schulz, C.: Recent advances in graph partitioning. Accepted as Chapter in AlgorithmEngineering, Overview Paper concerning the DFG SPP 1307 (2016). Preprint available at http://arxiv.org/abs/1311.3144
3. Clauset, A., Newman, M.E.J., Moore, C.: Finding community structure in very large networks. Phys. rev. E **70**(6), 066111 (2004)
4. Cramer, C.J.: Essentials of Computational Chemistry. Wiley, New York (2002)
5. Delling, D., Fleischman, D., Goldberg, A.V., Razenshteyn, I., Werneck, R.F.: An exact combinatorial algorithm for minimum graph bisection. Math. Program. **153**(2), 417–458 (2015)
6. Fedorov, D.G., Kitaura, K.: Extending the power of quantum chemistry to large systems with the fragment molecular orbital method. J. Phys. Chem. A **111**, 6904–6914 (2007)
7. Fedorov, D.G., Nagata, T., Kitaura, K.: Exploring chemistry with the fragment molecular orbital method. Phys. Chem. Chem. Phys. **14**, 7562–7577 (2012)
8. Fiduccia, C., Mattheyses, R.: A linear time heuristic for improving network partitions. In: Proceedings of the 19th ACM/IEEE Design Automation Conference, Las Vegas, NV, pp. 175–181, June 1982
9. Guerra, C.F., Snijders, J.G., te Velde, G., Baerends, E.J.: Towards an order-N DFT method. Theor. Chem. Acc. **99**, 391 (1998)
10. Garey, M.R., Johnson, D.S., Stockmeyer, L.: Some simplified NP-complete problems. In: Proceedings of the 6th Annual ACM Symposium on Theory of Computing (STOC 1974), pp. 47–63. ACM Press (1974)
11. Ghaddar, B., Anjos, M.F., Liers, F.: A branch-and-cut algorithm based on semidefinite programming for the minimum k-partition problem. Ann. OR **188**(1), 155–174 (2011)
12. Gordon, M.S., Fedorov, D.G., Pruitt, S.R., Slipchenko, L.V.: Fragmentation methods: a route to accurate calculations on large systems. Chem. Rev. **112**, 632–672 (2012)
13. He, X., Zhu, T., Wang, X., Liu, J., Zhang, J.Z.H.: Fragment quantum mechanical calculation of proteins and its applications. Acc. Chem. Res. **47**, 2748–2757 (2014)
14. Hendrickson, B., Leland, R.: A multi-level algorithm for partitioning graphs. In: Proceedings Supercomputing 1995, p. 28. ACM Press (1995)
15. Jacob, C.R., Neugebauer, J.: Subsystem density-functional theory. WIREs Comput. Mol. Sci. **4**, 325–362 (2014)
16. Jacob, C.R., Visscher, L.: A subsystem density-functional theory approach for the quantumchemical treatment of proteins. J. Chem. Phys. **128**, 155102 (2008)
17. Jensen, F.: Introduction to Computational Chemistry, 2nd edn. Wiley, Chichester (2007)
18. Kiewisch, K., Jacob, C.R., Visscher, L.: Quantum-chemical electron densities of proteins and of selected protein sites from subsystem density functional theory. J. Chem. Theory Comput. **9**, 2425–2440 (2013)

19. Lanyi, J.K., Schobert, B.: Structural changes in the l photointermediate of bacteriorhodopsin. J. Mol. Biol. **365**(5), 1379–1392 (2007)

20. Ochsenfeld, C., Kussmann, J., Lambrecht, D.S.: Linear-scaling methods in quantum chemistry. In: Lipkowitz, K.B., Cundari, T.R., Boyd, D.B. (eds.) Reviews in Computational Chemistry, vol. 23, pp. 1–82. Wiley-VCH, New York (2007)

21. Olsen, J.G., Flensburg, C., Olsen, O., Bricogne, G., Henriksen, A.: Solving the structure of the bubble protein using the anomaloussulfur signal from single-crystal in-house CuKα diffractiondata only. Acta Crystallogr. Sect. D **60**(2), 250–255 (2004)

22. Ormö, M., Cubitt, A.B., Kallio, K., Gross, L.A., Tsien, R.Y., Remington, S.J.: Crystal structure of the aequorea victoria green fluorescent protein. Science **273**(5280), 1392–1395 (1996)

23. Pavlopoulos, G.A., Secrier, M., Moschopoulos, C.N., Soldatos, T.G., Kossida, S., Aerts, J., Schneider, R., Bagos, P.G.: Using graph theory to analyze biological networks. BioData Min. **4**(1), 1–27 (2011)

24. Ramage, R., Green, J., Muir, T.W., Ogunjobi, O.M., Love, S., Shaw, K.: Synthetic, structural and biological studies of the ubiquitin system: the total chemical synthesis of ubiquitin. Biochem. J. **299**(1), 151–158 (1994)

25. Sanders, P., Schulz, C.: Think locally, act globally: highly balanced graph partitioning. In: Bonifaci, V., Demetrescu, C., Marchetti-Spaccamela, A. (eds.) SEA 2013. LNCS, vol. 7933, pp. 164–175. Springer, Heidelberg (2013)

26. Staudt, C., Sazonovs, A., Meyerhenke, H.: NetworKit: an interactive tool suite for high-performance networkanalysis. CoRR, abs/1403.3005 (2014)

27. Tronrud, D.E., Allen, J.P.: Reinterpretation of the electron density at the site of the eighth bacteriochlorophyll in the fmo protein from pelodictyon phaeum. Photosynth. Res. **112**(1), 71–74 (2012)

28. von Looz, M., Wolter, M., Jacob, C.,Meyerhenke, H.: Better partitions of protein graphs for subsystem quantum chemistry. Technical Report 5, Karlsruhe Institute of Technology (KIT), 3 (2016). http://digbib.ubka.uni-karlsruhe.de/volltexte/1000052814

29. Wesolowski, T.A., Weber, J.: Kohn-Sham equations with constrained electron density: an iterative evaluation of the ground-state electron density of interaction molecules. Chem. Phys. Lett. **248**, 71–76 (1996)

30. Zhang, D.W., Zhang, J.Z.H.: Molecular fractionation with conjugate caps for full quantummechanical calculation of protein-molecule interaction energy. J. Chem. Phys. **119**, 3599–3605 (2003)

Online Algorithm for Approximate Quantile Queries on Sliding Windows

Chun-Nam Yu[1]([✉]), Michael Crouch[2], Ruichuan Chen[3], and Alessandra Sala[2]

[1] Bell Labs, Murray Hill, USA
chun-nam.yu@nokia.com
[2] Bell Labs, Dublin, Ireland
[3] Bell Labs, Stuttgart, Germany

Abstract. Estimating statistical information about the most recent parts of a stream is an important problem in network and cloud monitoring. Modern cloud infrastructures generate in high volume and high velocity various measurements on CPU, memory and storage utilization, and also different types of application specific metrics. Tracking the quantiles of these measurements in a fast and space-efficient manner is an essential task in monitoring the health of the overall system. There are space-efficient algorithms for estimating approximate quantiles under the "sliding window" model of streams. However, they are slow in query time, which makes them less desirable for monitoring applications. In this paper we extend the popular Greenwald-Khanna algorithm for approximating quantiles in the unbounded stream model into the sliding window model, getting improved runtime guarantees over the existing algorithm for this problem. These improvements are confirmed by experiment.

1 Introduction

Existing cloud monitoring systems, e.g. Openstack Ceilometer, Openstack Monasca, and Ganglia, all adopt a similar architecture: an agent deployed at each cloud node collects local performance metrics, which are then sent via a message bus to a backend database for analytics. The database I/O and the bus bandwidth to the database, however, are the primary obstacles to scalability.

By developing lighter-weight algorithms for on-the-fly statistical summaries of large volumes of data, we hope to enable improved anomaly detection and system monitoring applications. We first attempted implementing the Arasu and Manku (AM) algorithm [1] for storing approximate rank information on windows of each stream of system metrics; however, this algorithm was designed primarily to minimize memory space usage. In testing, the amount of processing overhead for each element was prohibitive, particularly for query-intensive workloads.

In this paper, we design and test a more suitable algorithm for approximate quantile/rank reconstruction on sliding time windows. Inspired by the "Greenwald-Khanna (GK) algorithm" [5] for unbounded streams, we design a sliding window algorithm that can answer queries about the last W time units for any W up to a configurable threshold. We perform explicit analysis of the

© Springer International Publishing Switzerland 2016
A.V. Goldberg and A.S. Kulikov (Eds.): SEA 2016, LNCS 9685, pp. 369–384, 2016.
DOI: 10.1007/978-3-319-38851-9_25

time required by the GK algorithm and AM algorithm for processing input elements and answering queries; in the literature these algorithms have typically been analyzed for space performance. We present an algorithm which provides asymptotically improved processing and query time over the AM algorithm.

Both the AM algorithm and our algorithm use the GK algorithm as a subroutine; on "natural" streams, previous experiments [1] have demonstrated that the GK algorithm typically uses asymptotically less space and processing time than its worst-case analysis would suggest (it maintains $O(\frac{1}{\epsilon})$ elements instead of $O(\frac{1}{\epsilon} \log \epsilon n)$). We reanalyze our algorithm and the AM algorithm under this assumption, and find that the query time improvement is still significant, but that the insertion time improvement is much more modest.

We then perform an experimental comparison of our algorithm and the AM algorithm on two real data sets and four synthetic data sets, testing a range of approximation accuracies and window lengths. The experimental data confirmed that our algorithm offers a significant improvement on per-query performance. The improvement on the time required for processing each inserted element is more modest, supporting the version of our analysis performed assuming the experimentally-supported behavior of the GK algorithm.

Our algorithm yields significant improvements in processing time for query-intense workloads. This comes at the expense of the provable worst-case accuracy and space guarantees of the AM algorithm. However, our experimental tests indicate that we achieve comparable accuracy and space performance for every data set we have tested.

2   Streaming Quantile Algorithms

We begin by defining the problem of ϵ-approximate quantiles. We then briefly define the Greenwald-Khanna (GK) algorithm [5] for approximate quantiles, because our algorithm adapts the GK algorithm to the sliding window setting. We define the *rank* of an element x in a stream S as

$$\text{rank}(x) \triangleq \left| \{s \in S \mid x \leq s\} \right|. \tag{1}$$

Definition 1. *In the ϵ-approximate quantiles problem, we are given an accuracy parameter ϵ and then we read a stream of n numeric elements; after this stream, we are given a query parameter $\phi \in [0, 1]$, and we must output some x such that:*

$$\text{rank}(x) \in (\phi \pm \epsilon)n \tag{2}$$

(Equivalently, we could think of the query parameter as a "target rank" r_0, in which case our goal is $\text{rank}(x) \in r_0 \pm \epsilon n$.)

The survey [2] contains an excellent summary of the history of quantile approximation algorithms; we are interested in those with small memory usage. The most relevant approach for our analysis is a deterministic approximation algorithm using space $O(\frac{1}{\epsilon} \log n)$, called the "GK algorithm" for its authors, Greenwald and Khanna [5]. Our work adapts the GK algorithm to the setting of sliding window streams (Sect. 3), so we review the algorithm here.

The GK algorithm stores an ordered list of values from the stream; for each value v stored, we can quickly calculate two values $r^-(v)$, $r^+(v)$ which bound the true rank $\text{rank}(v)$ within a range ϵn:

$$r^-(v) \leq \text{rank}(v) \leq r^+ \leq r^-(v) + \epsilon n \tag{3}$$

In order to reconstruct good rank approximations, we will also require that the set $\{v_1, \ldots, v_i, \ldots, v_k\}$ of stored values is "sufficiently dense": for all i we require

$$r^-(v_i) \leq r^-(v_{i+1}) \leq r^-(v_i) + \epsilon n \tag{4}$$

It is shown in [5] that maintaining such a list, and deleting as many elements as possible while preserving invariants (3) and (4), suffices to answer the ϵ-approximate quantile problem using storage equal to $O(\frac{1}{\epsilon} \log \epsilon n)$ stream elements.

It would be possible to store the rank bounds of each v_i directly; however, whenever a new element x was read, we would need to update these values for every $v_i > x$. The main idea of the GK algorithm is to instead store "local" information about rank.

2.1   GK Algorithm

The GK algorithm maintains a sorted list S of triples $s_i = (v_i, g_i, \Delta_i)$ where:

- v_i is a value from the input stream; $v_1 \leq v_2 \leq \cdots \leq v_{|S|}$.
- g_i is $r^-(v_i) - r^-(v_{i-1})$ (the difference in minimum rank between neighbors).
- Δ_i is $r^+(v_i) - r^-(v_i)$ (the uncertainty in the rank of v_i).

Input. When a new value x is read, if x is the smallest or largest value seen so far, we insert the bucket $(x, 1, 0)$ at the beginning or end of S respectively. Otherwise, we locate the largest $v_i < x$, and insert the bucket $(x, 1, g_i + \Delta_i - 1)$ as s_{i+1}.

Compression. After every $\lfloor \frac{1}{2\epsilon} \rfloor$ insertions, there is a compression pass; we iterate through tuples in increasing order of i, "merging" every pair of adjacent tuples where $g_i + g_{i+1} + \Delta_{i+1} \leq 2\epsilon n$ according to the rule:

$$\text{merge}\big((v_i, g_i, \Delta_i), (v_{i+1}, g_{i+1}, \Delta_{i+1})\big) = (v_{i+1}, g_i + g_{i+1}, \Delta_{i+1}) \tag{5}$$

The first and last tuples are never merged.

Query. Note that

$$r^-(v_i) = \sum_{j \leq i} g_j \qquad r^+(v_i) = \Delta_i + \sum_{j \leq i} g_j \tag{6}$$

To query for an element of approximate rank r, we return v_i for the smallest i such that $r^-(v_i) \geq r - \epsilon n$. The merge criterion guarantees that such an element always exists, and that $r^+(v_i) \leq r + \epsilon n$.

Fact 2 ([5]). *The GK Algorithm solves the ϵ-approximate quantile problem using the memory space required to store $O(\frac{1}{\epsilon} \log \epsilon n)$ stream elements.*

The GK algorithm is very efficient in terms of space and processing time in practice, but only works for calculating quantiles on the entire history of the data stream. Several later works further developed quantile algorithms in this model [3, 10, 11]; however, for system monitoring purposes (and many other applications) we are interested in keeping an analysis of *recent* items up to date, while discarding the effects of older items.

3   Sliding Windows and Exponential Histograms

The sliding window model, introduced by Datar et al. [4], allows us to model infinite data streams when our goal is to compute properties of the most recent data. We are given a "window length" W before reading the stream, and queries are always intended to reflect the W most recently read elements.[1]

Quantile approximation in the sliding window model was first addressed by Lin et al. [6], who achieved space usage $O(\frac{\log(\epsilon W)}{\epsilon} + \frac{1}{\epsilon^2})$. Arasu and Manku [1] (AM) improved this to $O(\frac{1}{\epsilon} \log^2 \frac{1}{\epsilon} \log(\epsilon W))$, but in our application testing the time it required to process each element was too large; briefly, it replicates a window of length W into $O(\log 1/\epsilon)$ copies with subwindows of different sizes. Both of these algorithms are based on splitting the window into smaller subwindows and using GK sketches to summarize each subwindow. Mousavi and Zaniolo were able to provide improved runtime for the related problem of estimating equi-depth histograms [7, 8], but that problem is more restricted [7].

We instead test an adaptation of the original GK algorithm, using a sliding window algorithm called *exponential histograms* [4]. An exponential histogram is a data structure for the problem of maintaining the number of non-zero elements seen in the last W elements of a stream.

3.1   Exponential Histograms

An exponential histogram (EH) maintains a sorted list A of contiguous intervals; each interval $a_i = (t_i, d_i, k_i)$ where:

- t_i is the timestamp when the interval begins; $t_1 \leq t_2 \leq \cdots \leq t_{|A|}$.
- a_i covers the interval of observations from t_i to $t_i + d_i$.
- $t_i + d_i < t_{i+1}$.
- k_i is the count of nonzero observations over the interval (and a power of 2).

Input. Whenever a new observation is seen at time t, the interval $(t, 0, 1)$ is added as the last entry in A.

---

[1] For convenience of analysis we treat W as fixed; however, like many algorithms in this model, ours is easily adapted to answer queries about any window size $w \leq W$. For applications, W can thus be thought of as the *maximum* history length of interest.

Compression. Whenever there are more than $1/\epsilon$ intervals containing a particular count k, pairs of intervals are merged by replacing $(t_i, d_i, k), (t_{i+1}, d_{i+1}, k)$ with $(t_i, t_{i+1} - t_i + d_{i+1}, 2k)$. These merges occur from "most recent" to "least recent"; intuitively, older elements are thus kept in intervals of progressively larger counts (and coarser time resolution).

When $t_2 \leq t - W$ for the current time t, the first interval (t_1, d_1, k_1) is discarded. This ensures that at any time t, we are storing a set of intervals which fall entirely within the "active window" of the last W elements, plus possibly an interval which falls partly inside and partly outside the active window.

Query. Let C be the number of nonzero elements seen during the last W observations. We then have that C satisfies

$$\sum_{i=2}^{|A|} k_i \leq C \leq \sum_{i=1}^{|A|} k_i \tag{7}$$

The merge rule guarantees that the fraction of elements in the partial box (and thus the approximation uncertainty) is $O(1/\epsilon)$ fraction of the total. We then have

Fact 3 ([4]). *The exponential histogram algorithm maintains a $(1+\epsilon)$ approximation of the number of nonzero elements seen in the last W inputs, using space $O(\frac{1}{\epsilon} \log W)$.*

4   Algorithm

The GK algorithm stored tuples (v_i, g_i, Δ_i), where v_i was a stream element and g_i, Δ_i were integers. We wish to replace g_i and Δ_i, so that we know the rank of v_i only *considering the window of active elements*. We do this below, replacing the g_i and Δ_i counters with exponential histograms.

We also wish to ensure that each v_i value corresponds to an element *still within the active window*. We will exploit the connection between g_i and v_i: in the original algorithm, the g_i counter stores the number of observations which have been merged into the tuple (v_i, g_i, Δ_i), and v_i is the largest of those observations.

4.1   EH-with-max

We first describe a modified version of the EH structure, which will not only maintain a count of how many elements have been seen by the g_i structure, but be able to return one of the largest of these elements.

Let an EH-with-max structure be an exponential histogram (Sect. 3.1), modified in the following ways:

- Each interval a_i also stores a value v_i, which is the numerically largest value seen during that interval. This value is updated in the obvious ways.
- The structure supports an additional query operation, returning an *approximate maximum* $\max_{i=2}^{|A|} v_i$.

Fact 4. *The "approximate maximum" operation returns an element larger than all but some ϵ fraction of the nonzero elements in the active window.*

4.2   Our Algorithm

Our algorithm is a list of pairs (G_i, D_i), where G_i is an EH-with-max structure and D_i is an EH structure. At time t, we let:

- $v_i(t)$ denote the approximate maximum value from G_i at time t;
- $g_i(t)$ denote the count of G_i at time t;
- $\Delta_i(t)$ denote the count of D_i at time t.

Here we use ϵ_1 for the approximation factor of the underlying GK algorithm, and approximation factor ϵ_2 for all G_i and D_i.

We define two helper operations, merge and tail.

merge: The merge operation combines two EH structures. We follow Papapetrou et al. [9]'s technique for merging exponential histogram structures – before merging, we replace each interval (t_i, d_i, k_i) with appropriately weighted endpoints $(t_i, 1, k_i/2)$ and $(t_i + d_i, 1, k_i/2)$. We take the union of the resulting lists, and perform EH compression as normal.

For worst-case inputs the resulting aggregation can have additively increasing error [9]; however, as in that reference, we found that this problem did not arise in practice.

tail: The tail operation takes an EH structure and removes the last bucket $a_{|A|}$ (containing the most recent non-zero observation).

Input. When reading a value v, we add a new pair after the largest i such that $v_i(t) < v$. The G structure of the new tuple is an EH-with-max sketch containing the single point v at time t. The D structure of the new tuple is merge(D_i, tail(G_i)).

Compression. The compression operation is described in Algorithm 1.

Query. To query for an element of approximate rank r, we find the smallest i such that $\sum_j g_j(t) \geq r + \epsilon_1 W$, and return $\max_{j \leq i} v_i(t)$.

Algorithm 1. COMPRESS(), s the size of sketch

---

   for i from $s - 2$ to 0 **do**
     Let j be the index of larger of v_i and v_{i+1}
     if $g_i(t) + g_{i+1}(t) + \Delta_j(t) < 2\epsilon_1 W$ **then**
       $g_{i+1} = \texttt{merge}(g_i, g_{i+1})$
       $v_{i+1} = v_j$
       $\Delta_{i+1} = \Delta_j$
       Delete tuple (v_i, g_i, Δ_i)
     end if
   end for

---

5   Correctness

The simplifications in the merging condition (and our definition of EH merging) allowed us improved runtime performance and decreased development and maintenance cost, but they come at some cost of the ease of formal analysis of our algorithm. In particular, the approximation error for the EH sketches G_i, D_i can increase after each merge, from ϵ to $2\epsilon + \epsilon^2$ in the worst case. In practice we observe that the merge operation we use for EH sketches perform excellently, and the approximation error of the EH sketches hardly increases at all after multiple merges during the run of the algorithm. In the correctness analysis below we give error bounds on quantile queries conditional on the approximation error of the EH sketches at query time t. If an improved merge procedure with better approximation guarantee is available, it can be directly applied to our sliding window GK algorithm to improve the approximation bound.

We use $\mathbb{G}_i$, $\mathbb{D}_i$ to refer to the sets of elements being tracked by the EH sketches G_i and D_i, and $\mathbb{G}_i(t)$, $\mathbb{D}_i(t)$ to refer to their values at time t. We make the following assumption on the approximation quality of the EH sketches. At query time t, for all i

$$(1 - \epsilon_2')g_i(t) \leq |\mathbb{G}_i(t)| \leq (1 + \epsilon_2')g_i(t) \tag{8}$$

$$|\mathbb{D}_i(t)| \leq (1 + \epsilon_2')\Delta_i(t) \tag{9}$$

Here $g_i(t)$ and $\Delta_i(t)$ refer to the approximate counts from EH sketches while $|\mathbb{G}_i(t)|$ and $|\mathbb{D}_i(t)|$ are the exact answers. Notice we use ϵ_2' to differentiate against the precision parameter ϵ_2 for G_i and D_i, as the actual approximation error ϵ_2' can become bigger than ϵ_2 after multiple merging of EH sketches (although it rarely happens in practice).

We also assume $v_i(t)$ is the approximate maximum of the elements tracked by the EH sketch G_i, such that at least $1 - \epsilon_2'$ fraction of elements in $\mathbb{G}_i(t)$ are less than or equal to $v_i(t)$.

Theorem 5. *Correctness of Quantile: The query procedure returns a value v with rank between $(q - (\epsilon_1 + 2\epsilon_2'))W$ and $(q + (\epsilon_1 + 2\epsilon_2'))W$.*

Proof. See Appendix.

This result shows that our algorithm incurs only an extra error of $2\epsilon'_2$ from the use of approximate counting, compared to the GK algorithm in the unbounded stream model.

6   Analysis

Our key design consideration was improving on the per-update time complexity of the AM algorithm. To the authors' knowledge there is not an explicit amortized run-time analysis of the AM algorithm or even of the GK algorithm in the research literature; these are thus included.

6.1   Arasu-Manku Time Analysis

The GK algorithm [5] maintains approximate quantile summaries on an unbounded stream of elements. The AM algorithm [1] uses GK algorithm sketches as building blocks to perform sliding window quantile summaries. We thus begin by analyzing the GK algorithm.

Lemma 6. *The GK algorithm [5], reading N elements, has amortized update time complexity $O(\log N)$ and query complexity $O(\log \frac{1}{\epsilon} + \log \log \epsilon N)$.*

Proof. The GK algorithm maintains a sorted list of tuples, and iterates through the list performing a compression operation once every $\frac{1}{2\epsilon}$ insertions. If the list maintained contains s tuples, insertion can thus be done in time $O(\log s)$; the compression operation requires amortized time $O(\epsilon s)$.

For worst-case inputs, the list may contain $s = O(\frac{1}{\epsilon} \log \epsilon N)$ tuples [5], yielding an amortized time complexity of $O\big(\log(\frac{1}{\epsilon} \log \epsilon N) + \log \epsilon N\big) = O(\log N)$.

Queries are performed by accessing the sorted list of s tuples, and thus require space $O(\log s) = O(\log \frac{1}{\epsilon} + \log \log \epsilon N)$.   □

Note that for inputs seen in practice, experiments indicate that the GK list typically contains $s = O(\frac{1}{\epsilon})$ tuples [1], yielding amortized input time and query time of $O(\log \frac{1}{\epsilon})$. In an attempt to understand how the algorithms are likely to perform in practice, we will analyze the quantile-finding algorithms using both the worst-case bound and the experimentally supported bound.

Theorem 7. *The Arasu-Manku Algorithm [1] with error parameter ϵ and window length W has amortized update time $O(\log \frac{1}{\epsilon} \log W)$ and query time $O(\frac{1}{\epsilon} \log \frac{1}{\epsilon})$.*

Proof. The reader should see [1] for a full description of their algorithm. Briefly, it keeps sketches of the active window at each of $L = \lceil \log_2 \frac{4}{\epsilon} \rceil + 1$ levels. (Over the range of $0.001 \le \epsilon \le 0.05$ tested in our experiments, L ranges from 8 to 13; avoiding this factor is the key to our runtime improvement).

For each level ℓ in $0 \leq \ell < L$, the active window is partitioned into $\frac{4}{\epsilon 2^\ell}$ blocks, each containing $n_\ell = \frac{\epsilon W}{4} 2^\ell$ elements. Within the most recent block at each level, a GK sketch is maintained with error parameter $\epsilon_\ell = \frac{\epsilon}{4L} 2^{L-\ell}$. Using the analysis of the GK algorithm above and ignoring constant factors, we find worst-case amortized update time

$$\sum_{\ell=0}^{L-1} \log n_\ell = O(\sum_{\ell=0}^{L-1} \log \epsilon W 2^\ell) = O(L^2 + L \log \epsilon W)$$

$$= O(\log^2 \tfrac{1}{\epsilon} + \log \tfrac{1}{\epsilon} \log \epsilon W) = O(\log \tfrac{1}{\epsilon} \log W) \qquad \square$$

Blocks which are not the newest on each level are maintained only as summary structures (simple lists of quantiles), each requiring space $O(\frac{1}{\epsilon})$. Performing a query can require accessing one block at each level, and merging the lists can be done in linear time; we thus find a worst-case query time of $O(\frac{L}{\epsilon}) = O(\frac{1}{\epsilon} \log \frac{1}{\epsilon})$.

Via a similar analysis we find

Theorem 8. *Assuming the experimentally derived [1] GK space usage of $O(\frac{1}{\epsilon})$, the Arasu-Manku algorithm [1] has amortized update time $O(\log \frac{1}{\epsilon} \log \log \frac{1}{\epsilon})$ for randomly ordered inputs.*

6.2   Our Time Analysis

Our algorithm takes advantage of the $O(1)$ amortized update time of exponential histograms [4].

Theorem 9. *Our algorithm has worst-case update time $O(\log \log W + \epsilon \log W)$ and query time $O(\log \log W)$.*

Proof. Again, assume the algorithm maintains s tuples. On reading an input, we must find the appropriate EH structure, in time $O(\log s)$, and update it, in time $O(1)$. We also perform a linear-time compression sweep every $O(\frac{1}{\epsilon})$ inputs. Our algorithm thus has total amortized update time $O(\log s + \epsilon s)$.

The top level of our algorithm is a GK structure on the W live elements, plus possibly some expired elements. Any expired elements which are part of the structure will be removed when found; thus, any slowdown due to expired elements is limited to amortized $O(1)$. For the remaining elements, substituting $s = O(\log W)$ from Lemma 6, we find amortized time $O(\log \log W + \epsilon \log W)$.

Querying our structure requires a simple binary search through the sorted list of EH structures, which is time $O(\log s) = O(\log \log W)$, and a query of the value of the EH structure, which is time $O(1)$ [4]. $\qquad \square$

Similarly, we find

Theorem 10. *Assuming the experimentally derived [1] GK space usage of $O(\frac{1}{\epsilon})$, our algorithm has amortized update time $O(\log \frac{1}{\epsilon})$ and query time $O(\log \log \frac{1}{\epsilon})$ for randomly ordered inputs.*

	Worst-case		Experimental	
	Insertion	**Query**	**Insertion**	**Query**
Arasu-Manku	$\log \frac{1}{\epsilon} \log W$	$\frac{1}{\epsilon} \log \frac{1}{\epsilon}$	$\log \frac{1}{\epsilon} \log \log \frac{1}{\epsilon}$	$\frac{1}{\epsilon} \log \frac{1}{\epsilon}$
Our Algorithm	$\log \log W + \epsilon \log W$	$\log \log W$	$\log \frac{1}{\epsilon}$	$\log \log \frac{1}{\epsilon}$

Fig. 1. Summary of the time complexity of operations in the two algorithms. The "experimental" column performs the analysis assuming the experimental observation of [1] that the GK algorithm appears to use space $O(\frac{1}{\epsilon})$ for randomly-ordered inputs.

Figure 1 summarizes the time complexity results of this section. Our algorithm does provide improved asymptotic complexity in insertion operations, but particularly striking is the removal of the $\frac{1}{\epsilon}$ factor in the query complexity. The improvement in insertion complexity is significant in the worst-case analysis, but amounts to only a factor of $\log \log \frac{1}{\epsilon}$ under experimentally-supported assumptions about the behavior of the GK algorithm; thus, we would expect the gain in insertion times to be modest in practice.

7   Experiments

We implemented our sliding window GK algorithm (SW-GK) and compared it against the approach from [1] (AM). We expected our algorithm to demonstrate improved processing and query time, particularly for small values of ϵ. Both algorithms were implemented in C++; experiments were run on an AMD Opteron 2.1 GHz CPU server with 128 Gb of memory running Redhat Linux Server 6.4.

The algorithms were compared on two real datasets and four synthetic datasets. The first real dataset was a set of Cassandra database object read time data; it notably had a long tail and a lot of outliers. The second dataset is from the set of numbered minor planets from the Minor Planet Center[2] and contains the right ascension of different celestial objects. The four synthetic datasets were generated from a normal distribution, from a uniform distribution on the interval [0,1], and from Pareto distributions with tail index $\alpha = 1$ and $\alpha = 3$, for evaluating the algorithms under a diverse set of distributions. All datasets contained 10^7 entries.

We confirmed our expectation that our algorithm's runtime would grow more slowly than AM's as ϵ decreases. Figure 2 shows the CPU time of maintaining SW-GK and AM on the Cassandra dataset (inserts only, no queries) for varying values of W and ϵ. For $W = 10^4$ and $W = 10^5$, SW-GK runs faster than AM for most settings of ϵ. For $W = 10^6$, AM is faster except for the highest precision $\epsilon = 0.001$. Overall, both algorithms are reasonably fast and use on average less than 200 s to process 10^7 stream elements, i.e., less than 0.02 ms to process each element. Time performance was similar on the other data sets (omitted for lack of space).

---

[2] http://www.minorplanetcenter.net/iau/ECS/MPCAT-OBS/MPCAT-OBS.html.

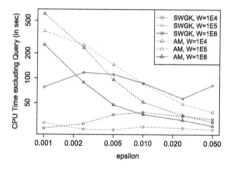

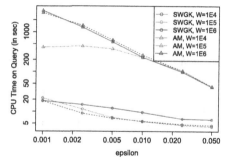

Fig. 2. CPU time without query of SW-GK and AM on the Cassandra dataset

Fig. 3. Query time of SW-GK and AM on the Cassandra dataset

Our algorithm showed an even more dramatic improvement in query time. We simulated a query-heavy workload by running quantile queries from 1 % up to 100 % in 1 % increments every 2000 stream elements. SW-GK is an order of magnitude faster than AM on query time, and close to 2 orders of magnitude faster on the highest precision 0.001 (Fig. 3). AM is slow because it needs to pool the quantile query results from many small GK sketches and then combine them by sorting. The query time complexity of SW-GK is similar to standard GK because the only major difference in the query function is the replacement of counts with EH counting (which answers count queries in constant time). The query time is fairly independent of window size for both algorithms. SW-GK is clearly the method of choice if query time is a concern. Again, performance was similar on the other data sets.

We also verified that our algorithm offers comparable space and accuracy performance to the AM algorithm. Figure 4 shows the average space usage on the Cassandra dataset, in terms of the number of tuples stored (sampled every

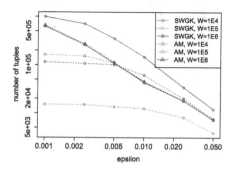

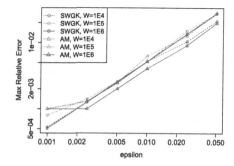

Fig. 4. Space usage of SW-GK and AM on the Cassandra dataset

Fig. 5. Maximum relative error of SW-GK and AM on the Cassandra dataset

2000 iterations). Each tuple contains 3 integer/float values (4 bytes each). We can see that except for the largest window size 10^6, SW-GK uses less or roughly the same amount of space as AM. Figure 5 shows the maximum error of SW-GK and AM for different precision and window sizes. We achieve better accuracy for small ϵ, though both algorithms are well within the ϵ bound.

Figure 6 shows the space usage, maximum relative error, CPU time and query time of SW-GK and AM on all 6 datasets, for $\epsilon = 0.01$ and $W = 10^5$. We can see that the results of both SW-GK and AM are very stable across the different datasets. The same is true for other settings of ϵ and W (not shown here due to space limitations).

7.1   Sorted Inputs

Because of the GK algorithm's improved space behavior on randomly ordered inputs, we tested the effect of ordered inputs on both algorithms. Figure 7 shows the space usage, maximum relative error, CPU time and query time on the Cassandra dataset sorted in ascending, descending and original order. SW-GK uses slightly more space in the sorted datasets compared to the original order, while AM stays constant. The maximum error stays constant for SW-GK, while AM has slight smaller error for the sorted datasets. Both algorithms run a little bit faster on the sorted datasets than the original ordering. Overall both algorithms have fairly consistent performance across different ordering of observations.

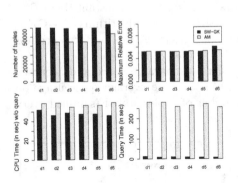

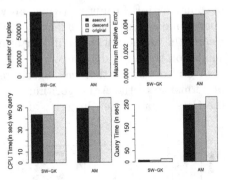

Fig. 6. Space, maximum relative error, CPU time and query time of SW-GK and AM on several datasets. d1:Cassandra, d2:Normal, d3:Pareto1, d4:Pareto3, d5:Uniform, d6:Planet

Fig. 7. Space, maximum relative error, CPU time and query time of SW-GK and AM on Cassandra data in ascending, descending, and original order

8   Conclusions

We have designed a sliding window algorithm for histogram maintenance with improved query time behavior, making it more suitable than the existing algorithm for applications such as system monitoring. Experimental comparison confirms that our algorithm provides competitive space usage and approximation error, while improving the runtime and query time.

Appendix: Correctness Analysis

Proof of Theorem 5

We use $\mathbb{G}_i$, $\mathbb{D}_i$ to refer to the sets of elements being tracked by the EH sketches G_i and D_i. We can define $\mathbb{G}_i$ and $\mathbb{D}_i$ as the value-timestamp pairs $\{(v, t), (v', t'), \ldots\}$ of all elements ever added to the EH sketches G_i and D_i. We use $\mathbb{G}_i(t)$ and $\mathbb{D}_i(t)$ to denote the set of value-timestamp pairs in $\mathbb{G}_i$ and $\mathbb{D}_i$ that has not expired at time t. We can think of $\mathbb{G}_i(t)$ and $\mathbb{D}_i(t)$ as exact versions of the EH sketches G_i and D_i, and they are useful in establishing our correctness claims.

We state without proof the following two claims:

Claim 1: At all time t, the set of $\mathbb{G}_i(t)$'s partition the set of all observations in the current window $[t - W, t]$.

Claim 2: At all time t, for all i, $\mathbb{D}_i(t) \subseteq \cup_{j>i}\mathbb{G}_j(t)$.

Claim 1 is true because all elements inserted started out as singleton sets of $\mathbb{G}_i(t)$, and subsequent merging in COMPRESS always preserves the disjointness of the $\mathbb{G}_i(t)$ and never drops any elements. Claim 2 is true because by the insertion rule, at the time of insertion $\mathbb{D}_i(t)$ is constructed from merging some $\mathbb{G}_{i+1}(t)$ and $\mathbb{D}_{i+1}(t)$. By unrolling this argument, $\mathbb{D}_{i+1}(t)$ is constructed from $\mathbb{G}_j(t)$ and $\mathbb{D}_j(t)$ with $j > i + 1$. Since $\mathbb{D}_i(t)$ starts as an empty set initially and none of the insertion and merge operations we do reorder the sets $\mathbb{G}_i(t)$, the elements in $\mathbb{D}_i(t)$ have to come from the sets $\mathbb{G}_j(t)$ for $j > i$.

Lemma 11. *At all t, for all i, all elements in $\cup_{j>i}\mathbb{G}_j(t)\setminus\mathbb{D}_i(t)$ have values greater than or equal to $v_i(t)$.*

Proof. We prove this by induction on t, and show that the statement is preserved after INSERT, expiration, and COMPRESS. As the base case for induction, the statement clearly holds initially before any COMPRESS operation, when all $\mathbb{G}_j$ are singletons and $\mathbb{D}_j$ are empty.

We assume at time t, an element is inserted, then an expiring element is deleted, then the timestamp increments.

INSERT: Suppose an observation v is inserted at time t between (v_{i-1}, G_{i-1}, D_i) and (v_i, G_i, D_i). We insert the new tuple $(v, EH(v, t), \mathtt{merge}(D_i, \mathtt{tail}(G_i)))$ into our summary. Here $EH(v, t)$ refers to the EH sketch with a single element v added at time t. In the set notation, this corresponds to inserting $(v, \mathbb{G} = \{(v, t)\}, \mathbb{D} = (\mathbb{G}_i\setminus\{v_i\}) \cup \mathbb{D}_i)$.

We assume the statement holds before insertion of v. For $r < i$, before insertion we know elements in $\cup_{j>r}\mathbb{G}_j(t)\setminus\mathbb{D}_r(t)$ are all greater than or equal to v_r by the inductive hypothesis. After insertion the new set becomes

$(\cup_{j>r}\mathbb{G}_j(t)\backslash\mathbb{D}_r(t))\cup\{(v,t)\}$, which maintains the statement because by the insertion rule we know $v_r \leq v$ for all $r < i$.

For $r \geq i$, insertion of v does not change the set $\cup_{j>r}\mathbb{G}_j(t)\backslash\mathbb{D}_r(t)$ at all, so the statement continues to hold.

At the newly inserted tuple v, we know $v < v_i$, and all elements in $\cup_{j>i}\mathbb{G}_j(t)\backslash\mathbb{D}_i(t)$ are greater than or equal to v_i by the inductive hypothesis. So all elements in $\cup_{j>i}\mathbb{G}_j(t)\backslash\mathbb{D}_i(t)$ are greater than v.

At v, the set in the statement becomes

$$\cup_{j\geq i}\mathbb{G}_j(t)\backslash((\mathbb{G}_i(t)\backslash\{v_i\})\cup\mathbb{D}_i(t))$$
$$=(\cup_{j>i}\mathbb{G}_j(t)\backslash\mathbb{D}_i(t))\cup\{v_i\}$$

All elements in this set are greater than or equal to v, so the statement holds for v as well.

EXPIRE: When the timestamp increments to $t + 1$, one of the elements v expires. Pick any v_i, the expiring element v can be in any one of the following 3 sets:

1. $\cup_{j\leq i}\mathbb{G}_j(t)$
2. $\mathbb{D}_i(t)$
3. $\cup_{j>i}\mathbb{G}_j(t)\backslash\mathbb{D}_i(t)$

By Claims 1 and 2, these 3 sets are disjoint and contain all observations in the current window. Assuming $v \neq v_i$, if v comes from set 1, then $\cup_{j\leq i}\mathbb{G}_j(t+1)$ decrease by 1 but does not affect the set $\cup_{j>i}\mathbb{G}_j(t+1)\backslash\mathbb{D}_i(t+1)$ in our statement. If v comes from set 2, then $\cup_{j>i}\mathbb{G}_j(t+1)\backslash\mathbb{D}_i(t+1)$ remains unchanged as v is contained in both $\mathbb{D}_i(t)$ and $\cup_{j>i}\mathbb{G}_j(t)$ (Claim 2). If v comes from set 3, then $\cup_{j>i}\mathbb{G}_j(t+1)\backslash\mathbb{D}_i(t+1)$ decreases by 1, the number of elements greater than v_i decreases by 1. The statement still holds in all these cases.

If $v = v_i$ is the expiring element, then at $t + 1$ there is another observation v' in the EH G_i that becomes the maximum element in G_i. But we know $v' \leq v_i$ as v_i is the maximum element in G_i before expiration, so the elements in $\cup_{j>i}\mathbb{G}_j(t+1)\backslash\mathbb{D}_i(t+1)$ which are greater than v_i are also greater than v', and the statement holds.

COMPRESS: Suppose the COMPRESS step merges two tuples $(v_{i-1}, G_{i-1}, D_{i-1})$ and (v_i, G_i, D_i). For $r > i$, this does not affect the set $\cup_{j>r}\mathbb{G}_j(t)\backslash\mathbb{D}_r(t)$. For $r < i - 1$, this does not affect the set $\cup_{j>r}\mathbb{G}_j(t)\backslash\mathbb{D}_r(t)$ as the deletion of $\mathbb{G}_{i-1}$ is compensated by setting $\mathbb{G}_i = \mathbb{G}_{i-1}\cup\mathbb{G}_i$. For $r = i$, if $v_i = \max(v_{i-1}, v_i)$ then the set $\cup_{j>i}\mathbb{G}_j(t)\backslash\mathbb{D}_i(t)$ does not change. Since v_i does not change either the statement holds after merging.

If $v_{i-1} = \max(v_{i-1}, v_i)$ (which is possible with inversion), then by inductive hypothesis we know $\cup_{j>i-1}\mathbb{G}_j(t)\backslash\mathbb{D}_{i-1}(t)$ contains elements that are greater than or equal to v_{i-1}. After merging by setting $v_i = v_{i-1}, \mathbb{G}_i = \mathbb{G}_{i-1}\cup\mathbb{G}_i, \mathbb{D}_i = \mathbb{D}_{i-1}$, the set in the statement becomes $\cup_{j>i}\mathbb{G}_j(t)\backslash\mathbb{D}_{i-1}(t)$, which is a subset of $\cup_{j>i-1}\mathbb{G}_j(t)\backslash\mathbb{D}_{i-1}(t)$. Therefore all elements in it are greater than or equal to v_{i-1} after merging.   □

Lemma 12. *At all time t, for all i, at least $1 - \epsilon'_2$ fraction of elements in the set $\cup_{j \le i} \mathbb{G}_j(t)$ have values less than or equal to $\max_{j \le i} v_j(t)$.*

Proof. For each individual $\mathbb{G}_j(t)$, by the property of tracking approximate maximum by our EH sketch G_j, $1 - \epsilon'_2$ fraction of the elements in $\mathbb{G}_j(t)$ are less than $v_j(t)$.

Taking union over $\mathbb{G}_j(t)$ and maximum over $v_j(t)$, we obtain the lemma. □

Theorem 5. *Correctness of Quantile: The query procedure returns a value v with rank between $(q - (\epsilon_1 + 2\epsilon'_2))W$ and $(q + (\epsilon_1 + 2\epsilon'_2))W$.*

Proof. We maintain the invariant: at all time t, for all i

$$g_i(t) + \Delta_i(t) \le 2\epsilon_1 W. \tag{10}$$

The function QUANTILE returns $v = \max_{j \le i} v_i(t)$, where i is the minimum index such that $\sum_{j \le i} g_j(t) \ge (q - \epsilon_1)W$. Suppose $v = v_p$, $p \le i$.

By Lemma 12, there are at least $(1 - \epsilon'_2) \sum_{j \le i} |\mathbb{G}_j(t)|$ elements less than or equal to $v_i(t)$ (and hence v). Now

$$
\begin{aligned}
&(1 - \epsilon'_2) \sum_{j \le i} |\mathbb{G}_j(t)| \\
&\ge \sum_{j \le i} |\mathbb{G}_j(t)| - \epsilon'_2 W && [\text{as } \sum_{j \le i} |\mathbb{G}_j(t)| \le W] \\
&\ge (1 - \epsilon'_2) \sum_{j \le i} g_j(t) - \epsilon'_2 W && [\text{by Eq. 8}] \\
&\ge \sum_{j \le i} g_j(t) - 2\epsilon'_2 W && [\text{as } \sum_{j \le i} g_j(t) \le W] \\
&\ge (q - \epsilon_1)W - 2\epsilon'_2 W \\
&= (q - (\epsilon_1 + 2\epsilon'_2))W
\end{aligned}
$$

Therefore v has minimum rank of $(q - (\epsilon_1 + 2\epsilon'_2))W$.

By Lemma 11, there are at least $\sum_{j > p} |\mathbb{G}_j(t)| - |\mathbb{D}_p(t)|$ elements greater than or equal to $v = v_p$. The maximum rank of v is

$$
\begin{aligned}
&W - (\sum_{j > p} |\mathbb{G}_j(t)| - |\mathbb{D}_p(t)|) \\
&= \sum_{j \le p} |\mathbb{G}_j(t)| + |\mathbb{D}_p(t)| && [\sum_j |\mathbb{G}_j(t)| = W] \\
&= \sum_{j < p} |\mathbb{G}_j(t)| + |\mathbb{G}_p(t)| + |\mathbb{D}_p(t)| \\
&\le (1 + \epsilon'_2) \sum_{j < p} g_j(t) \\
&\quad + (1 + \epsilon'_2)(g_p(t) + \Delta_p(t)) \\
&\le (1 + \epsilon'_2)(q - \epsilon_1)W + (1 + \epsilon'_2)(2\epsilon_1 W) \\
&\le (q + (\epsilon_1 + \epsilon'_2 + \epsilon_1 \epsilon'_2))W \\
&\le (q + (\epsilon_1 + 2\epsilon'_2))W && [\text{since } \epsilon_1 < 1]
\end{aligned}
$$

The inequality from the third last line comes from the invariant in Eq. 10 and the fact that $i \ge p$ is the minimum index with $\sum_{j \le i} g_j(t) \ge (q - \epsilon_1)W$, so $\sum_{j < p} g_j(t)$ has to be strictly less than $(q - \epsilon_1)W$. Therefore $v = v_p$ has maximum rank of $(q + (\epsilon_1 + 2\epsilon'_2))W$. Together with the minimum rank of v, this shows v gives an $(\epsilon_1 + 2\epsilon'_2)$-approximation to the quantile query problem on the qth quantile. □

References

1. Arasu, A., Manku, G.S.: Approximate counts and quantiles over sliding windows. In: PODS, pp. 286–296. ACM (2004)
2. Buragohain, C., Suri, S.: Quantiles on streams. In: Liu, L., Özsu, M.T. (eds.) Encyclopedia of Database Systems, pp. 2235–2240. Springer, New York (2009)
3. Cormode, G., Muthukrishnan, S.: An improved data stream summary: the count-min sketch and its applications. J. Algorithms **55**(1), 58–75 (2005)
4. Datar, M., Gionis, A., Indyk, P., Motwani, R.: Maintaining stream statistics over sliding windows. SIAM J. Comput. **31**(6), 1794–1813 (2002)
5. Greenwald, M., Khanna, S.: Space-efficient online computation of quantile summaries. In: ACM SIGMOD Record, vol. 30, pp. 58–66. ACM (2001)
6. Lin, X., Hongjun, L., Jian, X., Yu, J.X.: Continuously maintaining quantile summaries of the most recent n elements over a data stream. In: ICDE, pp. 362–373. IEEE (2004)
7. Mousavi, H., Zaniolo, C.: Fast and accurate computation of equi-depth histograms over data streams. In: EDBT, pp. 69–80. ACM (2011)
8. Mousavi, H., Zaniolo, C.: Fast computation of approximate biased histograms on sliding windows over data streams. In: SSDBM, p. 13. ACM (2013)
9. Papapetrou, O., Garofalakis, M., Deligiannakis, A.: Sketch-based querying of distributed sliding-window data streams. Proc. VLDB Endowment **5**(10), 992–1003 (2012)
10. Shrivastava, N., Buragohain, C., Agrawal, D., Suri, S.: Medians and beyond: new aggregation techniques for sensor networks. In: SenSys, pp. 239–249. ACM (2004)
11. Zhang, Q., Wang, W.: A fast algorithm for approximate quantiles in high speed data streams. In: SSDBM, p. 29. IEEE (2007)

Author Index

Printed in the United States
By Bookmasters